Energy Technology 2014
Carbon Dioxide Management and Other Technologies

TMS2014

143rd Annual Meeting & Exhibition

Energy Technology 2014
Carbon Dioxide Management and Other Technologies

Proceedings of a symposium, Energy Technologies and Carbon Dioxide Management, sponsored by the Energy Committee of the Extraction & Processing Division and the Light Metals Division of The Minerals, Metals & Materials Society (TMS) with papers contributed by two symposia, High-temperature Material Systems for Energy Conversion and Storage and Solar Cell Silicon, sponsored by the Energy Conversion and Storage Committee of the Electronic, Magnetic & Photonic Materials Division of The Minerals, Metals & Materials Society (TMS)

held during

TMS2014
143rd Annual Meeting & Exhibition

February 16-20, 2014
San Diego Convention Center
San Diego, California, USA

Edited by:

Cong Wang

Jan de Bakker • Cynthia K. Belt

Animesh Jha • Neale R. Neelameggham

Soobhankar Pati • Leon H. Prentice

Gabriella Tranell • Kyle S. Brinkman

WILEY | TMS

Wiley also publishes books in a variety of electronic formats. Some content that appears in print may not be available in electronic formats. For more information about Wiley products, visit the web site at www.wiley.com. For general information on other Wiley products and services or for technical support, please contact the Wiley Customer Care Department within the United States at (800) 762-2974, outside the United States at (317) 572-3993 or fax (317) 572-4002.

Library of Congress Cataloging-in-Publication Data is available.

ISBN 978-1-118-88820-9

Printed in the United States of America.

10 9 8 7 6 5 4 3 2 1

WILEY | TMS

TABLE OF CONTENTS
Energy Technology 2014

Alternative Green Processes

Energy in Iron and Steel

Carbon Dioxide Management

Novel Technologies and Life Cycle Assessment

Energy Efficiency and Furnace Technologies

Poster Session

High-temperature Material Systems for Energy Conversion and Storage

Solid Oxide Fuel Cells II

Solar Cell Silicon

Silicon Production and Solidification

Silicon Refining I

Silicon Refining II

PREFACE

This volume contains selected papers presented at the Energy Technologies and Carbon Dioxide Management symposium organized in conjunction with the TMS 2014 Annual Meeting & Exhibition in San Diego, California, USA, and organized by the TMS Energy Committee. As the title of the symposium suggests, the papers intend to address the issues, intricacies, and challenges relating to energy and environmental science. This volume also contains selected articles from the High-temperature Material Systems for Energy Conversion and Storage Symposium and the Solar Cell Silicon Symposium.[1]

This is the seventh year of the Energy Technologies and Carbon Dioxide Management symposium, which was initiated for TMS2008. During the first two years, emphasis was given to topics on CO_2 minimization by chemical reduction of oxides or physical minimization by other methods. It was called CO_2 Reduction Metallurgy. Starting in 2010, the proceedings have been renamed as *Energy Technology* with articles from the symposium on carbon dioxide and other greenhouse gas reduction metallurgy, energy efficiency, and waste heat recovery in metallurgical processes. In 2012 it was decided to encompass all these topics in a symposium titled "Energy Technologies and Carbon Dioxide Management".

Because energy is an emerging issue in any branch of the materials industry, the symposium intends to address the needs for sustainable technologies with reduced energy consumption and pollutants. Given the spread of topics in energy among numerous journals, making the work less accessible to many researchers, we decided to compile information on research activities in the area of energy, and this book is the result. The availability of focused scientific information into a few accessible resources should be attractive and gratifying to many researchers. The authors have contributed to this book, and provided a summary of current research on energy technology and CO_2 management.

The Energy Technologies and Carbon Dioxide Management symposium was open to participants from both industry and academia and focused on energy efficient technologies including innovative ore beneficiation, smelting technologies, recycling, and waste heat recovery. This volume also covers various technological aspects of sustainable energy ecosystems, processes that improve energy efficiency, reduce CO_2 and other greenhouse emissions. It also includes contributions from all areas of non-nuclear and non-traditional energy sources.

---

[1] Papers from the High-temperature Material Systems for Energy Conversion and Storage Symposium and the Solar Cell Silicon Symposium were also contributed to the TMS journal *Metallurgical and Materials Transactions E: Materials for Energy Systems*.

We hope this book will serve as a reference for both new and current materials beginners and metallurgists, particularly those who are actively engaged in exploring innovative technologies and routes that lead to more efficiency and sustainability.

This book could not materialize without contributions from the authors of included papers, the time and effort of the reviewers dedicated to the manuscripts during the review process, and the help received from the publisher. We thank them all! We wish to acknowledge the efforts of Energy Committee Chair Jaroslaw Drelich for enhancing the *Energy Technology 2014* proceedings. We also acknowledge the organizers of the other two symposia represented in this book.

Energy Technologies and Carbon Dioxide Management symposium organizers
Cong Wang
Jan de Bakker
Cynthia K. Belt
Animesh Jha
Neale R. Neelameggham
Soobhankar Pati
Leon H. Prentice

EDITORS

Cong Wang – Lead Organizer

Cong Wang is currently Senior Research Associate of the Department of Materials Science and Engineering of Northwestern University. Prior to joining Northwestern University, Dr. Wang was with the Saint-Gobain High Performance Materials Research and Development Center and the Alcoa Technical Center, respectively. He is a seasoned metallurgist/materials scientist with rich experience in aluminum alloys, copper alloys, iron and steels, magnesium alloys, and super-abrasives. He obtained his Ph.D. from Carnegie Mellon University; his M.S. from the Institute of Metal Research, Chinese Academy of Sciences; and his B.S. from Northeastern University with distinctions, respectively.

Dr. Wang is the 2011 TMS Young Leader Award winner and is a Key Reader for *Metallurgical and Materials Transactions*. He is the 2014 Outstanding Young Manufacturing Engineer Award winner conferred by the Society of Manufacturing Engineers. He serves the editorial committees of *the International Journal of Refractory Metals and Hard Materials*, *Journal of Materials Engineering and Performance*, and *Advanced Materials and Processes*. He is a frequent organizer of international conferences, including REWAS 2013 (Global Conference on Sustainability), Energy Technologies and Carbon Dioxide Management Symposium (2013 and 2014), and Materials Processing Fundamentals Symposium (2012, 2013, and 2014).

Jan de Bakker is a metallurgist with BBA, Inc., a consulting engineering firm based in Montreal, Canada. His roles include optimization and development of process flowsheets, plant audits, and OPEX reduction. He specializes in the design and optimization of process plants for iron ore and base metal mines. Dr. de Bakker received his Ph.D. in Mining Engineering from Queen's University in 2011, for which he developed a novel flowsheet for HCl recovery from $MgCl_2$ solutions. Prior to this, he obtained a master's degree in Environmental Engineering from Columbia University, where he also did his undergraduate studies. Dr. de Bakker has authored and coauthored a number of technical papers on topics in mineral processing and extractive metallurgy, and has taught several courses.

Cynthia K. Belt has managed energy programs at Superior Industries, Aleris International, and Kaiser Aluminum. She has published multiple papers in the area of energy management in the metals industry and has co-edited several proceedings volumes on energy and recycling.

Cindy is Past-Chair of the TMS Energy Committee and has been involved in numerous energy groups within TMS, AFS, and ASME. She earned her B.S. in Mechanical Engineering from Ohio Northern University.

Animesh Jha obtained his bachelor's (from the University of Roorkee, India) and master's degrees in Engineering (Indian Institute of Science, Bangalore, India) in 1979 and 1981, respectively. He obtained his Ph.D. and Diploma of Imperial College in October 1984 from the University of London. Dr. Jha has more than 30 years of research experience in the fields of materials processing, chemical thermodynamics, kinetics, glass science and photonic optical materials. His current research areas are on critical materials for the energy sector, light-matter interaction, and integrated photonic devices for energy-efficient light sources, and chemical and biological sensing. He has been a TMS member since 1992, and is also a member of the Institute of Physics, Optical Society of America, and IEEE. He is a fellow of the Institute of Physics.

Neale R. Neelameggham is "The Guru" at IND LLC, involved in technology marketing and international consulting in the field of light metals and associated chemicals (boron, magnesium, titanium, lithium, and alkali metals), rare earth elements, battery and energy technologies, etc. He has been an advisor in various environmental extractive processes for metal production and energy process firms.

Dr. Neelameggham has more than 38 years of expertise in magnesium production technology from the Great Salt Lake Brine in Utah, involved in process development of its startup company NL Magnesium through the present US Magnesium, LLC, from which he retired. He was involved in most of the process and equipment development of all areas of the plant from the raw material source—the Great Salt Lake Brine, which is concentrated by solar evaporation into a plant feed and further purified—followed by spray drying and preparation of anhydrous magnesium chloride cell feed to electrolytic cells and then into magnesium metal product and chlorine.

In addition, Dr. Neelameggham's expertise includes an in-depth and detailed knowledge of all competing technologies worldwide of magnesium production, both electrolytic and thermal processes, such as the Pidgeon Process, Murex Process, Zuliani Process, and others, as well as alloy development. This expertise is used in engineering a tailor-made magnesium process to fit any resources and has developed near zero-waste sustainable magnesium production process and equipment. He was a visiting expert at Beihang University of Aeronautics and Astronautics, Beijing, China.

Dr. Neelameggham holds 13 patents and a patent application on boron production and has several technical papers to his credit. As a member of TMS, AIChE, and a former member of American Ceramics Society, he is well versed in energy engineering, bio-fuels, rare-earth minerals and metal processing and related processes. Dr. Neelameggham has served in the Magnesium Committee of the TMS Light Metals Division (LMD) since its inception in 2000, chaired it in 2005, and in 2007 he was made a permanent co-organizer for the Magnesium Technology symposium. He has been a member of the Reactive Metals Committee, Recycling Committee, and Titanium Committee, and has been a Program Committee Representative of LMD. He received the LMD Distinguished Service Award in 2010.

In 2008, LMD and the Extraction & Processing Division (EPD) created the Energy Committee following the CO2 Reduction Metallurgy Symposium that he initiated. Dr. Neelameggham was the inaugural chair for the Energy Committee and has served as a co-organizer of the Energy Technology symposium since 2008. Dr. Neelameggham holds a doctorate in extractive metallurgy from the University of Utah. He has been selected as the Chair of the Hydrometallurgy and Electrometallurgy Committee of EPD for the 2013-2015 term.

 **Soobhankar Pati** is an Assistant Professor at IIT Bhubaneswar, India. His research interests includes clean energy, advance manufacturing process, and green metallurgy. Prior to joining IIT, he was with Infinium Inc., Natick, Massachusetts, where he was instrumental in developing a revolutionary process for electrolytic production of metals directly from their oxides. His innovations led to simpler and more robust gas handling in the process, and facilitated scale-up of the process from a few grams to kilogram scale. In addition, Dr. Pati was a visiting scientist at Boston University in the Department of Materials Science and Engineering, where he received a Ph.D. in 2010. At Boston University, his contributions led to breakthroughs that reduced the cost of pure oxygen production in this direct oxide electrolysis process. As part of his graduate research he developed a new

technology for using the energy in industrial and municipal waste to directly make hydrogen gas at high efficiency. His research work at Boston University won various clean energy awards. He is currently a member of TMS and actively takes part in Energy Committee and Magnesium Committee activities.

Leon H. Prentice is a Chartered Professional (Chemical) Engineer working in Melbourne, Australia. Dr. Prentice started his career in hydrometallurgical R&D, followed by six years in biomaterials development for a small Australian manufacturer. Since 2007 he has been a Senior Research Engineer with the Commonwealth Scientific and Industrial Research Organisation (CSIRO). He led the research team working on a new magnesium process (MagSonic™), for which he received from TMS the Vittorio de Nora Prize in 2013. Dr. Prentice leads or consults to a number of projects in titanium production, post-combustion capture, and manufacturing informatics, mostly focused on the scale-up and proof-of-operation of new technologies.

Gabriella Tranell received her Ph.D. in materials science and engineering from the University of New South Wales, Australia, in 1999. After 10 years as senior scientist and research manager with SINTEF in Trondheim, Norway, she is currently an associate professor and Deputy Head of the Department of Materials Science and Engineering, NTNU, Norway. She is also acting director of the Norwegian Centre for Renewable Energy – a collaboration between NTNU, SINTEF, the University in Oslo (UiO), and the Institute for Energy Technology (IFE). Her research interests include among others: production and refining of silicon, GHG and fume emissions from metallurgical industry, and recovery of REE from waste materials.

Kyle S. Brinkman is currently a principal engineer and program manager for renewable energy activities in the Science and Technology Directorate of the Savannah River National Laboratory (SRNL) and an adjunct professor at the University of South Carolina (USC) in the Department of Mechanical Engineering's Solid Oxide Fuel Cell Center of Excellence. Prior to working at SRNL, Dr. Brinkman was a fellow of the Japanese Society for the Promotion of Science working in a Japanese national laboratory, the National Advanced Institute of Science and Technology in Tsukuba, Japan, as part of the Research Center for Hydrogen Industrial Use and Storage "Hydrogenius". Dr. Brinkman received his Ph.D. in materials science and engineering from the Swiss Federal Institute of Lausanne in Switzerland (EPFL). He obtained an M.S. in ceramic and materials engineering and a B.S. in chemical engineering from Clemson University. His recent work and focus at SRNL has been in the areas of hydrogen storage and purification, electronic ceramic materials for gas separation in commercial (H_2, and CO_2) and nuclear domains (hydrogen isotopes), structure/property relations in solid oxide fuel cell systems, mixed ionic and electronic conductive ceramics for energy conversion, crystalline waste form development as a transformational waste management solution for a closed nuclear fuel cycle, and thin film material synthesis and characterization for sensors and enhanced non-proliferation objectives. He is the SRNL principal investigator for the DOE-BES Energy Frontier Research Center, "HeteroFoam".

Energy Technology 2014
Carbon Dioxide Management and Other Technologies

SYMPOSIUM:
Energy Technologies and Carbon Dioxide Management

Energy Technology 2014
Carbon Dioxide Management and Other Technologies

SYMPOSIUM: ENERGY TECHNOLOGIES AND
CARBON DIOXIDE MANAGEMENT

Alternative Green Processes

Session Chairs:

Cong Wang
Adam C. Powell

Energy Technology 2014: Carbon Dioxide Management and Other Technologies
Edited by: Cong Wang, Jan de Bakker, Cynthia K. Belt, Animesh Jha, Neale R. Neelameggham,
Soobhankar Pati, Leon H. Prentice, Gabriella Tranell, and Kyle S. Brinkman
TMS (The Minerals, Metals & Materials Society), 2014

A Review: Solar Thermal Reactors for Materials Production

Ben M. Ekman, Geoffrey Brooks and M. Akbar Rhamdhani

Faculty of Engineering and Industrial Science
Swinburne University of Technology
Hawthorn, VIC 3122, Australia.

Keywords: Solar Energy, Solar Reactors and Heat Transfer.

Abstract

Concentrated solar collectors including solar simulators have been studied to concentrate solar radiation to flux levels capable of reaching temperatures of 3000K. Currently, there are no industrial scale solar reactors used for material processing and only small research units have been tried. Various laboratory scale solar reactor designs are reviewed including their operating temperature and heat transfer efficiency. Solar reactor designs can be classed into directly or indirectly irradiated and single or double cavity construction. Thermal efficiencies of up to 40% have been achieved in experimental units while operating at 1500K. Problems are encountered that relate to the transfer of high heat flux. Hybrid reactor system should be considered in the future to overcome some of the practical issues associated with solar thermal reactors.

I.   INTRODUCTION

In the high temperature processing of materials, such as the production of metals, cement, refractory or glass, the reactions are typically performed in the range of 500° to 2000°C. Metals are produced from minerals, which maybe oxides, sulphides, chlorides or silicates, and to recover the metals, these compounds have to undergo chemical reaction to liberate the metal. The thermodynamic stability of these compounds found in minerals necessitates the use of high temperatures to break down the bonds between the metals and other elements. Normally, the raw metals produced from smelting require refining, which is often carried out in the molten state. As new energy sources have been established, new furnace designs have evolved to incorporate the latest technology. Although solar energy has existed longer than all the current energy sources, it has not yet been incorporated into any industrial smelting; refining or melting processes.

The solar radiation reaching the earth's atmosphere, called the "Solar Constant" (SC), is defined as the rate at which solar energy is incident on a surface normal to the sun's rays at the outer edge of the atmosphere when the earth is at its mean distance from the sun. The currently accepted value is 1367 W/m^2 [1]. The earth intercepts only a small amount of this and 30% is reflected back into space, 47% is converted to low-temperature heat and reradiated into space and 23% powers the evaporative and precipitation cycle of the biosphere [2]. As a result, the solar energy reaching the earth's surface is weakened considerably to about 1000 W/m^2 or one sun, on a clear day and much lower on cloudy or smoggy days [3]. The thermodynamic limit for solar concentration is determined from the inverse square law [4] by the factor $\sin^2\theta = 46,764$ sun's [5] where θ is the angular size of the sun and equals 0.0093 radians [7]. This would produce a steady state temperature equal to that of the sun's surface of approximately 5800°C. In practice however much lower concentrations are achieved due to geometrical and optical imperfections, shading effects and tracking inaccuracies. Optical configurations based on parabolic-shaped mirrors are commercially available for large-scale collection and concentration of solar energy for the generation of electrical power. The most common configurations used for concentration of the sun's energy in solar thermal applications are parabolic troughs, solar towers and parabolic dishes [8,9].

Although concentrated high temperature solar technology is currently applied successfully in the power generation industry, there are no commercial applications that use the intense heat generated for direct material processing and although a number of small laboratory scale reactors have been tried, these have not been progressed to full scale commercial size units. In the paper, we will review the current state of development of solar thermal reactors for materials production, highlighting the important thermodynamic and heat transfer limits of the reactors developed to date, as well as assessing the practicality of the designs being proposed.

II.   THERMODYNAMICS

Solar reactors for highly concentrated solar systems usually feature the use of a cavity-receiver type configuration. At temperatures above about 1000 K, the net power absorbed is diminished mostly by radiative losses through the aperture. For a perfectly insulated cavity-receiver (no convection or conduction heat losses), the energy absorption efficiency [10] in equation 1.

$$\eta_{absorption} = \frac{\alpha_{eff} Q_{aperture} - \varepsilon_{eff} A_{aperture} \sigma T^4}{Q_{solar}} \qquad (1)$$

where Q_{solar} is the total power coming from the concentrator, $Q_{aperture}$ the amount intercepted by the aperture of area $A_{aperture}$, α_{eff} and ε_{eff} are the effective absorptance and emittance of the solar cavity-receiver, respectively, T is the nominal cavity-receiver temperature, and σ the Stefan-Boltzmann constant (5.6705×10^{-8} $Wm^{-2}K^{-4}$). The incoming solar power is determined by the normal beam insolation I, by the collector area, and by taking into account the optical imperfections of the collection system. The capability of the collection system to concentrate solar energy is often expressed in terms of its mean flux concentration ratio $\hat{C}$ over an aperture normalized with respect to the incident normal beam insolation as follows:

$$\tilde{C} = \frac{Q_{aperture}}{I \times A_{aperture}} \qquad (2)$$

Smaller apertures reduce re-radiation losses but the intercepted solar radiation is reduced. The optimum aperture size is a compromise between maximising the radiation captured and minimising the radiation loss. The heat emitted by bodies with emissivities of less than one and if the surface emissivity is independent of wavelength body is given by:

$$\dot{Q} = \varepsilon A \sigma (T^4 - T_e^4) \qquad (3)$$

Where Q has units of Watts, A is the radiating area, ε is the surface emissivity, T is the temperature of the gray body and T_e is the temperature of the blackbody. For simplification, an aperture size that captures all incoming power is assumed so that $Q_{aperture} = Q_{solar}$. With this assumption and for a perfectly insulated isothermal blackbody cavity-receiver ($\alpha_{eff} = \varepsilon_{eff} = 1$), equations, 1 and 2 are combined to yield

6

$$\eta_{absorption} = 1 - \left(\frac{\sigma T^4}{I\tilde{C}} \right) \qquad (4)$$

A plot which shows the variation of the energy absorption efficiency with cavity temperature for various concentration ratios is shown in Figure 1, [4]. With a solar intensity of 1 kW per square meter, one solar intensity (C=1), on the very best of days at the Earth's surface, the absorption efficiency of a black flat surface at 300 K is about 0.54. The highest temperature achievable with one sun is 364 K; at 364 K the efficiency with which energy can be collected for a useful purpose is zero. The measure of how well solar energy is converted into chemical energy for a given process is the exergy efficiency, defined as;

$$\eta_{exergy} = \frac{- \dot{n}\Delta G_{rxn@298K}}{Q_{solar}} \qquad (5)$$

where ΔG_{rxn} is the maximum possible amount of work that may be extracted from the products as they are transformed back to reactants at 298 K.

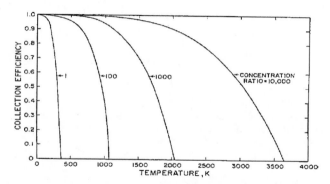

Fig. 1-Energy collection (absorption) efficiency of a black body cavity with its temperature at various solar concentration ratios [4].

Since the conversion of solar process heat to ΔG_{rxn} is limited by the solar absorption and Carnot efficiencies [5, 11], the maximum overall efficiency is;

$$\eta_{exergy,ideal} = \eta_{absorption} \times \eta_{Carnot}$$
$$= \left[1 - \left(\frac{\sigma T_H^4}{I\tilde{C}} \right) \right] \times \left[1 - \left(\frac{T_L}{T_H} \right) \right] \qquad (6)$$

where T_H and T_L are the upper and lower operating temperatures of the equivalent Carnot heat engine. $\eta_{exergy,ideal}$ is plotted in Figure 2. Because of the Carnot limit, one should try to operate thermochemical processes at the highest upper temperature possible; however, from a heat-transfer perspective, the higher the temperature, the higher the re-radiation losses.

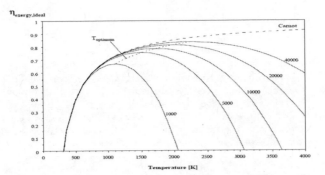

Fig. 2-The ideal exergy efficiency $\eta_{exergy,ideal}$ shown as a function of the operating temperature T_H. Also plotted is the Carnot efficiency and the locus of the optimum cavity temperature $T_{optimum}$ [4].

The highest temperature an ideal solar cavity-receiver is capable of achieving, defined as the stagnation temperature $T_{stagnation}$, is calculated by setting $\eta_{exergy,ideal}$ equal to zero [5,6], which yields;

$$T_{stagnation} = \left(\frac{IC}{\sigma}\right)^{0.25} \quad (7)$$

At this temperature, $\eta_{exergy,ideal} = 0$, because energy is being re-radiated as fast as it is absorbed. Concentration ratios of 10,000 and more have been achieved using paraboloidal primary reflectors and non-imaging secondary concentrators, which translate to stagnation temperatures above 3600 K. However, an energy-efficient process must run at temperatures that are substantially below $T_{stagnation}$. Assuming a uniform power-flux distribution, this relation yields the following equation for $T_{optimum}$;

$$T_{optimum}^5 - (0.75 T_L) T_{optimum}^4 - \left(\frac{\alpha eff T_L IC}{4\varepsilon_{eff}\sigma}\right) = 0 \quad (8)$$

Equation 8 was solved numerically by Steinfeld and Schubnell [12], and the locus of $T_{optimum}$ is shown in Figure 2. The optimal temperature for maximum efficiency varies between 1100 and 1800 K for uniform power-flux distributions with concentrations between 1000 and 13,000. For example, for a solar concentration ratio of 5000, as shown in Figure 2, the optimum temperature of a solar receiver is 1500 K giving a maximum theoretical efficiency of 75%. The maximum efficiency [13] must be equal to that of a Carnot heat engine [14] operating between T_H and T_L and is shown in equation 9, called Carnot Efficiency.

$$\eta_{Carnot} = \eta_{max} = 1 - \frac{T_L}{T_H} \quad (9)$$

The thermal dissociation and electro-thermal reduction of metal oxides proceeds without reducing agents (equation 10), while the carbothermal reduction of metal oxides uses solid carbon C (equation 11), or hydrocarbons (e.g., CH_4 equation 12), as reducing agents [6]. The corresponding overall chemical reactions may be represented as follows;

$$M_xO_y \rightarrow X_m + y/2\ O_2 \quad (10)$$

8

$$M_xO_y + yC \rightarrow xM + yCO \qquad (11)$$

$$M_xO_y + yCH_4 \rightarrow xM + y\,(2H_2 + CO) \qquad (12)$$

Except for the thermal dissociation of ZnO, the required temperature for effecting reaction 10 for oxides of Fe, Mg, Ti, Al and Si exceeds 3050 K [5]. Although it was possible to attain such stagnation temperatures with high-flux solar concentrating systems that deliver concentration ratios above 10,000, practical engineering and heat-transfer considerations suggest operation of solar reactors at substantially lower temperatures, especially when the process is to be conducted with high energy absorption efficiency. Under these circumstances, solar process heat alone would not make the reaction proceed; some amount of high-quality energy was required in the form of work. It may be supplied in the form of electrical energy in electrolytic processes or in the form of chemical energy by introducing a reducing agent in thermochemical processes. If one wishes to decompose metal oxides thermally into their elements without the application of electrical work, a chemical reducing agent is necessary to lower the dissociation temperature [5]. Coal as coke and natural gas are preferred reducing agents because of their availability and relatively low price. In the presence of carbon, the uptake of oxygen by the formation of CO brings about reduction of the oxides at much lower temperatures.

III.   SOLAR REACTORS

A number of studies have been published which describe the developments and advances in the field of solar thermochemical processes and their application for producing chemical fuels and material commodities [5, 6, 4, 15, 16, 11]. The choice of material to subject to solar reducton has been based on the temperature of thermal dissociation as in the case of ZnO or on the carbothermal temperature of dissociation including MgO, SiO_2 and Fe_2O_3 The zinc oxide reaction cycle is most commonly researched and studied extensively by Steinfeld [6, 15, 23], Palumbo [5, 20, 21] and others. Solar reactors may feature direct irradiation of reactants to provide a very efficient means of heat transfer directly to the reaction site. However, a major drawback when working with reducing or inert atmospheres is the requirement for a transparent window, which is a critical and troublesome component in high-pressure or severe gas environments. The introduction of an additional protecting cavity between the window and the reaction chamber can eliminate this problem. This cavity is directly irradiated by the concentrated solar radiation entering through its windowed aperture. As its temperature increases, its walls radiate to all directions. Thus, this inner cavity serves as the solar receiver, radiant absorber, and radiant emitter [11].

Schaffner et al.[17] describe the design, fabrication, and testing of a novel solar chemical reactor for recycling electric arc furnace fume dust (EAFD). The main objective of the process was the extraction of metal oxides, especially ZnO and PbO, by carbothermic reduction of their metal oxides using solar process heat. The main metals present in the fume dust were Zn at $37.8\%_{wt}$, Fe at $13.5\%_{wt}$, and Pb at $10.1\%_{wt}$ that are present primarily in their oxidized state. Carbon is used as the reducing agent and SiO_2 is employed as the main vitrification agent. The solar reactor concept proposed featured two cavities in series. The inner cavity is a graphite enclosure with a small windowed opening, the aperture, to let in concentrated solar power. The 10 kW solar reactor prototype was tested at the high flux solar furnace of the Paul Scherrer Institute which consisted of a 120 m^2 sun-tracking heliostat with an 8.5 m diameter paraboloidal concentrator. The reactor was subjected to mean solar flux intensities of 2000 kW m^{-2} with peak concentrations exceeding 5000 suns. Solar experiments were performed by Schaffner's et al.[17], in the temperature range of 1120 to 1400 K with both batch and continuous modes of operation. They concluded that the condensed off-gas products consisted mainly of Zn, Pb, and Cl and that extraction of up to 99% and 90% of the Zn originally contained in the EAFD was achieved in the residue for the batch and continuous solar experiments, respectively. An energy balance performed under approximate steady-state conditions resulted in a thermal efficiency as high as 8.2% although it was concluded that commercial applications will have to exclude the use of a purge gas, thus eliminating convection heat losses and reducing costs.

Solid metal zinc can be used either in a fuel cell or battery, or in a water-splitting reaction producing hydrogen and regenerating ZnO which can be recycled to the solar step. A specific feature of ZnO/Zn cycle is that the products of the dissociation reaction exiting the reactor are gaseous. ZnO is decomposed at about 2000 K and Zn is recovered after quenching the product gases. Table I summarises the studies and the reactors used. The reactors incorporate either a windowed aperture cover with a directly irradiated (single cavity) receiver shown in Figure 3, or Figure 4 with an indirect (two cavity) outer cavity receiver and inner cavity reactor design. A rotating or stationary cavity and with either a packed bed or gas/particle feed stream can be incorporated into this design.

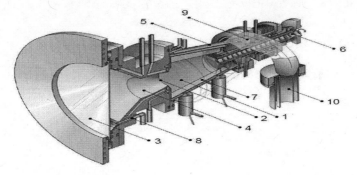

Fig. 3-Schematic of a "rotating-cavity" solar chemical reactor for the thermal dissociation of ZnO to Zn and O_2 at 2300 K [11].

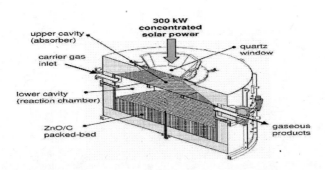

Fig. 4-Schematic of the solar chemical reactor, featuring two cavities in series containing a ZnO/C packed bed [21].

TABLE I. Summary of Zinc Oxide Reduction and Reactor Designs.

Reference	Material	Reactor Type And Power	Operating Temperature	Efficiency And Yield
Schaffner et al.[17]	Furnace Fume Dust ZnO, PbO	Two CavityRotating Indirect Windowed 2000kWm^{-2}	1120-1400k	Thermal 8.2% Zn 99%, Pb 90%
Steinfeld [11]	ZnO	Windowed Rotating Cavity Peak Flux 3955 kWm	2300k	Exergy 35%
Palumbo et.al. [18]	ZnO	Solar Flux 6000 To 10,000 W		Exergy 50%
Steinfeld,Meier[6], Steinfeld [11]	ZnO	Two Cavity Indirect Rotating	2000K	
Osinga,Steinfeld[20] Frommherz,,Weickert	ZnO	Two Cavity 5kW	1550k	Thermal 20%
Steinfeld [11]	ZnO	Vortex Windowed Reactor	1300k	
Weickert [21]	ZnO	Two Cavity 200 & 300 kW Indirectly Radiated	1350°c	Thermal 40% Zn 50kg/Hr
Steinfeld, Weiner [15]	ZnO	Windowed Rotating Cavity10kWReactor	1807-1907k	Zn 90%
Abanades,Charvin, Flamant [22]	ZnO	Direct Horizontal Solar Reactor	2300k	-

In 1999 Murray [23] reported, that two types of processes have great potential for the production of aluminium using high temperature solar process heat. A direct reduction process to Al or Al–Si alloy, or a reduction process to an intermediate such as AlN or Al_2S_3 that could be electrolysed more easily than Al_2O_3 using non-consumable electrodes. The direct reduction using high-temperature solar heat on an Al–Si ore to an alloy for further processing to pure Al or use as a casting alloy solves the problems encountered in a 'pure carbothermal' process to produce Al from Al_2O_3 and allows the use of non-bauxite ore sources if the problems with Fe and Ti can be solved. At IMP-Odeillo [24], the reactant pellets were directly irradiated using highly concentrated sunlight from a 2 Kw vertical axis solar furnace, which achieved a peak radiative flux as high as 1500 W/cm^2 over a diameter of several millimetres. Further qualitative tests have been performed in a 2 kW solar furnace, where the onset of production of both aluminium by direct carbothermal reduction, and Al-Si alloy via carbothermal reduction of a mixture of alumina, silica and carbon could be directly observed. Murray [24] concluded that solar process heat is a very promising replacement for electro thermal process heat now used in other high-temperature processes worldwide where heat energy is the real process requirement.

The thermal decomposition of calcium carbonate (limestone) to calcium oxide (lime), as represented in equation 13, is the main endothermic step in the production of lime and cement at 1300 K [6].

$$CaCO_3 \rightarrow CaO + CO_2 \tag{13}$$

Substituting concentrated solar energy in place of carbonaceous fuels, as the source of high-temperature process heat, is a means to reduce the dependence on conventional energy resources and to reduce emissions of CO_2 and other pollutants. Figures 5 show a solar reactor concepts for the solar calcination process based on direct irradiation and consists of a refractory-lined conical rotary kiln operated in a horizontal position.

Meier et al., [25] designed and tested a direct irradiated 10-kW solar rotary kiln reactor to effect the calcination reaction. The degree of calcination and the reactivity both depended on the reactant's decomposition temperature (1323 K to 1423 K), residence time (3–7 min), and feed rate (10–50 g/min). The reactor's energy efficiency reached 20% for solar flux inputs of about 1200 kW/m^2 and for quicklime production rates of about 1.3 kg/h. Figure 5 shows the 10kW solar lime reactor mounted in the solar furnace test facility. The solar concentrating system used for the solar calcination experiments delivered solar power close to 15 kW with a peak concentration of about 3000 Suns on a focal spot of 8 cm diameter. For some experiments, the degree of calcination exceeded 98%, which is beyond the 90–95% that is often accepted by industrial standards. The most important observation made from the experimental data was that a typical solar energy to chemical energy conversion efficiency was about 13%, and the maximum efficiency reached 20%. Meier et al., [25] concluded from the results of the initial solar experiment, that a solar reactor can produce completely calcined lime with virtually any quality ranging from low to high reactivity.

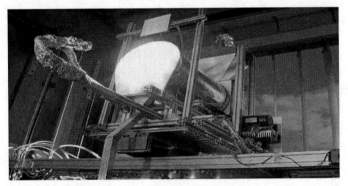

Fig. 5-Test of the 10-kW solar lime reactor prototype in the solar furnace [25].

Meier et al., [26], developed the solar chemical reactor technology to demonstrate endothermic calcination reaction at 1200 K to 1400 K. The indirectly heated 10kW$_{th}$ multi-tube rotary kiln prototype processed 1 to 5 mm limestone particles and produced high purity lime that was not contaminated with combustion by-products. The quality of the solar produced quicklime met the industrial standards in terms of reactivity (low, medium, and high) and degree of calcination (exceeding 98%). The reactor's efficiency, (defined as the enthalpy of the calcination reaction at ambient temperature (3184 kJ kg−1) divided by the solar energy input), reached 30–35% for quicklime production rates up to 4 kg/hr. The solar lime reactor prototype operated reliably for more than 100 h at solar flux inputs of about 2000 kW/m^2, withstanding the thermal shocks that occur in solar high temperature applications. The performance of one particular experimental run was measured at 1395 K, the degree of calcination was 98.2%, and the CaO production rate was 64.2 g min^{-1}. The solar energy to chemical energy conversion efficiency was 34.8%. Meier et al., [26], believed that the indirectly heated rotary kiln may also be operated in hybrid mode. In place of concentrated solar radiation, any conventional external heat source such as oil or gas burners could be introduced through the aperture into the cavity. Scaled up industrial sized units have been proposed including a 20MW$_{th}$ sized unit with a production rate of 20t/hr peak output [26]. It was expected that the efficiency of industrially sized reactors would be higher than that of the small-scale experimental version if conduction losses were reduced and sensible heat were recovered for preheating the limestone particles [28]. Such a reactor with a thermal efficiency near 45–55% that also produces high quality quicklime would demonstrate that an Industrial Solar Lime Plant could be an economically viable path for reducing CO_2 emissions in specific market sectors of the lime industry. It must be remembered that the solar lime processes discussed would require hybridisation to operate continuously if it is to be adopted by industry.

IV.   CONCLUSIONS

Reactor-Furnace vessels that apply to Smelting, Converting and Refining are designed to operate at high temperatures in aggressive and highly polluting environment and it is difficult to see how they would be amenable to apply solar thermal energy On the other hand melting, reheating and kiln firing operate in a relatively dust and fume free environment, both as a batch or continuous operations, on a wide variety of materials including ferrous and non-ferrous metals, precious and semi-precious metals, ceramics and refractory materials. These types of furnaces are used in the metal casting industry and have a thermal efficiencies ranging from 7% for a crucible furnace to 76% for an induction furnaces. The solar reactors reviewed have shown thermal efficiencies in this range and as high as 55%. It can be envisaged that some commercial operations could take advantage of solar thermal or be solar hybridized for continuous operations.

Concentrated solar thermal technology has reached a stage where the high temperatures required for material processes are achievable. Reactor design has endeavored to contain the thermal energy by incorporating window covers and gas purging to optimize the entry of solar energy into the aperture. Reactor designs have used either a direct irradiated single cavity or indirectly irradiated two cavity configurations. It is clear that where there are no gases emitted the single cavity design is preferred, however, a two stage design is often required to maintain solar energy effectiveness. The efficiency of these reactors have been shown to be significantly affected by the particle size of the feed. However, it is difficult to see how the rotation of the cavity could be practical in commercial large scale operations. To date there has only been limited research on hybrid design but this would appear to be a worthwhile avenue of research and development, as hybrid designs can help overcome some of the weaknesses associated with solar thermal reactors.

ACKNOWLEDGMENTS

The authors gratefully acknowledge the financial support of Australian Renewable Energy Agency (ARENA) formerly the Australian Solar Institute (ASI).

REFERENCES

1.   Fröhlich, C., and R. W. Brusa (1981), "Solar Radiation and its Variation in Time", Solar Physics 74, 209.
2.   Goswami, D. Yogi, Kreith, Frank, and Kreider, Jan F., Principles of Solar Engineering, 2nd edition. Taylor and Francis, Philadelphia, PA, 2000.
3.   Y. A. Cengel "Heat And Mass Transfer." Mc Graw Hill Publish 2006
4.   Fletcher E.A. 2001 Solarthermal Processing: A Review. Journal of Solar Energy Engineer May 2001 vol 123
5.   A. Steinfeld &   R. Palumbo "Solar Thermochemical Process Technology" Solar Thermochemical Process Technology Academic Press, Vol. 15, pp. 237-256, 2001.
6.   Steinfeld, A., Meier A "Solar Fuels And Materials". Encyclopedia of Energy Vol. 5 2004.
7.   Pitz-Paal .R. High Temperature Solar Concentrators.Vol.1 Solar Energy Conversion and Photoenergy Systems.
8.   Lovegrove K. "Solar Thermal Energy Systems in Australia". The international Journal of environ studies 2006.
9.   Mills A.A. "Reflections of the Burning Mirrors of Archimedes with a Consideration of the Geometry and Intensity of Sunlight Reflected from Plane Mirrors" European Journal of Physics. 2004
10.   Fletcher, E. A, and Moen, R. L.(1977). Hydrogen and Oxygen from Water. Science 197.
11.   Steinfeld A., Solar thermochemical production of hydrogen–a review. Solar Energy 78 2005.
12.   Steinfeld, A., and Schubnell, M. (1993). Optimum Aperture Size and Operating Temperature of a Solar.
13.   Martinek J., Channel M., Lewandowski A., and Weiner A.W., Considerations for the Design of Solar-Thermal Chemical Processes. Journal of Solar Energy Engineering Aug. 2010 Vol.132.
14.   De Vos, A. Thermodynamics of Solar Energy Conversion. Wiley-Vch., Pub., 2005.
15.   Steinfeld A., Weiner A.W., Thermochemical Production of Fuels with Concentrated Solar Energy. Optics Express Vol.18 2010.
16.   Kodama t., High-Temperature Solar Chemistry for Converting Solar Heat to Chemical Fuels. Progress in Energy and Comb. Sci. 29 2003.
17.   Schaffner B., Meier A. , Wuillemin D. ,Hoffelner W., Steinfeld A., Recycling of Hazardous Solid Waste Material Using High-Temperature Solar Process Heat. 2. Reactor Design and Experimentation. VOL. 37, NO. 1, 2003 Environmental Science & Technology.
18.   R. Palumbo, J. Ledè, O. Boutin, E. Elorza Ricart, A. Steinfeld, S. Muller, A. Weidenkaff, E.A. Fletcher, J. Bielicki The production of Zn from ZnO in a high-temperature solar decomposition quench process. I. The scientific framework for the process Chemical Engineering Science, 53 (14) (1998), p. 2503.
19.   R. Palumbo, M. Keunecke, S. Moller, A. Steinfeld Reflections on the design of solar thermal chemical reactors: thoughts in transformation Energy, 29 (2004), pp. 727–744.
20.   Osinga T., Frommherz U., Steinfeld A., Weickert C., Experimental Investigation of the Solar Carbothermic Reduction of ZnO using a Two-Cavity Solar Reactor. J. of Solar Energy Eng. Vol 126 2004.
21.   Weickert C., Frommherz U., Kraupl S., Guillot E., Olalde G., Epstein M., Santen S., Osinga T., Steinfeld A., A 33 kW Solar Chemical Pilot Plant for the Carbothermic Production of Zinc. J. of Solar Energy Eng. Vol 129 May. 2007.
22.   Abanades S., Charvin P., Flamant G., Design and Simulation of a Solar Chemical Reactor for the Thermal Reduction of Metal Oxides: Case Study of Zinc Oxide Dissociation. Chem. Eng. Sci. 62 2007.
23.   Murray J., Aluminium production using high-temperature solar process heat. Solar Energy Vol. 66 No.2 1999.
24.   Murray J., Solar Production of Aluminum by Direct Reduction: Preliminary Results for Two Processes. Journal of Solar Energy Engineering. Vol. 123 May 2001.
25.   Meier A., Bonaldi E., Cella G.M., Lipinski W., Wuillemin D., Palumbo R., Design and experimental investigation of a horizontal rotary reactor for the solar thermal production of lime. Energy 29. 2004. p 811.
26.   Meier A., Bonaldib E., Cellab G.M., Lipinskia W., Wuillemina D., Solar chemical reactor technology for industrial production of lime. Solar Energy Vol. 80, Issue 10, Oct. 2006, Pages 1355.

14

Energy Technology 2014: Carbon Dioxide Management and Other Technologies
Edited by: Cong Wang, Jan de Bakker, Cynthia K. Belt, Animesh Jha, Neale R. Neelameggham,
Soobhankar Pati, Leon H. Prentice, Gabriella Tranell, and Kyle S. Brinkman
TMS (The Minerals, Metals & Materials Society), 2014

Ferroelectric-Enhanced Photocatalysis with TiO$_2$/BiFeO$_3$

Yiling Zhang[1], Paul A. Salvador[1], Gregory S. Rohrer[1]

[1]Department of Materials Science and Engineering,
Carnegie Mellon University, Pittsburgh, PA 15213, USA

Keywords: Photocatalysis, Ferroelectrics, Heterostructure, Functional semiconductor, Thin Film

Abstract

Photocatalysis utilizes solar energy to produce hydrogen by splitting H$_2$O. When a thin film of photocatalyst TiO$_2$ is supported on a ferroelectric BiFeO$_3$ substrate (band gap ~2.5 eV), the TiO$_2$/BiFeO$_3$ heterostructure is capable of photochemically reducing Ag$^+$ to Ag^0 in aqueous solutions under blue light with enhanced efficiency. The observation of spatially selective silver patterns on surface of the heterostructure after reaction suggests that photogenerated electrons and holes in ferroelectric BiFeO$_3$ are driven in opposite directions to reduce their chance of recombination. Comparisons of the amounts of reduced silver on TiO$_2$ film grains with distinct phases/orientations show that the reactivity is mildly preferential on anatase TiO$_2$ phase and does not strongly depend on crystallographic orientation of BiFeO$_3$.

Introduction

Titania (TiO$_2$) is capable of photocatalytically converting H$_2$O to hydrogen and oxygen under the irradiation of ultraviolet light. [1, 2] One of the limiting factors for titania to be used as an effective photocatalyst is its band gap. Because the absorption edges of rutile and anatase titania are at 3.0 eV and 3.2 eV, respectively, they can only absorb light in the UV portion of the spectrum, which is about 3% of the available solar energy. [3-5] This limitation has motivated attempts to modify the absorption edge of titania so that it can absorb visible light.

One strategy is to substitute a portion of the titanium and/or the oxygen with other atoms. There are numerous reports on visible light absorption of titania by doping and co-doping of Cr, V, Mo, Sb, Fe, N, S, C, and F. [6-34] Another strategy is to add an adsorbed molecular species that absorb visible light and donate an electron or hole to the titania, such as organic dyes [33, 35, 36] and Ce^{3+}/Ce^{4+}. [37, 38]

The strategy in the present work takes the approach of depositing a pure titania thin film on a visible light absorbing substrate BiFeO$_3$ to induce visible light activity of the heterostructure. BiFeO$_3$ is a semiconductor with a band gap of about 2.5 eV [39-41] and is photocatalytically active in visible light. [42, 43] It is also ferroelectric with spontaneous polarization along pseudo-cubic <111> directions of 6.1 μC cm^{-2}. [44] The spontaneous polarization gives rise to internal electric fields that drive photogenerated charge carriers, namely electrons and holes, in opposite directions to reduce their chance of recombination. [45-47] Additionally, the separation of electrons and holes leads to the spatial separation of reduction and oxidation half reactions so that the back reaction of intermediates is also suppressed. Such charge carrier separation has been reported in TiO$_2$/BaTiO$_3$ heterostructures with UV light illumination. [48-51] The TiO$_2$/BiFeO$_3$ heterostructures in this work combine both advantages of the relatively narrow band gap and the internal fields in BiFeO$_3$ to enhance the overall photocatalytic activity.

In this paper, the $TiO_2/BiFeO_3$ heterostructures are shown to be photochemically active in blue light (~460nm). Instead of reduction of protons (H^+), reduction of aqueous silver cations to silver metal [52-55] was used as a marker reaction for demonstration. The patterns of silver reduction on surface closely match the structures of ferroelectric domains in $BiFeO_3$, which is strong evidence supporting the effect of internal fields. The dependence of activity on titania phases and crystallographic orientations is also summarized.

Experimental Work

Polycrystalline $BiFeO_3$ pellet substrates were synthesized from equimolar amounts of Bi_2O_3 (Alfa Aesar 99.99%) and Fe_2O_3 (Alfa Aesar 99.945%) powders. The mixture of powders was ball milled in ethanol for 24 hr and dried at 85 °C. Then the powders underwent calcination at 700 °C for 3 h in air. The powder was again ball milled in ethanol for 24 h, dried, and uniaxially compressed in a cylindrical die to form disk-shaped pellets about 1 cm in diameter and 3 mm thick. The pellets were sintered at 850 °C for 3 h. X-ray diffraction indicated that the majority phase was $BiFeO_3$ and that there are small amounts of minority phases such as $Bi_2Fe_4O_9$, $Bi_{25}FeO_{40}$, and Fe_2O_3. The sintered pellets were lapped flat with an aqueous Al_2O_3 suspension (9 µm or 3 µm, Logitech) and polished with a 0.02 µm colloidal SiO_2 suspension (Mastermet 2, Buehler) using a Logitech autopolisher. The polished pellets were thermally etched at 600 °C for 3 h to heal the polishing damage.

Pulsed laser deposition (PLD) was used to deposit thin films of titania onto the $BiFeO_3$ pellet substrates. The pellet substrates were cleaned twice by sonication in a bath of methanol for 5 – 10 min before loaded in the deposition chamber. A KrF ($\lambda = 248$ nm) laser was pulsed at 3 Hz with an energy density of 2 J cm^{-2}. The target-to-substrate distance was maintained at ~ 6 cm. Initially, a base pressure of ~ 10^{-5} Torr was established in the chamber with the substrate at 120 °C. A constant flow of oxygen at 5 mTorr was introduced into the chamber when the substrates were heated to 500 – 700 °C at a rate of 25 °C min^{-1} and during deposition. Before deposition, the surface of the TiO_2 target was cleaned by laser ablation for 10 min. In deposition, 1500 laser pulses were fired to produce films of about 10 nm. After deposition, the chamber was cooled down to room temperature at a rate of 25 °C min^{-1} in a static atmosphere of 5 Torr oxygen.

The photochemical reduction of silver was used to test activity of the heterostructure samples. This marker reaction has a relatively close redox potential level to the reduction of H^+ to H_2. [56] But unlike gaseous H_2, reduced silver particles are adhered to sample surfaces to be detected by atomic force microscopic techniques. A 0.115 M $AgNO_3$ solution was prepared by dissolving appropriate amount of $AgNO_3$ crystallites in 50 mL of de-ionized H_2O. A viton O-ring was placed on top of the film. The $AgNO_3$ solution was added to fill the space bound by the O-ring and the film surface. A quartz slip was placed on top of the O-ring to seal the space. A blue LED with emission energy of 2.53 – 2.70 eV (LUXEON, Philips Lumileds, San Jose, CA) was used as the light source. The reaction assembly was brought as close as possible to the lens of the LED. The LED was operated at 750 mA to trigger the photochemical reduction reaction for 60 s. After reaction, the sample was rinsed by sequential immersion in two baths of de-ionized water to dilute the remnant solution adhered on the surface. After rinsing, a stream of clean nitrogen was directed at the surface to blow dry the remaining liquid.

An NTegra atomic force microscope (AFM) system (NT-MDT) was used to examine the surface topography of the samples. Piezoresponse force microscopy (PFM) images were taken to show the ferroelectric domain structures of the heterostructures. Electron backscatter diffraction (EBSD) was used to determine the crystallographic orientations of grains in the samples.

Results and Discussion

First, the photochemical activity of $BiFeO_3$, rutile, and anatase is evaluated by themselves under blue light irradiation. The AFM topographic images in Fig.1 show the surfaces of $BiFeO_3$ (a,b), rutile (c,d), and anatase (e,f), respectively. While the $BiFeO_3$ and rutile are bulk materials, the anatase is a 20 nm film deposited on $SrTiO_3$ (100) under conditions that produce anatase (001). [57, 58] The topographic contrast arises from a variety of sources including surface steps (S), residual polishing scratches (PS), boundaries between ferroelectric domains (DB), surface contamination (SC), and inclusions of minority phases (MP), which are labeled in the images. After the photochemical reaction, certain regions on the $BiFeO_3$ surface are covered with reduced silver particles (bright contrast) of heights in the range of 20 to 130 nm. On the rutile and anatase surfaces, only minor changes are found, corresponding to either few silver particles or surface contamination from the $AgNO_3$ solution. This is considered background activity and is presumably present in all of the experiments, but is overwhelmed by the activity of $BiFeO_3$. Based on Fig. 1, it is clear that $BiFeO_3$ is far more active under blue light illumination than both rutile and anatase phases.

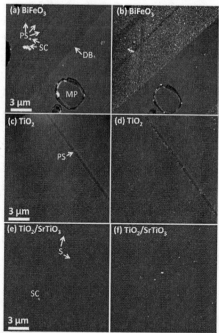

Figure 1 AFM topographic images of surfaces of (a) bare $BiFeO_3$ before reaction, (b) $BiFeO_3$ after reaction, (c) bare rutile TiO_2 before reaction, (d) rutile TiO_2 after reaction, (e) 20 nm anatase TiO_2 film on $SrTiO_3$ (100) before reaction, (f) 20 nm anatase TiO_2 film on $SrTiO_3$ (100) after reaction. The topographic contrast in all images is 100 nm from bright to dark.

The images in Fig. 2 show the correlation between the piezoresponse of the BiFeO$_3$ substrate (a) and of the TiO$_2$/BiFeO$_3$ and the photochemical activity of the titania film (c, d). The two PFM images indicate the same ferroelectric domain structure of BiFeO$_3$ before and after the 10 nm TiO$_2$ film was coated. The two AFM images show that there were more silver particles reduced on some domains than others after the sample was illuminated with blue light. The typical difference in height between domains of different activity is 100 nm. Comparison of the pattern of reduced silver with the domain structure indicates a close match and suggests that silver is preferentially reduced on certain domains, in this case, the domains with dark contrasts.

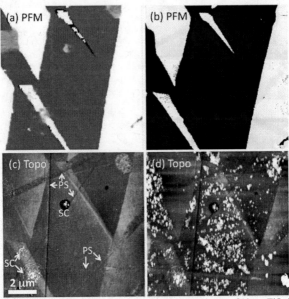

Figure 2 (a) PFM phase image of bare BiFeO$_3$ surface. (b) PFM phase image of 10 nm TiO$_2$/BiFeO$_3$. (c) Topographic image of 10 nm TiO$_2$/BiFeO$_3$ before reaction. (d) Topographic image of 10 nm TiO$_2$/BiFeO$_3$ after reaction. PFM phase contrast scale is -180° to 180° from bright to dark. Topographic contrast scale is 15 nm from bright to dark.

The above results indicate that the photochemical reduction of silver initiated by visible light is spatially selective on the BiFeO$_3$ surface. Assuming this is analogous to what occurs in BaTiO$_3$ and PZT [48, 59-61], then the reduction reaction is probably associated with domains that have polarizations pointed at the surface, where bands are bent downward and electrons are driven toward the surface. The main difference is that the same reaction can only be triggered by UV light for BaTiO$_3$ and PZT, while it is triggered with blue light for BiFeO$_3$.

The observations from the TiO$_2$/BiFeO$_3$ heterostructure indicate that the electrons that participate in the reactions on the TiO$_2$ film surface are created by absorption in the BiFeO$_3$ substrate. This is completely analogous to the conclusion by Burbure *et al.* [48, 50] in studies of TiO$_2$ films on BaTiO$_3$ excited by UV radiation. One observation that supports this conclusion is that when TiO$_2$ is not supported by BiFeO$_3$, it absorbs very little blue light and does not reduce much silver as in Fig.1(d, f). Another observation is that the reduced silver pattern matches the domain pattern in BiFeO₃ substrate as in Fig.2. Both observations suggest electrons created in the BiFeO₃ substrate travel through the TiO₂ thin film to react on the surface.

Based on the results, if the goal were only to have the highest reactivity for silver reduction, BiFeO₃ would be the best choice. However, to be useful for water splitting, the photocatalyst must be stable in aqueous solutions and its conduction band must be higher than the hydrogen reduction level. These two conditions are met by TiO₂, but not by BiFeO₃. [62-64]

To understand the mechanism of visible light activity, plausible energy level diagrams for the TiO₂/BiFeO₃ heterostructure are constructed in Fig.3. Assuming that the energy levels at the interface are strongly influenced by the polarization, then the bands in BiFeO₃ bend upward when the polarization is directed away from the surface (negative domains) and downward when directed toward the surface (positive domains), as shown in Fig. 3(a) and (b), respectively. Therefore, photogenerated electrons in negative domains are stopped by an energy barrier at the interface. In positive domains, electrons are driven to the interface. The experimental observation of spatially selective patterns of reduced silver supports this mechanism.

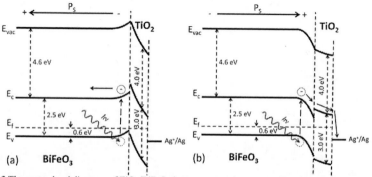

Figure 3 The energy level diagrams of TiO₂/BiFeO₃ heterostructure in contact with solution in the case of (a) negative domain and (b) positive domain. The bands at the TiO₂/BiFeO₃ interface bend upward in (a) and downward in (b) corresponding to the P_s. A photon with sufficient energy $h\nu$ excites an electron in the BiFeO₃ valence band near the interface. The excited electron in (a) is repelled from the interface. The excited electron in (b) is driven to the interface, then through the film to the surface.

The dependence of photochemical activity on titania phase and crystallographic orientation is also of interest for this heterostructured photocatalyst. The method to identify titania phase in the thin film and orientation of BiFeO₃ grains in the substrate using EBSD is described in detail elsewhere. [65]

Fig.4 is a stereographic orientation triangle for the pseudo-cubic BiFeO₃ that covers all possible orientations in a cubic crystallographic system. The three vertices indicate low index planes, namely (001), (101), and (111). There are in total 52 grains in two TiO₂/BiFeO₃ samples examined using AFM. Each grain is considered reactive if part of the grain surface is covered with silver particles of thickness greater than 20 nm, and unreactive otherwise. The red dash line approximately divides the triangle into two regions of BiFeO₃ orientations that allow the growth of anatase or rutile titania, a result from previous studies. [65] In the anatase growth region, 65% (15 grains) of the 23 grains are reactive. In the rutile growth region, 41% (12 grains) of the 29 grains are reactive. Based on the statistics, anatase phase is mildly more reactive than rutile phase in the TiO₂/BiFeO₃ heterostructure. This is slightly different from the results observed on the TiO₂/BaTiO₃ heterostructure, where both phases are of similar activity. [51] In regard of crystallographic orientations, it is found that there are no preferred orientations for reactive grains in both regions, which is a case dissimilar to bulk titania. [66-70]

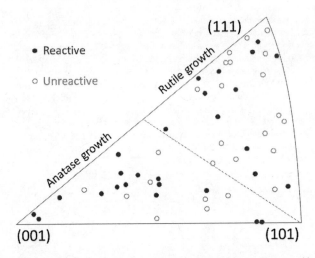

Figure 4 The stereographic orientation triangle for BiFeO$_3$. Reactive grains are represented by solid circles, and unreactive ones are by open circles. The red dash line roughly splits the triangle into anatase growth and rutile growth regions.

The observations of spatial selectivity of silver reduction and that no preferred orientation for reactive grains indicate that there are other factors, apart from polarization and orientation, influencing the photocatalytic behavior of the TiO$_2$/BiFeO$_3$, such as chemical adsorptions on the surface. Further investigation is required to elucidate those unknown factors. Compared with orientation, which influences the activity mostly through the growth of titania phases, the polarization is more significant in determining activity of the heterstructure.

Conclusion

TiO$_2$/BiFeO$_3$ heterostructure reduces silver cations from aqueous solutions when illuminated by visible blue light with energy less than TiO$_2$'s band gap. The patterns of reduced silver on the TiO$_2$ surface mimic the domain structures in the BiFeO$_3$ substrate. The observations indicate that electrons photogenerated in the BiFeO$_3$ substrate can travel to the surface of the TiO$_2$ film where they participate in the reduction of silver. Electron transport to the surface is speculated to be favored in positive domains with polarizations pointed toward the surfaces. The anatase TiO$_2$ films are mildly more reactive than the rutile ones. Both titania phases have little preference on crystallographic orientations for higher reactivity. The photocatalytic behavior of the TiO$_2$/BiFeO$_3$ heterostructures is mainly dictated by a combination of orientation and polarization associated with the BiFeO$_3$ substrates.

References:

1. A. Fujishima and K. Honda, "Electrochemical Photolysis Of Water at a Semiconductor Electrode," *Nature*, **238** [5358] 37-38 (1972).
2. F. E. Osterloh, "Inorganic Materials as Catalysts for Photochemical Splitting of Water," *Chem Mater*, **20** [1] 35-54 (2008).
3. R. G. Breckenridge and W. R. Hosler, "Electrical Properties Of Titanium Dioxide Semiconductors," *Phys Rev*, **91** [4] 793-802 (1953).
4. J. Pascual, J. Camassel, and H. Mathieu, "Fine-Structure In the Intrinsic Absorption-Edge Of TiO_2," *Phys Rev B*, **18** [10] 5606-14 (1978).
5. H. Tang, F. Levy, H. Berger, and P. E. Schmid, "Urbach Tail Of Anatase TiO_2," *Phys Rev B*, **52** [11] 7771-74 (1995).
6. R. Asahi, T. Morikawa, T. Ohwaki, K. Aoki, and Y. Taga, "Visible-light Photocatalysis in Nitrogen-doped Titanium Oxides," *Science*, **293** [5528] 269-71 (2001).
7. D. M. Chen, Z. Y. Jiang, J. Q. Geng, Q. Wang, and D. Yang, "Carbon and Nitrogen Co-doped TiO_2 with Enhanced Visible-light Photocatalytic Activity," *Ind Eng Chem Res*, **46** [9] 2741-46 (2007).
8. X. B. Chen and C. Burda, "The Electronic Origin of the Visible-light Absorption Properties of C-, N- and S-doped TiO_2 Nanomaterials," *J Am Chem Soc*, **130** [15] 5018-19 (2008).
9. Y. Cong, J. L. Zhang, F. Chen, M. Anpo, and D. N. He, "Preparation, Photocatalytic Activity, and Mechanism of Nano-TiO_2 Co-doped with Nitrogen and Iron (III)," *J Phys Chem C*, **111** [28] 10618-23 (2007).
10. A. Ghicov, B. Schmidt, J. Kunze, and P. Schmuki, "Photoresponse in the Visible Range from Cr Doped TiO_2 Nanotubes," *Chem Phys Lett*, **433** [4-6] 323-26 (2007).
11. J. L. Gole, J. D. Stout, C. Burda, Y. B. Lou, and X. B. Chen, "Highly Efficient Formation of Visible Light Tunable $TiO_{2-x}N_x$ Photocatalysts and Their Transformation at the Nanoscale," *J Phys Chem B*, **108** [4] 1230-40 (2004).
12. W. K. Ho, J. C. Yu, and S. C. Lee, "Low-temperature Hydrothermal Synthesis of S-doped TiO_2 with Visible Light Photocatalytic Activity," *J Solid State Chem*, **179** [4] 1171-76 (2006).
13. H. Irie, Y. Watanabe, and K. Hashimoto, "Carbon-doped Anatase TiO_2 Powders as a Visible-light Sensitive Photocatalyst," *Chem Lett*, **32** [8] 772-73 (2003).
14. H. Kato and A. Kudo, "Visible-light-response and Photocatalytic Activities of TiO_2 and $SrTiO_3$ Photocatalysts Codoped with Antimony and Chromium," *J Phys Chem B*, **106** [19] 5029-34 (2002).
15. S. Klosek and D. Raftery, "Visible Light Driven V-doped TiO_2 Photocatalyst and Its Photooxidation of Ethanol," *J Phys Chem B*, **105** [14] 2815-19 (2001).
16. D. Li, H. Haneda, S. Hishita, and N. Ohashi, "Visible-light-driven N-F-codoped TiO_2 Photocatalysts. 2. Optical Characterization, Photocatalysis, and Potential Application to Air Purification," *Chem Mater*, **17** [10] 2596-602 (2005).
17. D. Li, N. Ohashi, S. Hishita, T. Kolodiazhnyi, and H. Haneda, "Origin of Visible-light-driven Photocatalysis: A Comparative Study on N/F-doped and N-F-codoped TiO_2 Powders by Means of Experimental Characterizations and Theoretical Calculations," *J Solid State Chem*, **178** [11] 3293-302 (2005).
18. H. X. Li, X. Y. Zhang, Y. N. Huo, and J. Zhu, "Supercritical Preparation of a Highly Active S-doped TiO_2 Photocatalyst for Methylene Blue Mineralization," *Environ Sci Technol*, **41** [12] 4410-14 (2007).
19. T. Morikawa, R. Asahi, T. Ohwaki, K. Aoki, and Y. Taga, "Band-gap Narrowing of Titanium Dioxide by Nitrogen Doping," *Jpn J Appl Phys 2*, **40** [6A] L561-L63 (2001).

20. S. Nahar, K. Hasegawa, and S. Kagaya, "Photocatalytic Degradation of Phenol by Visible Light Responsive Iron-doped TiO_2 and Spontaneous Sedimentation of the TiO_2 Particles," *Chemosphere,* **65** [11] 1976-82 (2006).

21. R. Nakamura, T. Tanaka, and Y. Nakato, "Mechanism for Visible Light Responses in Anodic Photocurrents at N-doped TiO_2 Film Electrodes," *J Phys Chem B,* **108** [30] 10617-20 (2004).

22. K. Nishijima, B. Ohtani, X. L. Yan, T. Kamai, T. Chiyoya, T. Tsubota, N. Murakami, and T. Ohno, "Incident Light Dependence for Photocatalytic Degradation of Acetaldehyde and Acetic Acid on S-doped and N-doped TiO_2 Photocatalysts," *Chem Phys,* **339** [1-3] 64-72 (2007).

23. T. Ohno, M. Akiyoshi, T. Umebayashi, K. Asai, T. Mitsui, and M. Matsumura, "Preparation of S-doped TiO_2 Photocatalysts and Their Photocatalytic Activities under Visible Light," *Appl Catal a-Gen,* **265** [1] 115-21 (2004).

24. T. Ohno, T. Tsubota, M. Toyofuku, and R. Inaba, "Photocatalytic Activity of a TiO_2 Photocatalyst Doped with C^{4+} and S^{4+} Ions Having a Rutile Phase under Visible Light," *Catal Lett,* **98** [4] 255-58 (2004).

25. J. H. Park, S. Kim, and A. J. Bard, "Novel Carbon-doped TiO_2 Nanotube Arrays with High Aspect Ratios for Efficient Solar Water Splitting," *Nano Lett,* **6** [1] 24-28 (2006).

26. Y. Sakatani, J. Nunoshige, H. Ando, K. Okusako, H. Koike, T. Takata, J. N. Kondo, M, Hara, and K. Domen, "Photocatalytic Decomposition of Acetaldehyde under Visible Light Irradiation over La^{3+} and N Co-doped TiO_2," *Chem Lett,* **32** [12] 1156-57 (2003).

27. S. Sakthivel, M. Janczarek, and H. Kisch, "Visible Light Activity and Photoelectrochemical Properties of Nitrogen-doped TiO_2," *J Phys Chem B,* **108** [50] 19384-87 (2004).

28. N. Serpone, D. Lawless, J. Disdier, and J. M. Herrmann, "Spectroscopic, Photoconductivity, And Photocatalytic Studies Of TiO_2 Colloids - Naked And with the Lattice Doped with Cr^{3+}, Fe^{3+}, And V^{5+} Cations," *Langmuir,* **10** [3] 643-52 (1994).

29. M. Shen, Z. Y. Wu, H. Huang, Y. K. Du, Z. G. Zou, and P. Yang, "Carbon-doped Anatase TiO_2 Obtained from TiC for Photocatalysis under Visible Light Irradiation," *Mater Lett,* **60** [5] 693-97 (2006).

30. V. Stengl and S. Bakardjieva, "Molybdenum-Doped Anatase and Its Extraordinary Photocatalytic Activity in the Degradation of Orange II in the UV and Vis Regions," *J Phys Chem C,* **114** [45] 19308-17 (2010).

31. T. Tachikawa, S. Tojo, K. Kawai, M. Endo, M. Fujitsuka, T. Ohno, K. Nishijima, Z. Miyamoto, and T. Majima, "Photocatalytic Oxidation Reactivity of Holes in the Sulfur- and Carbon-doped TiO_2 Powders Studied by Time-resolved Diffuse Reflectance Spectroscopy," *J Phys Chem B,* **108** [50] 19299-306 (2004).

32. X. H. Wang, J. G. Li, H. Kamiyama, Y. Moriyoshi, and T. Ishigaki, "Wavelength-sensitive Photocatalytic Degradation of Methyl Orange in Aqueous Suspension over Iron(III)-doped TiO_2 Nanopowders under UV and Visible Light Irradiation," *J Phys Chem B,* **110** [13] 6804-09 (2006).

33. W. J. Youngblood, S. H. A. Lee, K. Maeda, and T. E. Mallouk, "Visible Light Water Splitting Using Dye-Sensitized Oxide Semiconductors," *Accounts Chem Res,* **42** [12] 1966-73 (2009).

34. W. J. Zhang, Y. Li, S. L. Zhu, and F. H. Wang, "Surface Modification of TiO_2 Film by Iron Doping Using Reactive Magnetron Sputtering," *Chem Phys Lett,* **373** [3-4] 333-37 (2003).

35. B. O'Regan and M. Gratzel, "A Low-Cost, High-Efficiency Solar-Cell Based on Dye-Sensitized Colloidal TiO_2 Films," *Nature,* **353** [6346] 737-40 (1991).

36. F. Sauvage, D. H. Chen, P. Comte, F. Z. Huang, L. P. Heiniger, Y. B. Cheng, R. A. Caruso, and M. Graetzel, "Dye-Sensitized Solar Cells Employing a Single Film of Mesoporous TiO_2 Beads Achieve Power Conversion Efficiencies Over 10%," *Acs Nano,* **4** [8] 4420-25 (2010).

37. E. A. Kozlova, T. P. Korobkina, A. V. Vorontsov, and V. N. Parmon, "Enhancement of the O_2 or H_2 Photoproduction Rate in a Ce^{3+}/Ce^{4+}-TiO_2 System by the TiO_2 Surface and Structure Modification," *Appl Catal a-Gen,* **367** [1-2] 130-37 (2009).

38. E. A. Kozlova, T. P. Korobkina, and A. V. Vorontsov, "Overall water splitting over Pt/TiO2 Catalyst with Ce3+/Ce4+ shuttle charge transfer system," *Int J Hydrogen Energ,* **34** [1] 138-46 (2009).

39. S. R. Basu, L. W. Martin, Y. H. Chu, M. Gajek, R. Ramesh, R. C. Rai, X. Xu, and J. L. Musfeldt, "Photoconductivity in $BiFeO_3$ Thin Films," *Appl Phys Lett,* **92** [9] (2008).

40. T. Choi, S. Lee, Y. J. Choi, V. Kiryukhin, and S. W. Cheong, "Switchable Ferroelectric Diode and Photovoltaic Effect in $BiFeO_3$," *Science,* **324** [5923] 63-66 (2009).

41. F. Gao, Y. Yuan, K. F. Wang, X. Y. Chen, F. Chen, and J. M. Liu, "Preparation and Photoabsorption Characterization of $BiFeO_3$ Nanowires," *Appl Phys Lett,* **89** [10] (2006).

42. C. M. Cho, J. H. Noh, I. S. Cho, J. S. An, K. S. Hong, and J. Y. Kim, "Low-Temperature Hydrothermal Synthesis of Pure $BiFeO_3$ Nanopowders Using Triethanolamine and Their Applications as Visible-Light Photocatalysts," *J Am Ceram Soc,* **91** [11] 3753-55 (2008).

43. F. Gao, X. Y. Chen, K. B. Yin, S. Dong, Z. F. Ren, F. Yuan, T. Yu, Z. Zou, and J. M. Liu, "Visible-light Photocatalytic Properties of Weak Magnetic $BiFeO_3$ Nanoparticles," *Adv Mater,* **19** [19] 2889-+ (2007).

44. J. R. Teague, R. Gerson, and W. J. James, "Dielectric Hysteresis in Single Crystal $BiFeO_3$," *Solid State Commun,* **8** [13] 1073-74 (1970).

45. P. S. Brody, "Large Polarization-Dependent Photovoltages in Ceramic $BaTiO_3$ + 5 wt Percent $CaTiO_3$," *Solid State Commun,* **12** [7] 673-76 (1973).

46. P. S. Brody, "High-Voltage Photovoltaic Effect in Barium-Titanate And Lead Titanate Lead Zirconate Ceramics," *J Solid State Chem,* **12** [3-4] 193-200 (1975).

47. V. M. Fridkin, "Review of Recent Work on the Bulk Photovoltaic Effect in Ferro And Piezoelectrics," *Ferroelectrics,* **53** [1-4] 169-87 (1984).

48. A. Bhardwaj, N. V. Burbure, A. Gamalski, and G. S. Rohrer, "Composition Dependence of the Photochemical reduction of Ag by $Ba_{1-x}Sr_xTiO_3$," *Chem Mater,* **22** [11] 3527-34 (2010).

49. N. V. Burbure, P. A. Salvador, and G. S. Rohrer, "Influence of Dipolar Fields on the Photochemical Reactivity of Thin Titania Films on $BaTiO_3$ Substrates," *J Am Ceram Soc,* **89** [9] 2943-45 (2006).

50. N. V. Burbure, P. A. Salvador, and G. S. Rohrer, "Photochemical Reactivity of Titania Films on $BaTiO_3$ Substrates: Origin of Spatial Selectivity," *Chem Mater,* **22** [21] 5823-30 (2010).

51. N. V. Burbure, P. A. Salvador, and G. S. Rohrer, "Photochemical Reactivity of Titania Films on $BaTiO_3$ Substrates: Influence of Titania Phase and Orientation," *Chem Mater,* **22** [21] 5831-37 (2010).

52. B. G. Ershov and A. Henglein, "Time-resolved Investigation of Early Processes in the Reduction of Ag^+ on Polyacrylate in Aqueous Solution," *J Phys Chem B,* **102** [52] 10667-71 (1998).

53. B. G. Ershov and A. Henglein, "Reduction of Ag^+ on Polyacrylate Chains in Aqueous Solution," *J Phys Chem B,* **102** [52] 10663-66 (1998).

54. P. D. Fleischauer, J. R. Shepherd, and H. K. A. Kan, "Quantum Yields of Silver Ion Reduction on Titanium-Dioxide And Zinc Oxide Single-Crystals," *J Am Chem Soc,* **94** [1] 283-85 (1972).

55. A. Henglein, "Reactions of Organic Free-Radicals at Colloidal Silver in Aqueous-Solution - Electron Pool Effect and Water Decomposition," *J Phys Chem-Us,* **83** [17] 2209-16 (1979).

56. A. J. Bard, R. Parsons, and J. Jordan, *Standard Potentials in Aqueous Solution;* Marcel Dekker: New York, 1985.

57. C. C. Hsieh, K. H. Wu, J. Y. Juang, T. M. Uen, J. Y. Lin, and Y. S. Gou, "Monophasic TiO_2 Films Deposited on $SrTiO_3(100)$ by Pulsed Laser Ablation," *J Appl Phys,* **92** [5] 2518-23 (2002).

58. S. Yamamoto, T. Sumita, T. Yamaki, A. Miyashita, and H. Naramoto, "Characterization of Epitaxial TiO_2 Films Prepared by Pulsed Laser Deposition," *J Cryst Growth,* **237** 569-73 (2002).

59. J. L. Giocondi and G. S. Rohrer, "Spatial Separation of Photochemical Oxidation and Reduction Reactions on the Surface of Ferroelectric $BaTiO_3$," *J Phys Chem B,* **105** [35] 8275-77 (2001).

60. J. L. Giocondi and G. S. Rohrer, "Spatially Selective Photochemical Reduction of Silver on the Surface of Ferroelectric Barium Titanate," *Chem Mater,* **13** [2] 241-42 (2001).

61. D. Tiwari and S. Dunn, "Photochemistry on a Polarisable Semiconductor: What do we understand today?," *J Mater Sci,* **44** [19] 5063-79 (2009).

62. R. Q. Guo, L. Fang, W. Dong, F. G. Zheng, and M. R. Shen, "Magnetically Separable $BiFeO_3$ Nanoparticles with a Gamma-Fe_2O_3 Parasitic Phase: Controlled Fabrication and Enhanced Visible-light Photocatalytic Activity," *J Mater Chem,* **21** [46] 18645-52 (2011).

63. S. Li, Y. H. Lin, B. P. Zhang, C. W. Nan, and Y. Wang, "Photocatalytic and Magnetic Behaviors Observed in Nanostructured $BiFeO_3$ Particles," *J Appl Phys,* **105** [5] (2009).

64. Z. J. Shen, W. P. Chen, G. L. Yuan, J. M. Liu, Y. Wang, and H. L. W. Chan, "Hydrogen-induced Degradation in Multiferroic $BiFeO_3$ Ceramics," *Mater Lett,* **61** [22] 4354-57 (2007).

65. Y. L. Zhang, A. M. Schultz, L. Li, H. Chien, P. A. Salvador, and G. S. Rohrer, "Combinatorial Substrate Epitaxy: A High-throughput Method for Determining Phase and Orientation Relationships and Its Application to $BiFeO_3/TiO_2$ Heterostructures," *Acta Mater,* **60** [19] 6486-93 (2012).

66. P. A. M. Hotsenpiller, J. D. Bolt, W. E. Farneth, J. B. Lowekamp, and G. S. Rohrer, "Orientation Dependence of Photochemical Reactions on TiO_2 Surfaces," *J Phys Chem B,* **102** [17] 3216-26 (1998).

67. M. Kobayashi, V. Petrykin, M. Kakihana, and K. Tomita, "Hydrothermal Synthesis and Photocatalytic Activity of Whisker-Like Rutile-Type Titanium Dioxide," *J Am Ceram Soc,* **92** [1] S21-S26 (2009).

68. J. B. Lowekamp, G. S. Rohrer, P. A. M. Hotsenpiller, J. D. Bolt, and W. E. Farneth, "Anisotropic Photochemical Reactivity of Bulk TiO_2 Crystals," *J Phys Chem B,* **102** [38] 7323-27 (1998).

69. T. Ohno, K. Sarukawa, and M. Matsumura, "Crystal Faces of Rutile and Anatase TiO_2 Particles and Their Roles in Photocatalytic Reactions," *New J Chem,* **26** [9] 1167-70 (2002).

70. T. Taguchi, Y. Saito, K. Sarukawa, T. Ohno, and M. Matsumura, "Formation of New Crystal Faces on TiO_2 Particles by Treatment with Aqueous HF Solution or Hot Sulfuric Acid," *New J Chem,* **27** [9] 1304-06 (2003).

Energy Technology 2014: Carbon Dioxide Management and Other Technologies
Edited by: Cong Wang, Jan de Bakker, Cynthia K. Belt, Animesh Jha, Neale R. Neelameggham,
Soobhankar Pati, Leon H. Prentice, Gabriella Tranell, and Kyle S. Brinkman
TMS (The Minerals, Metals & Materials Society), 2014

EFFECT OF Cu THIN FILMS' THICKNESS ON THE ELECTRICAL

PARAMETERS OF METAL-POROUS SILICON DIRECT HYDROGEN

FUEL CELL

Cigdem ORUC LUS[1] and Sevinc YILDIRIM[1]

[1]Yildiz Technical University, Department of Physics, Davutpasa, Istanbul 34210, Turkey

Keywords: Porous silicon, hydrogen fuel cell, cupper catalyst film, electrical parameters.

Abstract

Metal-Porous Silicon (PS) based structures dipped in water show the direct hydrogen fuel cell behavior. The effect of the thickness of the cupper (Cu) film on the electrical properties of direct hydrogen fuel cell is studied by fabricating Cu/Porous Silicon/n-Silicon/İndium structures. PS samples were produced electrochemical etching of n-type single crystalline silicon with a resistivity of 1,25 Ωcm and (111) orientation. The thin films of Cu with different thickness between 200 and 600 nm were deposited on PS surface by electron-beam technique. Investigation of the effect of Cu layer on the performance of the structure indicates that the performance parameters such as open circuit voltage (Voc) and short circuit current (Isc) strongly depend on the Cu layer thickness. It was observed that the divice with an 350 nm Cu layer exhibits the highest open circuit voltage (0,5 V), short circuit current (0,01 mA) and power density (10 W.m^{-2}).

Introduction

Hydrogen is the most attractive and ultimate candidate for a future fuel and an energy carrier. Hydrogen is recognized as the environmentally desirable clean fuel of the future since it can be used directly in different types of hydrogen fuel cells [1]. In the past years, the rise in portable electronics requires the development of miniature fuel cells compatible with standard silicon micro fabrication technology. Most silicon micro-fuel cells have used the polyperfluo-rosulfonic acid membranes (Nafion R) as a proton exchange membranes (PEM) that is not really integrated with standard micro fabrication techniques [2-5]. The composite materials such as inorganic-organic hybrid materials (Nafion-silica, Nafion-borosiloxane etc) are next generation of proton conducting membrane [4-8]. But, fuel cells run on pure hydrogen gas, which can be produced with other energy resources. Hydrogen smallest atom and thus the H_2 gas can leak easiliy. The application of hydrogen for power sources has been greatly restrained due to the lack of safe and convenient generation and storage methods.

A kind of hydrogen fuel cells Metal-PS-Si structures used as an energy water directly. There is no need pure hydrogen. Cu-PS-Si Schottky-type fuel cells with as proton-conducting membrane and the thin-film cupper catalist and water as fuel, which are interesting for fabrication of low-cost miniature fuel cells. Generation of the open-circuit voltage and short-circuit current has been observed by Cu-PS-Si structure dipping in different hydrogen-containing solution.

Experimental Procedures

The substrates used to prepare the Cu/PS/n-Si/In structure were n-type (111) oriented silicon wafers with resistivity in the $\rho = 1.25 \times 10^{-2}$ Ωcm range. After surface cleaning, the PS layers were prepared by electrochemical anodization in a HF:H$_2$O = 1:3 (by volume) mixture solution with 15 mA.cm^{-2} current density and 55 min etching time under white light illumination. After anodization proses Cu thin film with different thickness between 200 and 600 nm were deposited onto the PS layer by electron-beam technique. The thickness of the Cu films was measured with an oscillating quartz crystal thin film monitor (Inficon, Leybold). PS layers on the Si substrates and free-standing layers were characterised by porosity (60,5%) and thickness (10-15 μm) measurements. The average porosity, i.e. the void fraction in the porous layer, was measured by gravimetry technique [9]. The evaluation of the morphological characterizations of the Cu-PS surface were performed by scanning electron microscopy (SEM). The changes in device performance, including open circuit voltage, short circuit current and power density were investigated as function of Cu film thickness. The open circuit voltage (Voc) and short-circuit current (Isc) were measured directly by a Thurbly-1503 digital multimeter.

Results

Porous silicon, in this study, was determined to be the film resistivity $\rho = 2.88 \times 10^{+4}$ Ωcm and the PS layers of 60.5% porosity were analysed. To obtain different thicknesses of metal-PS, the Cu is coated on the PS surface by the e-beam method. The Cu thicknesses are in the range of 200-600 nm. The characteristics of the 250 nm thickness Cu-PS surface is shown in Figure 1 (a). Here, some regions, especially the peak points of the hills are covered with Cu film, but the pits area haven't got Cu film. Due to the fact that Cu thin film can split the hydrogen ions from the water, these samples have low values of open-circuit voltage , short-circuit current and power density because of Cu thin film being less coated on the porous silicon surface. The Cu-PS joint coating thickness is 350 nm compared to the others, due to the extreme porosity of the metal when covered with more homogen and the pit area also has Cu metals in Figure 1 (b). Because these samples have more area with Cu coated, their values with respect to open-circuit voltage, short-circuit current and power density are higher than the others.

(a)   **(b)**

Figure 1. (a) Joint of a film thickness of 250 nm Cu-PS (5000-fold magnification), (b) Joint of a film thickness of 350 nm Cu-PS (5000-fold magnification).

The hydrogen-voltaic effect generation of a open-circuit voltage between contacts to Cu film, and Si in water exposition is observed for a Cu-PS-Si Schottky-type structure. In other words, the Cu-PS-Si structure creates the potential differences when dipping in hydrogen based liquids which occur between Cu contact and Si contact without bias voltage, but not between the contacts to the PS layer and Si substrate. We suggest that in the Cu-PS-Si cell, similar to the Proton Exchange Membrane (PEM) fuel cell, the Cu film and PS layer play the role of the catalytic anode and electrolyte, respectively. The interface region between the porous and crystalline silicon (PS-Si), which is very impact and stressed, plays the role of the cathode. Electrons and protons formed in the Cu catalyst film (anode) after hydrogen splitting (Anode: $H_2 \rightarrow 2H^+ + 2e^-$), pass through the axternal circuit and along the pore surface of the PS layer (electrolyte), and reach the cathode (PS-Si interface) region. Water molecules and oxygen from the air can easily penetrate the PS-Si interface due to imperfections in this area. Here the hydrogen is recombined and reacts with oxygen to produce water molecules (Cathode: $2H^+ + 2e^- + (1/2)O^2 \rightarrow H_2O$) [10]. The hydrogen ions (protons) act as a donor and occur throughout the PS. The protons (H^+) concentration gradient occurs in the border region, resulting in dipoles. These dipoles give rise to open circuit voltage. Because of the Cu-PS Voc joint, the creation mechanism of the hydrogen ions (protons) is connected to the diffusion limit of the Cu-PS film, and the Cu film thickness is to be investigated. To determine this theory, the PS surface is covered with Cu and Voc exchange views on the various thicknesses of Au. Voc were found to be the caused by the Cu thicknesses in experiments.

The reasons of effect of Cu film thickness to the open circuit voltage:

1. The SEM surface analysis of the Cu-PS, shows us that the less thickness of Cu films the peak points of the hills are covered with Cu, but the pits have none . Due to the fact that Cu thin film can split hydrogen ions from water, these samples has low values for open-circuit voltage (Voc), short-circuit current (Isc) and power density. Because the Cu thin film is less coated on the porous silicon surface.

2. When the Cu-film thickness increases, more area of PS surface with Cu coated a larger portion of the surface and spreads becoming more homogeneous (as SEM analysis shows). This also forms more hydrogen ions, and increases the values of the open circuit voltage (Voc), short-circuit current (Isc) and power density, which are higher than the others.

3. When the Cu metal film thickness increased greatly in the open circuit voltage, Voc was observed in the fall. The Cu film thickness in this case, increased so much that it started to disappear, and the protons could not pass through the pore structure of the reduced spaces. Therefore, very thick films, and the open circuit voltage Voc values began to fall (Figure 2 a).

There the highest value of open circuit voltage is 0,5 V. The open circuit voltage of Cu-PS-Si structure is nearly zero in the air condition.

The open circuit voltage (Voc) of the Cu/PS/n-Si cell is thought to decrease with the annealing temperature. Samples include annealing in air at 200° C for specific periods. Temperature were repeated with an interval of 10 min. With the annealing, values began to fall. Oxygen in the air before the film's surface as a result of diffusion of Cu in the film consists of CuO_2. The Cu diffusion of oxygen in the film as time progresses, the deeper regions of the cupper arrives and CuO_2 the film thickness increases with time. Cu during annealing in air the film in the direction of the joint to prevent diffusion of hydrogen into the film CuO_2. For this reason, the open circuit voltage drops in Figure 2 b. There is used thickness of Cu film 350 nm.

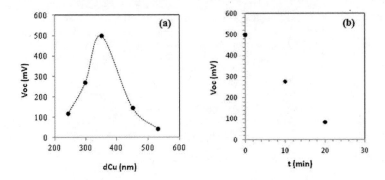

Figure 2. (a)The open circuit voltage-Cu film thickness for the Cu/PS/n-Si structure, (b) Annealing effect on the duration of the open circuit voltage (T= 200^0 C, d=350 nm).

After examining the effect of the open circuit voltage of the thickness of the metallic Cu film, the effect on the short-circuit current was also investigated. A similar change was observed on the short-circuit current. Voc, Isc values above the short-circuit shows that the flow causes the exchange of values for the same reasons, the change is shown Figure 3 (a). There the highest value of the short circuit current is 0,010 mA. The short circuit current of the Cu-PS-Si structure is nearly zero in air condition. The annealing effect of Cu film thickness o the Isc observed a similar change Figure 3 (b).

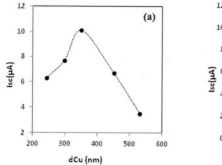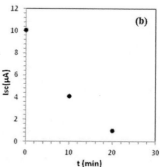

Figure 3. (a) The short circuit current Cu film thickness for the Cu/PS/n-Si structure, (b) Annealing effect on the duration of the short circuit current (T= 200^{0}C, d=350 nm).

Summary

In this study, the fabrication and characterization of Cu/PS/n-Si direct hydrogen fuel cell with porous silicon membrane and Cu catalyst thin film has been presented. The performance of Cu/PS/n-Si direct hydrogen fuel cell was measured at room temperature by water. Produced of Cu metal thin films thicknesses 200-600 nm by e-beam method and has the necessary contacts. With the film thickness increases, and more well-coated with PS and electrical properties are rised. SEM analysis of this effect in Cu-PS-Si structure is very open and supported by the observed. The best thickness of Cu thin film is 350 nm and this thickness test results confirm that about 0,5 V open circuit voltage (Voc), 0,01 mA short circuit current (Isc) and about 10 W.m^{-2} the power density. The open circuit voltage (Voc) of the Cu/PS/n-Si cell is thought to decrease by the annealing temperature. Samples include annealing in air at 200°C for specific periods. With the annealing, values began to fall. Finally, the thicknesses of the Cu film is very effective on the electrical parameters of Metal-PS-Si direct hydrogen fuel cells.

Acknowledgement

This work was partially supported by the research funds of Yildiz Technical University, Istanbul, Turkey (2011-01-01-YULAP02).

References

1. T. Dzhafarov and S.A.Yuksel, "Nano-Porous Silicon for Gas Sensor and Fuel Cell Applications", *Journal of Qafqaz University*, Number 25, (2009), 20-35.

2. N. Gyoko, I. Naohiro, T. Takaharu, Y.J. Rong, T. Koji and H. Michio, "Porous Silicon as a Proton Exchange Membrane for Micro Fuel Cells", *Electrochesmistry*, Vol 73, (2005), 939-941.

3. H. Presting, J. Konle, V. Starkov, A. Vyatkin and U. König, "Porous Silicon for Micro-sized Fuel Cell Reformer Units", *Materials Science and Engineering B*, Vol 108, (2004), 162-165.

4. T. Pichonat and B. Gauthier-Manuel, "Realization of Porous Silicon Based Miniature Fuel Cells", *Journal of Power Sources*, Vol 154, (2005), 198-201.

5. S. Gold, K.Chu, C. Lu, M.A. Shannon and R.I. Masel, Acid Loaded "Porous Silicon as a Proton Exchange Membrane for Micro-Fuel Cells", Journal of Power Sources, Vol 135, (2004), 198-203.

6. T.D. Dzhafarov, B.C. Omur, C. Oruc, Z.A Allahverdiev, "Hydrogen Characteristics of Cu-PS-Si Structures", *J. Phys. D: Appl. Phys.*, Vol 35, (2002), 3122-3126.

7. T.D. Dzhafarov, S.A. Yuksel and C.O. Lus, "Porous Silicon-Based Gas Sensors and Miniature Hydrogen Cells", *Japanese Journal of Appl. Phys.*, Vol 47, (2008), 8204-8207.

8. S. Basu, *Recent Trends in Fuel Cell Science and Technology,* (Springer, New York, 2007).

9. T.D. Dzhafarov, C. Oruc, S. Aydin, "Humidity-voltaic characteristics of Au-porous Silicon Interfaces", *J. Phys. D: App. Phy.*, 37 (2004), 404-408.

10. T.D. Dzhafarov, S. Aydin Yuksel, "Nano-Porous Silicon-Based Mini Hydrogen Fuel Cells", *Alternative Fuel*, August (2011), 309-333.

Energy Technology 2014: Carbon Dioxide Management and Other Technologies
Edited by: Cong Wang, Jan de Bakker, Cynthia K. Belt, Animesh Jha, Neale R. Neelameggham,
Soobhankar Pati, Leon H. Prentice, Gabriella Tranell, and Kyle S. Brinkman
TMS (The Minerals, Metals & Materials Society), 2014

Preparation of Silica Encapsulated Stearic Acid as Composite Phase Change Material via Sol-gel Process

Xueting Liu[1], Hao Bai[1*], Yuanyuan Wang[1], Kang Zhou[1], Hong Li[1]

[1]Department of ecological science and engineering, School of metallurgical and ecological engineering, University of science and technology 30# Xueyuan Road, Beijing, 100083, China
*Corresponding author: baihao@metall.ustb.edu.cn

Abstract

Phase change materials are widely used in efficient utilization of energy and new energy technology. Especially, more attention has been paid to composite phase material, which is applied on architectural energy-saving and efficient working fluid. In this paper, stearic acid-silicon dioxide composite material was prepared by sol-gel process from tetraethyl orthosilicate (TEOS) and stearic acid (SA). The material was investigated by Field emission scanning electron microscopy (FE-SEM), Fourier transform infrared spectroscopy (FT-IR) and Differential scanning calorimeter (DSC). The FE-SEM results show that 3D-grids of silicon dioxide were formed clearly and SA was encapsulated in the silicon dioxide shell. The FT-IR results show that there were no chemical reactions between silicon dioxide and SA, which is the chemical property of the composite material, is stable. The DSC results show that the composite phase change material has favorable heat capacity. This material is expected to be used to make a cooling working fluid.

Key words: composite phase change material; sol-gel process; stearic acid; silicon dioxide

1. Introduction

Energy thermal storage becomes increasingly prominent issue in recent years, because thermal storage has been used in many aspects, for example, it can be added into building materials to improve heat conditioning. Another potential application of thermal storage materials is that the intermittent waste heat can be collected and recycled effectively by using the thermal storage facility. Thus, energy thermal storage plays a significant role in conserving available energy and improving its utilization [1]. Energy thermal storage includes sensible heat storage and latent heat storage. The latent heat storage has great heat density with a constant temperature during the phase change process [2-5]. Large quantities of latent heat storage will be absorbed and released when phase change takes place. Based on the former researches, some phase change material (PCM), such as fatty acid and paraffin waxes, has been proved to have high latent heat storage capacity and appropriate thermal properties [5-6]. However, using the PCM in energy storage systems directly is restricted, e.g. some properties of organic PCM, like   form-stability and thermal conductivity, are not good enough to use. Meanwhile, almost all the inorganic PCMs are corrosive and most materials may cause sub cooling and segregation [7-9].The other problem is that the leakage when the temperature reaches the phase change point. Therefore, encapsulated composite phase change material (CPCM) emerged in order to solve the problem of leaking out, minimize the sub cooling and prevent phase segregation. [10-13] However, the easy flammability of fatty acid also has hindered the development of PCM. Accordingly, it has been found out that silicon compounds are environmentally friendly promising substitutes for halogen-containing flame retardants. Especially, silsesquioxanes are excellent in the flame retardancy. So CPCM can be encapsulated with SiO_2 acting supporting material. [14-16]

In this paper, stearic acid-silicon dioxide composite material was prepared by sol-gel process from tetraethyl orthosilicate (TEOS) and stearic acid (SA). In the CPCM, SA acted as the latent heat storage material. According to its proper phase change temperature, SA is a favorable organic phase change material for thermal energy storage. [17] SA melts at $(55\pm1)°C$ with a latent heat of 200.7J/g. Meanwhile, SiO_2 serves as the inorganic supporting material. The role of the SiO_2 was used to improve the shape-stability of CPCM, which has high surface area, porous structure and excellent thermal stability. [18-19] The material is expected to be used to make cooling working fluid.

Up to now, the silicon-stearic acid composite phase change material has not been widely reported. In this paper, the synthesis and properties will be discussed in detail.

2. Experimental
2.1 Materials:
Main materials: Tetraethyl silicate TEOS (analytically pure); Stearic acid SA (chemically pure).
Auxiliary materials: Absolute alcohol EtOH (analytically pure); Hydrochloric acid (analytically pure); DI water.

2.2 Preparation of composite phase change material
Silicon dioxide was prepared by using the sol-gel method. TEOS acted as the precursor. Ethanol (EtOH) was used as the solvent. Hydrochloric acid served as catalyst.

A beaker with volume of 500ml was used as a reactor. Firstly, 40g ethanol and 75g DI water were mixed together under magnetic stirring for about 5min. Then HCl was added into the mixture, adjusting the pH to 3. After that, 40g tetraethoxy silicate (TEOS) was added into the solution drop by drop under stirring (400rpm) for 60min at the temperature of 60°C in a water bath.

Table 1 The compositions of the phase change material

Samples	Compositions (g)			SA Addition (g)
	TEOS	EtOH	DI water	
M1	40	40	75	20
M2	40	40	75	30
M3	40	40	75	40
M4	40	40	75	50

When the solution turned milky, melted stearic acid was added into the mixture. Then stirring (300rpm) continued for 90min at the temperature of 75°C in a water bath.

Finally, the composite phase change material will be dried in an oven at 60°C until it is completely anhydrous.

The quantity of SA in samples were 20g,30g,40g and 50g and marked as M1~M4, respectively. The detailed composition of these samples were given in Table 1

2.3 Characterization methods of the composite phase change material

The morphology and microstructure of the silicon shell and CPCM was investigated by Field Emission Scanning Electron Microscopy (FE-SEM). The chemical structural analysis of the material was observed by Fourier Transform Infrared Spectroscopy (FT-IR). And the thermal ability was determined by Differential Scanning Calorimeter (DSC) at a heating rate of 5 °C per minute with the temperature range from room temperature to 80 °C.

3. Results and discussion

3.1 Morphology analysis of the structure of silicon dioxide and composite phase change

For the material, stearic acid was used as the phase change material. SiO_2 acted as shell, which prevented the leakage of SA. A series of samples were prepared at different mass fractions of SA (20, 30, 40 and 50g).

The morphologies of the samples were investigated by FE-SEM. Fig.1 presents FE-SEM images of structures of SiO_2. (Shown below)

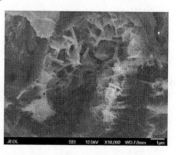

Fig.1 (a) is the FE-SEM image of SiO_2. It shows that the surface of SiO_2 is rough and porous. Hydrolysis reaction in the process of sol-gel results in the porous structure. In order to investigate the distribution of the pores, map analysis image of Si element was taken.

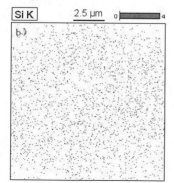

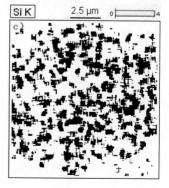

Fig. 1 FE-SEM images of SiO₂

Fig.1 (b) is the map analysis image of SiO_2. Si is characteristic element and the dots on the picture represent the distribution of Si element. It is found that the dots' distribution is not uniform and some blank areas exist, which suggests there are some pores distributed in the SiO_2. In this sense, in order to understand the size of the pores, it is necessary to measure the blank areas in the picture.

In this paper, a special statistical method is used to process Fig.1 (b) to obtain a new picture (shown as Fig.1(c)), in which the blank areas in the SiO_2 can be indentified clearly. It is found that there are many irregular pores distributed in the material. In the picture, more than 50 pores were measured and the largest one reaches 1.5 μ m in diameter, and the average diameter is about 0.7 μ m.

Fig.2 shows the FE-SEM images of composite phase change material. Fig.2 (a) ~ (d) shows the FE-SEM images of M1~M4. It is observed from the images that surface of the material is rough and porous.

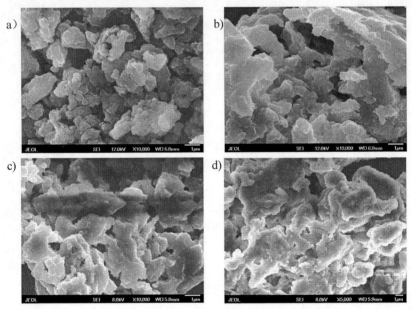

Fig.2 FE-SEM image of CPCM

3.1 FT-IR analysis of the composite phase change material

FT-IR spectroscopy can be used to reveal the interactions in the composite material. The FT-IR spectra of the SA and the composite material are shown in Fig.1. In the pure SA spectrum, there are adsorption peaks at the wave number of 2917.14cm⁻¹ and 2848.91cm⁻¹ which represent the symmetrical stretching vibration of C-H group. The peak at 1702.27cm⁻¹ is the typical absorption peak for the stretching vibration of C=O group. The peaks at 1432.46 cm⁻¹ and 1295.84 cm⁻¹ are assigned to the in-plant bending vibration of the functional group of the –OH in SA and the peak of 933.95 cm^{-1} represents the out-of-plant bending vibration of the –OH function group. Besides, there are some shoulder type absorption band ranging from 2500 cm^{-1} to 2700 cm^{-1} in wave number, which were caused by the stretching vibration of C-H, the swing vibration of O-H and the stretching vibration of C-O. The peak at 721 cm^{-1} represents the in-plant swing vibration of –OH functional group.

In the composite material spectrum, vibration bands for Si-O-Si asymmetric stretching at 1000-1100 cm^{-1}(see zone A in Fig. 3) and Si-O-Si symmetric at 782 cm^{-1} are clearly shown in Fig.3 (b). Besides, the absorption peaks at 2917 cm^{-1},2848 cm^{-1},1702 cm^{-1},1432 cm^{-1},1295 cm^{-1} and 933 cm^{-1} of the pure SA were also   found in the spectrum of CPCM. The infrared spectrum has no new peaks. It proves that there are no chemical reaction between SA and SiO$_2$. Hence, we can say, SA is well encapsulated in the structure of SiO$_2$.

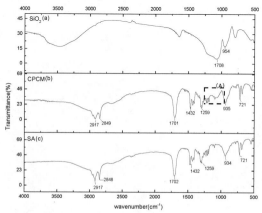

Fig.3 FTIR spectra of SiO$_2$; CPCM and SA

3.3 Heat storage ability of the composite phase change material

The phase change temperature and phase change heat were recorded by Differential Scanning Calorimeter (DSC). Ar was used as a standard for temperature calibration. The heating rate was 5℃/min from 25℃ to 80℃. The DSC curves of pure SA and CPCMs are shown in Fig.5 (a) ~ (d). And the data of pure SA and the samples of M1~M4 are recorded in Table 2. The phase change temperature of SA is 55℃ with the latent heat of 200.7J/g. But for the samples of M1~M4, the phase change temperature point changes a little,not far from 55 ℃.

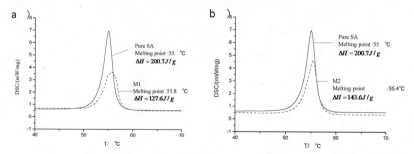

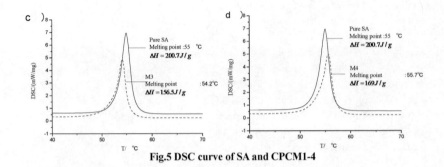

Fig.5 DSC curve of SA and CPCM1-4

The content of SA in the material has a significant effect on the heat storage ability of CPCM. The samples of M1~M4 represent that different amount of SA were added in the same content matrix material. The results show that the higher content of SA in the phase change material is, the higher phase change latent heat is.

Comparing the latent heat data of CPCM with SA, the encapsulation ratio of SA can be determined from formula (1) below. The value of η is the ratio of SA in CPCM. ΔH refers the latent heat of CPCM, and ΔH_{SA} refers to the latent heat of pure SA.

$$\eta = \frac{\Delta H}{\Delta H_{SA}} \times 100\% \tag{1}$$

The exact data of DSC are shown in Table 2. The ratio of SA in CPCM from M1 to M4 is 63.6%, 71.5%, 80% and 84.2%. The results show that the ratio of the SA in the samples increases with increase of SA adding amount in the synthesis process. The lowest level of the sample is M1, with the latent 127.6J/g and the highest level one is M4 with the latent heat of 169J/g. It still has good heat storage ability.

Table 2 DSC date of SA and M1-4

Sample	SA(g) added	Ratio (%)	Temperature(℃)	Latent(J/g)
SA	/	100	55	200.7
M1	20	63.6	55.8	127.6
M2	30	71.5	55.4	143.6
M3	40	80.0	54.2	156.5
M4	50	84.2	55.7	169

4. Potential application

The material is expected to be used to make a cooling working fluid, a kind of slurry, in which the milled fine particles of CPCM are dispersed in the liquid. The slurry can overcome traditional phase change material disadvantages of low thermal conductivity and weak fluidity. It is hopeful that the phase change slurry can be used in industry.

5. Conclusions

Stearic acid-silicon dioxide composite material was prepared by sol-gel process from tetraethyl orthosilicate (TEOS) and stearic acid (SA). The SA is the phase change material in the CPCM and the SiO_2 serves as the inorganic supporting material. There are no chemical changes between the two materials, and they are only physical integration. The melting point of CPCM is about $(55 \pm 1)°C$ with the latent heat range from 127.6 J/g to 169 J/g with the content of SA increased in the CPCM. The largest ratio of CPCM is 84.2%. The results show that the composite phase change material has favorable heat capacity. This material is expected to be used to make cooling working fluid.

References

1. M.M.Farid, A.M.Khuhair, S.A. Razack, et al, "A review on phase change energy storage: materials and application", Energy Conversion and Management, 45(2004): 1597-1615.

2. Agyenim Francis, Hewitt Neil, Eames Philip, "A review of materials, heat transfer and phase change problem formulation for latent heat thermal energy storage systems(LHTESS)", Renew Sustain Energy , 14(2010):615-28.

3. M.Li, Z.S.Wu,J., M.Tan, "Properties of form-stable paraffin/silicon dioxide/expanded graphite phase change composites prepared by sol－gel method", Applied Energy, 92 (2012) :456－461

4. Zalba B, Martin JM, Cabeza LF, Mehling H, "Review on thermal energy storage with phase change: materials, heat transfer analysis and applications.", Appl Therm Eng, 23(2003) :251－83

5. Tyagi VV, Buddhi D. PCM thermal storage in buildings: a state of art. Renew Sustain Energy 11(2007):1146－66.

6. GY.Fang, H.Li, X.Liu,"Preparation and properties of lauric acid/silicon dioxide composites as form-stable phase change materials for thermal energy storage", Materials Chemistry and Physis,122(2010):533-536.

7. Z.Chen, F.Shan, L.Cao, G.Y.Fang, "Synthesis and thermal properties of shape-stabilized lauic acid/activate carbon composites as phase change materials for thermal energy storage", Solar Energy Materials & Solar Cells, 102(2012):131-136.

8. Y.Wang, T.D.Xia, H.Zheng, H.X.Feng, "Stearic acid/silica fume composite as form-stable phase change material for thermal energy storage", Energy and Building, 43(2011)2365-2370.

9. H.Z.Zhang, X.D.Wang, D.Z.Wu, "Silicon encapsulation of n-octadecane via sol-gel progress: A novel microencapsulated phase-change material with enhanced thermal conductivity and performance", Journal of Colloid and Interface Science, 343(2010)246-255.

10. H.Ye, X.S.Ge, "Preparation of polyethylene paraffin compound as a form-stable solid–liquid phase change material", Solar Energy Materials and Solar cells, 64(2000)37-44.

11. Y.P. Zhang, K.P. Lin, R. Yang, H.F. Di, Y. Jiang," Preparation, thermal performance and

application of shape-stabilized PCM in energy efficient buildings", Energy and Buildings ,38 (2006) 1262–1269.

12. W.L. Cheng, R.M. Zhang, K. Xie, N. Liu, J. Wang, "Heat conduction enhanced shape-stabilized paraffin/HDPE composite PCMs by graphite addition: preparation and thermal properties, Solar Energy Materials and Solar Cells 94 (2010) 1636–1642.

13. Ahmet A. A, "Diesters of high-chain dicarboxylic acids with 1-tetradecanol as novel organic phase change materials for thermal energy storage", Solar Energy Materials and Solar cells104 (2012)102-108.

14. Y. Qian, P. Wei, P.K. Jiang, J.P. Liu, "Preparation of halogen-free flame retardant hybrid paraffin composites as thermal energy storage materials by in-situ sol-gel process." Solar Energy Materials and Solar cells, 107(2012)13-19.

15. P. Kiliaris, C.D. Papaspyrides, Polymer/layered silicate (clay) nanocomposites: an overview of flame retardancy, Progress in Polymer Science 35 (2010) 902–958.

16. D.K. Chattopadhyay, D.C. Webster, Thermal stability and flame retardancy of polyurethanes, Progress in Polymer Science 34 (2009) 1068–1133.

17. T.Zhang,Y.Wang, H.Shi, W. T. Yang, "Fabrication and performances of new kind microencapsulated phase change material based on stearic acid core and polycarbonate shell", Energy Coversion and Management 64(2012)1-7

18. X.Zhang, Y.Y.Wu, S.Y. He, D.Z.Yang, "Structural characterization of sol-gel composites using TEOS/MEMO as precursors", Surface and Coatings Technology 201(2007)6051-6058.

19. G.H.Zhang, C.Y.Zhao, "Thermal and rheological properties of microencapsulated phase change materials", Renewable Energy 36(2011)2959-2966.

Energy Technology 2014: Carbon Dioxide Management and Other Technologies
Edited by: Cong Wang, Jan de Bakker, Cynthia K. Belt, Animesh Jha, Neale R. Neelameggham,
Soobhankar Pati, Leon H. Prentice, Gabriella Tranell, and Kyle S. Brinkman
TMS (The Minerals, Metals & Materials Society), 2014

CELLULOSE ACETATE MEMBRANES FOR CO_2 SEPARATION FROM WGS REACTION PRODUCTS

Naidu V. Seetala[1], Upali Siriwardane[2], Tushar V. Kudale[2]

[1]Department of Mathematics and Physics, Grambling State University, Carver Hall 81,
RWE Jones Drive, Grambling, LA, 71245, USA
[2]Chemistry Program, Louisiana Tech University, Carson Taylor Hall, Room 316,
600 West Arizona Ave., Ruston, LA, 71272, USA

Keywords: Cellulose acetate, Triethyl citrate plasticizer, Gas separation membranes, Water-gas-shift reaction, Positron annihilation lifetime spectroscopy, Nano-porosity, CO_2 permeability

Abstract

Cellulose acetate (CA) membranes with 25% triethyl citrate (TEC) as plasticizer were prepared with varying thickness for CO_2 gas separation from water-gas-shift (WGS) reaction products. The AFM analysis of the CA membrane showed that the uniform coating had fewer and smaller pores as the film thickness increased, and corroborated by gas permeability studies. The CO_2 permeability has decreased faster than CO permeability with the CA/TEC membrane thickness, and findings support that the CA membrane could be used to entrap CO_2. Several CA/TEC membranes were also staked to increase the separation efficiency. Positron Annihilation Lifetime Spectroscopy (PALS) was used to estimate the nano-porosity (pore size and concentration) changes of CA/TEC films. PALS results show a decrease in pore size with increasing CA-TEC concentration in acetone, while the pore concentration increases. The PALS results are correlated with the CO_2/CO permeability variations as the CA-TEC concentration change.

Introduction

Hydrogen is one of the main energy sources available for the future advancement of clean fuel technologies especially for fuel cells. Generation of hydrogen and CO_2 from Water Gas Shift (WGS) reaction mixtures of CO and H_2O is of interest to many researchers since it allows a large scale production of hydrogen [1, 2]. Once hydrogen is produced using the WGS reaction, the output products contain mixture of CO, CO_2, H_2, and water vapor. Metallic membranes have been used as hydrogen separation membranes [3-6]. As H_2 is separated from the mixture using the metal membranes, we need to separate CO_2, and find ways of sequestration of CO_2 to avoid environmental problems by greenhouse gasses. Polymeric Membranes have been used for CO_2 separation [7]. Plasticizer can control the micro-structural properties such as pore size and concentration of polymer membranes [8], which influence the size/molecular weight dependent diffusion rates of the gas molecules. We have used cellulose acetate with triethyl citrate as plasticizer to separate CO_2 from the WGS reaction products. Positron annihilation lifetime spectroscopy (PALS) is very sensitive and able to identify different kinds of defects and their agglomerates such as vacancies, vacancy loops, and voids [9, 10]. PALS has been successful in determining the free volume in polymers that influence the properties of the polymers [11, 12]. Microscopic properties of the free volume obtained from PALS have been correlated to macroscopic properties [13, 14]. We used PALS to study the nanoporosity (pore size and concentration) and fractional free volume changes of CA/TEC films, and used to understand the variations observed in the CO_2/CO permeability as the CA-TEC concentration in the membranes change.

Experimental

Preparation of Cellulose Acetate Membranes

Cellulose acetate (CA) 75%, tri-ethyl citrate (TEC) 25% plasticizer, and acetone solution were used to drip coat a thin membrane on filter paper (S & S 100). All three chemicals were put into a conical flask and were stirred for about 15 min. until all the solid CA was dissolved into the acetone. Circular filter papers of 1.125" OD were used for CA film deposition. CA film is deposited on filter paper by simple dip solution method in which filter paper is soaked in the solution for 30 seconds and then allowed to dry overnight for permeability testing. The filter papers were also weighed before and after coating. Seven filter papers were prepared replicated 3 times for gas permeability testing. A separate batch of four membranes (Table 1) was also prepared by recoating 1, 2, 3, 4 times with (CA+TEC)/Acetone ratios: 0.2 (thicker membrane), 0.133, 0.1, and 0.067 g/mL (thinner membrane) for AFM study. Membranes prepared in this way were allowed to dry overnight.

Table 1: CA-TEC thin films prepared at different concentrations in acetone solution

CA (g)	TEC (ml)	TEC (g)	CA+TEC (g)	CA:TEC	Acetone (mL)	(CA+TEC)/Acetone (g/mL)
1.50	0.44	0.50	2.00	75%:25%	10	0.200
1.00	0.29	0.33	1.33	75%:25%	10	0.133
0.75	0.22	0.25	1.00	75%:25%	10	0.100
0.50	0.15	0.17	0.67	75%:25%	10	0.067

Another set (Table 2) of seven CA-TEC/filter-paper films with uniform thickness were prepared to study gas permeability as a function of thickness by stacking the films. They were prepared using 1g CA and 0.29 mL TEC, using a 75%:25% weight CA/TEC ratio, and 10 mL acetone solution.

Table 2: CA-TEC thin films prepared at different thickness/weight by stacking the films

Film Number	Filter-paper Weight (g)	CA+TEC deposited Paper Weight (g)	Each Film - Weight of CA+TEC (g)	Stacked Films - Net Weight of CA+TEC (g)
1	0.0552	0.0802	0.025	0.025
2	0.0550	0.0795	0.0245	0.0495
3	0.0576	0.0869	0.0293	0.0788
4	0.0577	0.0846	0.0269	0.1057
5	0.0545	0.0806	0.0261	0.1318
6	0.0516	0.0789	0.0273	0.1591
7	0.0534	0.0800	0.0266	0.1857

Atomic Force Microscope (AFM)

Agilent 5400 Atomic Force Microscope (AFM) in contact mode was used for scanning and characterization of CA/TEC films deposited on tissue paper by dip coating. The topology features of the fabricated membranes are observed to check whether the film is deposited uniformly on the filter paper using system software available [15]. The AFM tip is attached to the end of the cantilever (made up of Silicon Nitride) with a low spring constant (typically 0.001-5 nN/nm). The tip makes gentle contact with the sample, exerting from ~ 0.1 - 1000 nN force on the sample. The error signal is used as the input to a feedback circuit which, after amplification, controls the z-height piezo actuator. The initial scan size was kept to 10 × 10 microns with 256 data points per line with a scan rate of 2 lines per second.

Positron Annihilation Lifetime Spectrometer (PALS) Analysis

A standard positron annihilation lifetime spectrometer [16] was used to obtain lifetime spectra for each sample. Two identical samples were prepared for each condition. A ^{22}Na source was sandwiched between two identical samples under study and the lifetime spectrum was collected. A 16 ns delay was introduced for the time calibration of the spectrometer and found to be 0.0123 ns/ch. A ^{60}Co source that gives two gamma rays simultaneously was used to find the instrumental time resolution and found to be 35 ps. The time resolution was used to de-convolute the positron lifetime spectra of the CA+TEC polymer films into three lifetime components using POSFIT computer program [17]. The first and second lifetime components are related to positron annihilation with electrons with in polymer chains and between polymer chains, respectively, without forming a positronium (hydrogen like) atom. As the electron concentration at the annihilation site dectreases the corresponding positron lifetime increases. The electron concentration within the free volume (pores) of the polymer is so low such that a positron can find an isolated electron to combine to form a positronium atom that eventually decays to gamma rays. The third lifetime component is related to the positronium lifetime in nano-pores of these films. The positronium lifetime is proportional to the pore size. The relative intensities of these three lifetime components are proportional to the number of positrons annihilated in these three states described, which are directly related to the concentrations of these different sites available within the polymer. The third lifetime component was used to estimate the pore size, pore concentration, and fractional free volume using a simple model [14].

Results and Discussion

AFM Characterization of Cellulose Acetate Membranes

We used cellulose acetate membranes for CO_2 separation from CO/CO_2 mixture. Cellulose acetate (CA) with 25% triethyl citrate plasticizer was used to drip coat filter paper at different film thicknesses using different concentrations of CA/TEC in 10 ml acetone solution. The (CA+TEC)/acetone ratios are: 0.2 (thicker film), 0.133, 0.1, and 0.067 g/mL (thinner film). The CA/TEC film surfaces were examined by AFM and the view graphs of the surface topography are shown in Figure 1 for the 4 CA-TEC films (from the most concentrated i.e. thickest film to thinner film). The AFM pictures show that the thicker films have more uniform coating with fewer pores compared to the thinner films.

CO/CO₂ Permeability Dependence on CA/TEC Membrane Thickness

Cellulose acetate membranes with different thickness were analyzed by GC-TCD with 50%CO: 50%CO₂ gas mixture introducing in a pre-stream. The data is summarized in Table 3.

After the gas chromatogram (GC) run, the area under curve (AUC) was for the CO and CO₂ peaks and correlated with the mole fraction of CO and CO₂ gases permeated. The data (AUC) obtained is shown in Table 3 in a descending order of the CA/TEC film thickness for the (CA+TEC)/acetone ratios of 0.2 (thicker film), 0.133, 0.1, and 0.067 g/mL (thinner film).

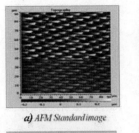

a) AFM Standard image

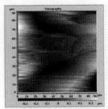

b) Membrane (0.07) image

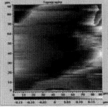

c) Membrane (0.133) image

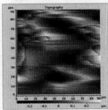

d) Membrane (0.2) image

Figure 1: AFM images of CA/TEC dip coated films

Table 3: Fractional concentrations of CO and CO₂ in the outlet gas mixture after passing 50%CO:50%CO₂ inlet gas mixture through CA-TEC membranes dip coated to different concentrations on filter paper.

No	(CA+TEC)/Acetone (g/mL)	CO (AUC)	CO₂ (AUC)	CO (conc.)	CO₂ (conc.)
1	0.200 (thicker membrane)	0.08	0.14	0.44	0.18
2	0.133	0.12	0.35	0.66	0.46
3	0.100	0.12	0.41	0.64	0.53
4	0.067 (thinner membrane)	0.08	0.38	0.44	0.50
5	0 (blank filter paper)	0.09	0.38	0.50	0.50

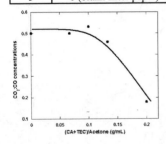

Figure 2: Permeability of CO₂ shown as CO₂:CO ratio after passing 50%CO₂:50%CO mixture through CA/TEC membranes.

42

The CO_2:CO concentration ratios from the GC analysis of outlet gas products after passing 50%CO_2+50%CO mixture through the CA/TEC membranes deposited on filter paper is shown in Figure 2. The results clearly indicate that the CO_2 permeability is reducing faster than CO permeability as the thickness of the CA/TEC membrane increases. From Figure 2 it can be inferred that, there is an increase in the CO permeability (i.e. CO enrichment in the product outlet) with the increase in the CA/TEC membrane thickness. So, we can easily conclude that there is decrease in CO_2 permeability. This proves that CA membranes can entrap CO_2.

N_2/CO/CO_2 Permeability as a Function of Number of CA-TEC Layers

Table 4: Fractional concentrations of N_2, CO, and CO_2 in the outlet gas mixture after passing equal amounts of N_2, CO, and CO_2 gasses through CA-TEC membrane layers.

Membrane layers	Weight of layers (g)	Conc. N_2	Conc. CO	Conc. CO_2
1	0.025	0.026	0.010	0.263
2	0.050	0.019	0.012	0.287
3	0.079	0.026	0.022	0.220
4	0.106	0.019	0.024	0.182
5	0.132	0.022	0.019	0.205
6	0.159	0.018	0.024	0.151
7	0.186	0.024	0.029	0.162

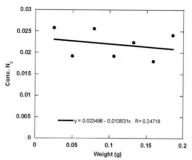

Figure 3: Permeability of N_2 vs. number of CA membranes (weight)

The results of N_2, CO, and CO_2 gas permeability through multiple layers of CA-TEC membranes are presented in Table 4 as a function of number of layers (weight). The permeability curves (Figures 3-5) show the variation of the post–stream concentration of N_2, CO, and CO_2 with time. CA-TEC membrane has almost negligible effect on N_2 with almost 100% permeable. It can be inferred from the chromatograms that there is CO enrichment and CO_2 entrapment as the number of membranes is increased.

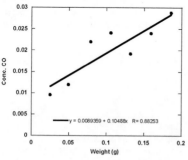

Figure 4: Permeability of CO vs. number of CA membranes (weight)

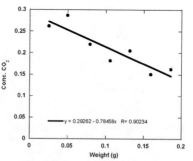

Figure 5: Permeability of CO_2 vs. number of CA membranes (weight)

Positron Lifetime/Nano-porosity Studies of CA-TEC Membranes

The nano-porosity and free volume changes of CA with 25% TEC as a function of its concentration in fixed amount of acetone were studied using Positron Annihilation Lifetime Spectroscopy (PALS) techniques. The third lifetime component associated with positronium annihilation in the polymer is used to estimate the pore size, pore concentration, and fractional free volume using a simple model [14]. Figure 6 shows the positron lifetime spectrum of CA-25%TEC compared with the lifetime spectrum for cellulose acetate butyrate (CAB) with 60%diacetin plasticizer. This shows that CA-TEC film is less porous compared to CAB film. The third lifetime component associated with positronium annihilation in the polymer

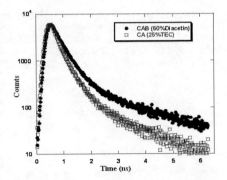

Figure 6: Positron lifetime spectra for (□) CA-25%TEC and (•) CAB-60%diacetin

is used to estimate the pore size, pore concentration, and fractional free volume using a simple model [14].

Table 5: Positron lifetime results of CA-TEC membranes dip coated on filter paper

CA-TEC/Acetone concentration	τ_1 (ns)	τ_2 (ns)	τ_3 (ns)	I_1 (%)	I_2 (%)	I_3 (%)
0.2	0.190	0.401	1.794	42.12	46.92	10.96
0.133	0.178	0.410	1.881	46.90	44.31	8.79
0.1	0.183	0.414	1.929	47.89	43.48	8.63
0.067	0.180	0.430	1.944	50.15	41.54	8.32

The POSFIT computer program analysis of the lifetime spectra for CA-TEC films provided three lifetimes. The variations in the lifetimes and related intensities with the CA-TEC/Acetone concentration are shown in Table 5. The third lifetime component is related to positronium annihilation in the pores, where τ_3 is proportional to the average pore size and I_3 is proportional to the pore concentration. The variations in these parameters listed in Table 4 show a decrease in pore size with increasing CA-TEC concentration in acetone, while the pore concentration increases. A simple model [14] is used to estimate the pore size (R) and the fractional free volume (F_v).

Pore size R is estimated by: $\tau_3^{-1} = 2[1-R/Ro + \{1/2\pi\}\sin(2\delta R/Ro)]$ (1)
Where $Ro = R + \delta R$ and $\delta R = 1.66$ Å
And the free volume fraction Fv by: $F_v = C * V_f * I_3$.   (2)
Where C (1/400) is a constant [18], and V_f is pore volume.

The estimates of pore size and fractional free volume are shown in Figure 7. The pore size decreases with the CA-TEC concentration in acetone, while the overall fractional free volume shows an increase with the CA-TEC concentration. These results may explain the variations observed in the CO_2/CO permeabilities. The CO_2 permeability decreased faster than CO permeability with the CA-TEC concentration, which agrees with the decrease in pore size shown by PALS results. Decrease in pore size might be reducing the CO_2 diffusion as the CA-TEC concentration increases.

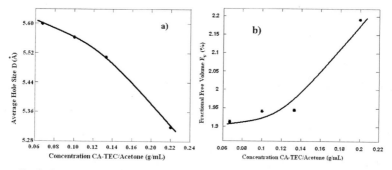

Figure 7: a) Average pore size and b) fractional free volume as a function of CA-TEC concentration in acetone.

Conclusions

Cellulose acetate (CA) films with 25wt% triethyl citrate (TEC) plasticizer were prepared for CO_2 separation with varying thickness of the films by mixing in 10 ml acetone at different concentrations and dip-coating onto filter papers. The Atomic Force microscope observations of these membranes show that the thicker films have more uniform coating with fewer pores compared to the thinner films. The CO/CO_2 permeability studies were performed using the prototype stainless steel reactor housing for the membranes with 50%CO:50%CO_2 gas mixture in the inlet side and a GC analyzer at the outlet side. The GC results clearly indicate that the CO_2 permeability is reducing faster than CO permeability as the thickness of the CA/TEC membrane increases. This may infer that CA membranes may be used to entrap CO_2. The N_2, CO, and CO_2 gas permeabilities through multiple layers of CA membranes as a function of number of layers (weight) show almost negligible effect on N_2 while there is CO enrichment and CO_2 entrapment as the number of membranes increased. The positron lifetime (PALS) analysis provided an estimate of fractional free volume in these films. The pore size decreased with the CA-TEC concentration in acetone, while the overall fractional free volume showed an increase with the CA-TEC concentration. These results may explain the variations observed in the CO_2/CO permeabilities. Decrease in pore size shown by PALS might be the cause for reducing the CO_2 diffusion as the CA-TEC concentration increased.

Acknowledgements

Work is supported by the Department of Energy (DOE), grant#DE-FG26-07NT43064.

References

1. H. M. Cheng, et al., "Hydrogen storage in carbon nano-tubes," *Carbon*, 39 (2001) 1447-1454.

2. O. Z. Ilsen, "Catalytic processes for clean hydrogen production from hydrocarbons," *Turkish Journal of Chemistry*, 31, (2007) 531-550.

3. H. Gao, et al., "Chemical Stability and Its Improvement of Palladium-Based Metallic Membranes," *Ind. Eng. Chem. Res.*, 43 (2004) 6920-6930.

4. J. M. Armor, "Applications of catalytic inorganic membrane reactors to refinery products," *J. Membr. Sci.* 147 (1998) 217-233.

5. T. Moss, et al., "Multilayer Metal Membranes for Hydrogen Separation," *Intl. J. Hydrogen Energy*, 23 (1998) 99-106.

6. T. Ozaki, et al., "Hydrogen permeation characteristics of V-Ni-Al alloys", *Int. J. Hydrogen Energy* 28 (2003) 1229-1235.

7. C. Scholes, S. Kentish, and G. Stevens, "Carbon Dioxide Separation through Polymeric Membrane Systems for Flue Gas Applications," *Recent Patents on Chemical Engineering*, 1 (2008) 52-66; "Effects of Minor Components in Carbon Dioxide Capture Using Polymeric Gas Separation Membranes", *Separation & Purification Reviews* 38 (2009) 1 – 44.

8. S. V. Naidu, et al., "Effect of Plasticizer on Free Volume and Permeability in Cellulose Acetate Pseudolatex Membranes Studies by Positron Annihilation and Tracer Diffusion Methods", *Mat. Sci. Forum*, 255-257 (1997) 333-336.

9. T. E. M. Staab, et al., "Irradiation experiment revisited – Stability and positron lifetime of large vacancy clusters in silicon," *Materials Science Forum*, 363-365 (2001) 135-137.

10. S. C. Sharma, et al., "Positron annihilation and conductivity measurements on poly(pyrrole tosylate) and poly(pyrrole fluoride)," *Phys. Rev.*, B41 (1990) 5258-5265.

11. T. Hiradea and T. Kumadab, "The effect of γ-irradiation on positronium formation in polyethylene," *Radiation Physics and Chemistry*, 60 (2001) 541–544.

12. Y. Y. Wang, et al., "Positron annihilation in amine-cured epoxy polymers - pressure dependence," *J. Polym. Sci.*, B28 (1990) 1431- 1441.

13. Y. C. Jean, et al., "Anisotropy of free-volume hole dimensions in polymers probed by positron annihilation spectroscopy," *Phys. Rev.* B42 (1990) 9705-9708.

14. Y. C. Jean, "Free-volume properties of polymers probed by positron annihilation spectroscopy," *Mater. Sci. Forum*, 105-110 (1992) 309-316.

15. D. F. Stamatialis, C. R. Dias, and M. N. de Pinho, "Atomic force microscopy of dense and asymmetric cellulose-based membranes", *Journal of Membrane Science*, 160 (1999) 235-242.

16. N. V. Seetala, et al., "Positron lifetime and SEM studies of porous silica," *World Journal of Engineering* 10 (2013) 199-204.

17. P. Kirkegaard, N. J. Pedersen, and M. Eldrup, POSFIT computer program from Risø National Laboratory (1989).

18. Y. C. Jean, "PAS for chemical analysis: a novel probe for microstructural analysis of polymers", *Microchem. J.*, 42 (1990) 72-102.

Energy Technology 2014
Carbon Dioxide Management and Other Technologies

SYMPOSIUM: ENERGY TECHNOLOGIES AND
CARBON DIOXIDE MANAGEMENT

Energy in Iron and Steel

Session Chairs:

Wanlin Wang
Il Song

Energy Technology 2014: Carbon Dioxide Management and Other Technologies
Edited by: Cong Wang, Jan de Bakker, Cynthia K. Belt, Animesh Jha, Neale R. Neelameggham,
Soobhankar Pati, Leon H. Prentice, Gabriella Tranell, and Kyle S. Brinkman
TMS (The Minerals, Metals & Materials Society), 2014

DETERMINATION OF ENERGY REQUIREMENTS FOR IRONMAKING PROCESSES: IT'S NOT THAT STRAIGHTFORWARD

Hong Yong Sohn and Miguel Olivas-Martinez

Department of Metallurgical Engineering, University of Utah;
135 S. 1460 E., Rm 412; Salt Lake City, UT 84112-0114, USA

Keywords: Energy Requirement, Flash Ironmaking, Blast Furnace

Abstract

Different approaches are used for calculating the 'energy requirement' of a process. The determination of energy requirement becomes confusing for a process in which a fuel is also a reactant. The key issue is which heat of chemical reaction should be included in the 'energy requirement' value. We will use the example of ironmaking processes to illustrate the problem. The different approaches essentially boil down to the question: does one include the combustion heat of the reductant used in the reduction reaction or just the heat of the reduction reaction? Depending on the viewpoint, either approach can be accepted; it is a matter of convention. There is a need, however, to select a standard approach because the absolute value of 'energy requirement' of a process depends on the choice. The energy requirement of a novel flash ironmaking process will be compared with that of an average blast furnace operation.

Introduction

Different approaches are used for performing energy balance calculations and perhaps more significantly in presenting the 'energy requirement' of a process. The question becomes more involved for a process in which one or more reactants are also used to generate the process heat by combustion. Often the key issue is which heat of chemical reaction to include in presenting the 'energy requirement' of the process.

It is worthwhile to note at the outset that the difference in energy requirements between different processes for converting similar raw materials to the same desired product is largely unaffected by the choice of chemical reactions to be included. However, the absolute value presented as the 'energy requirement' of a process can be different by a large amount depending on the approach. We use the example of ironmaking processes to illustrate the problem. In these processes, as currently practiced or under development, carbon and/or hydrocarbons (including hydrogen) are used as the fuel as well as the reactant for reducing the iron oxide mineral.

The different approaches in this case essentially boil down to the following question: does one include the combustion heat of the reductant used in the reduction reaction or treat the reductant portion as just a reactant? Depending on the viewpoint, either approach can be considered acceptable. It is a matter of convention. There is a need, however, to select a standard approach because the absolute value of 'energy requirement' of an ironmaking process (e.g., the currently dominant blast furnace process) depends on such a choice. It is particularly important to clearly state the specific approach used when the energy requirement of ironmaking is compared with those of other industrial processes such as petrochemicals production.

Determination of Energy Requirement in Ironmaking Processes: Chemical Aspects

Definitions of Chemical Energy Terms

To reduce iron oxides to iron, the energy requirement for the reduction reaction can be defined in two ways: One is to consider the reduction as decomposing iron oxide to iron and oxygen (reverse of the energy of formation of iron oxide); the other is to consider it as a reaction of iron oxide with a reductant. [As shown below, the former is equivalent to including the heat of combustion of the reactant acting as a reductant in the total energy requirement, whereas the latter is equivalent to not including it.] The following is a simplified demonstration of these equivalences.

The demonstration is made just for the energy involved with chemical changes at a reference temperature (usually 298 K), excluding sensible heat which does not affect the argument discussed here.

Basis of the Simplified Demonstration

1. The solid charge is composed of pure iron oxide. No flux materials are considered.
2. Carbon is used as the fuel/reductant. However, this methodology can easily be extended to the case in which hydrogen or a hydrocarbon is used as the fuel/reductant.
3. The total carbon input is divided into two groups: fuel carbon and reductant carbon.
4. The heat balance is performed over the entire system. The process outputs are iron and the complete combustion product (CO_2) of fuel carbon plus the reduction product (CO_2), all at 298 K.

Chemical Reactions

The overall chemical reaction taking place to reduce iron oxide to iron using carbon as the fuel/reductant is

$$[Fe_3O_4 + 2(C)_r] + [x(C)_f + x(O_2)_f] = 3Fe + 2(CO_2)_r + x[CO_2]_f \qquad (1)$$

where $(C)_r$ is the carbon consumed by the reduction reaction, $(C)_f$ is the carbon burned to generate heat, $(O_2)_f$ is the oxygen for reaction with the fuel carbon, $(CO_2)_f$ is the carbon dioxide produced by the reduction reaction, $(CO_2)_f$ is the carbon dioxide produced by combustion of the fuel carbon, and x is the number of moles of the fuel portion of carbon required to generate the heat for the reduction process.

The enthalpy change of the overall reaction (1) is a combination of the energy required to reduce iron oxide and the energy produced by combustion of the fuel/reductant. As mentioned in the Introduction, different definitions of 'energy requirement' are possible depending on what one considers as the reduction reaction between the following two reactions:

$$Fe_3O_4 = 3Fe + 2(O_2)_r \qquad (2)$$

and

$$Fe_3O_4 + 2(C)_r = 3Fe + 2(CO_2)_r \qquad (3)$$

Approach 1: Energy Requirement for Reduction Reaction Based on Oxide Decomposition

When the reduction of iron oxide is attributed to the decomposition of the iron oxide [Equation (2)], the energy requirement for reduction process is the difference between the heats of reactions (1) and (2), which is equivalent to the heat of the following combustion reaction:

$$[2(C)_r + x(C)_f] + [2(O_2)_r + x(O_2)_f] = [2(CO_2)_r + x(CO_2)_f] \qquad (4)$$

The heat of this reaction corresponds to the chemical heating value of the total amount of carbon supplied to the system. A number of reports [1-3] on energy use in the ironmaking industry follow this approach.

Approach 2: Energy Requirement Based on Oxide Reaction with Reductant

When the reduction of iron oxide is attributed to the reaction of iron oxide with carbon [Equation (3)], the energy requirement for reduction reaction is the difference between the heats of reactions (1) and (3), which is equivalent to the heat of the following combustion reaction:

$$x(C)_f + x(O_2)_f = x(CO_2)_f \tag{5}$$

The heat of this combustion reaction corresponds to the chemical heating value of just the fuel portion of carbon $(C)_f$. Some energy balance calculations [4] use this approach.

In practice, the reduction of an iron oxide occurs by contacting it with a reductant (e.g., C), that is, reaction (3). In such a case, $(C)_r$ may be considered just as another reactant regardless of its chemical energy content and thus, the energy requirement for such a process includes only the energy required by reaction (3), the sensible heat of iron and off-gas, and the process heat losses.

Results

The application of the above two approaches to the calculation of energy requirements is illustrated using a novel flash ironmaking process under development at the University of Utah and an average blast furnace operation. A material and energy flow diagram for the novel flash ironmaking process using natural gas, prepared using the commercially available software METSIM [5], is shown in Figure 1.

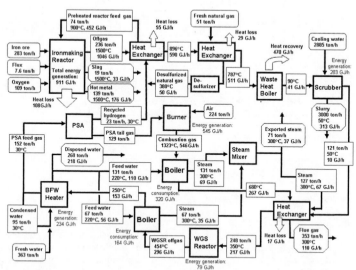

Figure 1. Material and energy flows for a commercial-scale reformerless flash ironmaking process producing 1 million tons of iron per year at 1773 K (300 days operation in 1 year). Adapted from Pinegar and coworkers [5].

51

Material and Energy Flows: System Boundary

Energy balance calculations start with material balance around a clearly defined system boundary. A system boundary can be drawn around different parts of a process, depending on the purpose of the calculation. An illustration is given in Figure 2. The solid lines represent material flow with associated sensible and latent heats, and the open arrows indicate the flow of heat only. Once the boundary is defined, any open-ended streams that cross it are the input and output streams. Any streams entirely inside a boundary (e.g., the off-gas stream in Figure 2 in the case of the outer boundary) are not included in the balance calculations around that boundary.

It is noted that the heat contents of any streams that can be recovered for a useful purpose such as steam could be credited to reduce the input amount of energy or 'energy requirement'. Heat losses and the sensible heat of a material stream that are not recovered are output items, and increase the input amount and thus the 'energy requirement'. Further, all heats of chemical reaction (chemical heat contents), including heat of combustion, are calculated at 298 K and all sensible heat in input or output streams is calculated relative to this temperature.

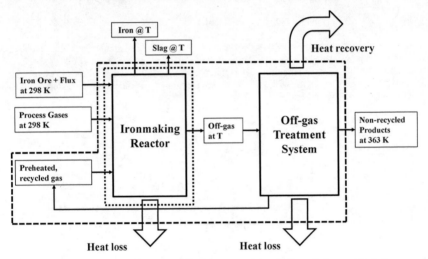

Figure 2. Examples of system boundaries for material and energy balance calculations for an ironmaking process.

Energy Balances for Ironmaking Processes

Tables I and II present energy inputs and outputs, respectively, in terms of Approaches 1 and 2 for the flash ironmaking process using natural gas or hydrogen and for an average blast furnace operation. The annual production rates in all the cases are 1 million tons of iron from a single unit of ironmaking reactor. The energy balance for the flash ironmaking process is based on the flow sheet presented in Figure 1. The energy balance for an average blast furnace operation was calculated using published material balance data and applying the same method of calculating energy values as for the flash ironmaking process.

Table I. Energy Input Items for a Commercial-Scale Flash Ironmaking Process Using Natural Gas or Hydrogen Compared with Those for an Average Blast Furnace Process. [Production rate is equal to 1 million tons of iron] Adapted from Pinegar and coworkers [5, 6]

	Process	Reformerless Natural Gas		Hydrogen		Blast Furnace[a]	
	Approach	1	2	1	2	1	2
ITEMIZED INPUT (GJ/mtFe)	Fuel combustion[b]	19.22	13.45	14.05	8.28	13.60	8.33
	Heat recovery (sum of next 2)	- 4.77		-2.80		- 1.32	
	(Waste heat boiler)	(-3.39)					
	(Steam not used)	(-1.38)					
	Sub-total	14.45	8.68	11.25	5.48	12.28	7.01
	Ore/Coke preparation[c]					5.68	
	$CaCO_3$ and $MgCO_3$ calcination (external)	0.26		0.26			
	Total	14.71	8.94	11.51	5.74	17.96	12.69

[a] Energy balance was calculated by METSIM based on the published material balance.
[b] Fuel combustion energy input was calculated by adding the energy (subtracting reaction enthalpy) for iron ore reduction from the difference of heats of formation of all output components and input components [Equation (1)]. This amount is equivalent to the heat of combustion (4) for Approach 1 and reaction (5) for Approach 2.
[c] From Fruehan and coworkers [2] and Stubbles [3].

For the blast furnace process, the energy consumed by sintering, pelletizing and cokemaking operations, which are not required in flash ironmaking, is part of the energy requirement. For the flash ironmaking process, the cases of using two different reductants/fuels – natural gas or hydrogen – are presented. The results of applying Approach 2 were adapted from Pinegar and coworkers [5, 6] and those of applying Approach 1 were calculated by modifying the 'Fuel combustion' term in the input category and the 'Reduction' term in the output category, which are the only numbers that are different between Approaches 1 and 2.

In terms of the descriptions of Approaches 1 and 2 formulated above, the 'fuel combustion' energy input was calculated by adding the energy (subtracting reaction enthalpy) for iron ore reduction from the difference of heats of formation of all output components and input components [Equation (1)]. This amount is equivalent to the heat of combustion of reaction (4) for Approach 1 and reaction (5) for Approach 2.

The 'reduction' energy on the output category corresponds to the decomposition of magnetite (Fe_3O_4), reaction (2) (6.68 GJ/mtFe), for Approach 1 for flash ironmaking process. For the blast furnace, it corresponds to the heat of decomposition of hematite (Fe_2O_3) (7.37 GJ/mtFe). [Magnetite concentrate must be pelletized or sintered to be used in a blast furnace during which it is converted to hematite, generating heat. This heat generation is reflected in the energy requirement for pelletization. Thus, it is appropriate to use hematite as the feed material in blast furnace operations.] In Approach 2, 'reduction' energy corresponds to the heat of reduction reaction (3) (0.91 GJ/mtFe) for flash ironmaking and it is the heat of the reaction $Fe_2O_3 + 3/2C = 2Fe + 3/2CO_2$ for the blast furnace.

The energy requirement calculated using Approach 2 is the 'process energy requirement' and the difference between the total input energy by Approach 1 and the process energy requirement may be termed 'reductant energy' ('feedstock energy' in petrochemicals production [7]).

Table II. Energy Output Items for a Commercial-Scale Flash Ironmaking Process Using Natural Gas or Hydrogen Compared with Those for an Average Blast Furnace Process. [Production rate is equal to 1 million tons of iron] Adapted from Pinegar and coworkers [5, 6]

	Process	Reformerless Natural Gas		Hydrogen		Blast Furnace[a]	
	Approach	1	2	1	2	1	2
ITEMIZED OUTPUT (GJ/mtFe)	Reduction[b]	6.68	0.91	6.68	0.91	7.37	2.10
	Sensible heat of iron	1.27 (1773 K)				1.35 (1873 K)	
	Sensible heat of slag	0.24 (1773 K)				0.47 (1873 K)	
	Slurry (H_2O (l))	2.25 (323 K)		1.93			
	Hot water not used	1.57 (493 K)					
	Flue gas	0.79 (573 K)				0.26 (363 K)	
	Removed water vapor			0.01			
	$CaCO_3$ decomposition					0.33	
	Slagmaking					-0.17	
	Heat loss in the reactor	0.78		0.78		2.60	
	Heat loss in the heat exchangers (sum of next 3)	0.73		0.34		0.07	
	(Reactor feed gas heater)	(0.40)					
	(Natural gas heater)	(0.21)					
	(WGS reactor feed gas heater)	(0.12)					
	Steam not used (363 K)	0.14					
	Sub-total	**14.45**	**8.68**	**11.25**	**5.48**	**12.28**	**7.01**
	Pelletizing[c]					3.01	
	Sintering[c]					0.65	
	Cokemaking[c]					2.02	
	$CaCO_3$ and $MgCO_3$ calcination (external)	0.26		0.26			
	Total	**14.71**	**8.94**	**11.51**	**5.74**	**17.96**	**12.69**

[a] Energy balance was calculated by METSIM based on the published material balance.
[b] For flash ironmaking process, the reduction energy corresponds to the decomposition of magnetite (2) (6.68 GJ/mtFe) for Approach 1. For the blast furnace, it corresponds to the heat of decomposition of Fe_2O_3 (7.37 GJ/mtFe). In Approach 2, reduction energy corresponds to the heat of reduction reaction (3) (0.91 GJ/mtFe) for flash ironmaking and it is the heat of the reaction $Fe_2O_3 + 3/2C = 2Fe + 3/2CO_2$ for the blast furnace.
[c] From Fruehan and coworkers [2] and Stubbles [3].

Discussion

As can be seen in Tables I and II, the difference in the energy requirements between each ironmaking process and the blast furnace remains largely the same regardless of the treatment given to the fuel used as a reductant.

The 'energy requirements' of these processes can be obtained directly from the above energy inputs (Table I) and outputs (Table II) from the total amounts. This is possible because in Table I the recovered heat is presented as a negative input item. The energy balance results can be presented with such an item listed as a positive output item in Table II, in which case the 'balance' is still achieved. The presentation of the 'energy requirement' value, however, may be somewhat confusing. Thus, it may be clearer to present the result of energy requirement calculation following the format shown in Table III, as is used by some authors (e.g., Remus and coworkers [1]).

Table III. Energy Balance for a Commercial-Scale Flash Ironmaking Process Using Natural Gas or Hydrogen Compared with Energy Balance for an Average Blast Furnace Process

	Process	Reformerless Natural Gas		Hydrogen		Blast Furnace	
	Approach	1	2	1	2	1	2
Energy Input (GJ/mtFe)	Fuel combustion	19.22	13.45	14.05	8.28	13.60	8.33
	Pelletizing					3.01	
	Sintering					0.65	
	Cokemaking					2.02	
	Limestone calcination (external)	0.26					
	Total	19.48	13.71	14.31	8.54	19.28	14.01
Energy Output (GJ/mtFe)	Heat recovery	4.77		2.80		1.32	
	Total	4.77		2.80		1.32	
Net Requirement = Input – Output (GJ/mtFe)		**14.71**	**8.94**	**11.51**	**5.74**	**17.96**	**12.69**

In some reports [2, 8], the energy required for decomposing iron oxide to iron and oxygen, reaction (2) is used as the 'theoretically lowest' or 'theoretical minimum' energy requirement. It is noted that this reaction, which does not actually take place under ironmaking processes, involves the largest enthalpy change (energy requirement) of any chemical reactions involved in the production of iron from iron oxide. This is used as an item of energy requirement in Approach 1 above. In Approach 2, the endothermic heat of the reduction reaction [i.e., reaction (3)] is used for calculating energy requirement.

A simple food-for-thought question: If we consider a process in which the reaction
$$2/3Fe_2O_3 + C = 4/3Fe + CO_2 \qquad (6)$$
occurs at room temperature, what do we consider as the energy requirement for the process? Is it the heat of reaction (6), or should it include the heat of combustion of the carbon used as a reactant because it has a heating value?

Concluding Remarks

When calculating 'energy requirements' and presenting the results, it is essential to clearly state the approach used. Specifically and most importantly, it should be stated whether or not the chemical heat content (the heat of combustion) of a reactant that can also be used as a fuel is included in the calculated value of energy requirement. Even with these clarifications of approaches, it will make it much clearer to present distinct values for 'process energy requirement' that includes only the heating value of the fuel and 'reductant energy' ('feedstock energy' in petrochemicals production) that represents the chemical heating value of the material used as a reactant.

In addition, a statement regarding whether the higher heating value (HHV; assumes liquid water in the combustion products) or the lower heating value (LHV; assumes water vapor in the combustion products at 298 K) should be included when the fuel contains hydrogen. Another item of information often neglected that must be provided is the system boundary around which the material and energy balances are performed; that is, clear definitions of input and output streams and conditions.

Acknowledgments

The authors thank Dr. D. R. Forrest, Technology Manager, Advanced Manufacturing Office, U.S. Department of Energy for raising the issues regarding differences in calculating energy requirements for different ironmaking processes. We also thank J. Cresko of DOE for suggesting that we prepare a written record of this issue. This material contains results of work supported by the U.S. Department of Energy under Award Number DE-EE0005751.

References

1. R. Remus, M.A. Aguado Monsonet, S. Roudier, and L. Delgado Sancho, "Best Available Techniques (BAT) Reference Document for Iron and Steel Production. Industrial Emissions Directive 2010/75/EU (Integrated Pollution Prevention and Control)," Joint Research Centre (JRC) of the European Union: Spain, 2013, 304-305.

2. R.J. Fruehan, O. Fortini, H.W. Paxton, and R. Brindle, "Theoretical Minimum Energies to Product Steel for Selected Conditions," March 2000, U.S. Department of Energy, http://www1.eere.energy.gov/industry/steel/pdfs/theoretical_minimum_energies.pdf (accessed April 30, 2011).

3. J. Stubbles, "Energy Use in the U.S. Steel Industry: An Historical Perspective and Future Opportunities," September 2000, U.S. Department of Energy, http://www1.eere.energy.gov/manufacturing/resources/steel/pdfs/steel_energy_use.pdf (accessed September 17, 2013).

4. J.A. Burgo, "The Manufacture of Pig Iron in the Blast Furnace," *The Making, Shaping and Treating of Steel*, 11th ed., D.H. Wakelin Ed. (Pittsburgh, PA: The AISE Steel Foundation, 1999) 712-713.

5. H.K Pinegar, M.S. Moats, and H.Y. Sohn, "Flow Sheet Development, Process Simulation and Economic Feasibility Analysis for a Novel Suspension Ironmaking Technology Based on Natural Gas: Part I. Flow Sheet and Simulation for Ironmaking with Reformerless Natural Gas," *Ironmaking Steelmaking*, 39 (2012), 398-408.

6. H.K. Pinegar, M.S. Moats, and H.Y. Sohn, "Process Simulation and Economic Feasibility Analysis for a Hydrogen-Based Novel Suspension Ironmaking Technology," *Steel Res. Int.*, 82, (8) (2011), 951–963.

7. N. Santero, "Feedstock Energy in Bitumen," http://www.ucprc.ucdavis.edu/P-LCA/pdf/04_feedstock_web.pdf (accessed May 7, 2013).

8. J de Beer, E. Worrell, and K. Blok, "Future Technologies for Energy-Efficient Iron and Steel Making,' *Annu. Rev. Energy*, 23 (1998), 123–205.

Energy Technology 2014: Carbon Dioxide Management and Other Technologies
Edited by: Cong Wang, Jan de Bakker, Cynthia K. Belt, Animesh Jha, Neale R. Neelameggham,
Soobhankar Pati, Leon H. Prentice, Gabriella Tranell, and Kyle S. Brinkman
TMS (The Minerals, Metals & Materials Society), 2014

BLAST FURNACE IRONMAKING: PROCESS ALTERNATIVES AND CARBON INTENSITY

P. Chris Pistorius, Jorge Gibson and Megha Jampani

Center for Iron and Steelmaking Research, Department of Materials Science and Engineering, Carnegie Mellon University, 5000 Forbes Avenue, Pittsburgh PA 15213, USA

Keywords: Blast furnace, Ironmaking, Carbon intensity

Abstract

The main reason for the significant carbon intensity of integrated steelmaking (approximately 1.8 tons of CO_2 per ton of steel) is the use of carbon-based reductants in blast furnace ironmaking. Several changes to blast furnace ironmaking have been proposed as ways to reduce carbon intensity and cost. A conceptually simple two-zone blast furnace mass and energy balance can be used to quantify the expected impact of such process changes on coke rate, carbon intensity, and productivity. Examples to be discussed include top gas recycling, combining lower blast temperatures with increased coal injection to produce a rich top gas, using prereduced feed, and tuyère injection of natural gas. The use of the operating window – which illustrates how allowable flame temperatures and top-gas temperatures constrain feasible oxygen enrichment levels – is illustrated. Increased use of natural gas would lower the carbon intensity of blast furnace ironmaking.

Introduction

Because of the significant carbon intensity of steelmaking (approximately 1.8 tons of CO_2 emitted per ton of steel produced) and the high production rate of steel (more than 1.4 billion tonnes per annum), steelmaking is responsible for a large percentage, approximately 6.7%, of world anthropogenic CO_2 emissions [1]. By far the largest input of the carbon into conventional steelmaking is as coke, which is essential to blast furnace ironmaking. A typical fuel rate for blast furnace ironmaking is 450 kg of fuel per tonne of hot metal (THM). ("Fuel" includes coke and injectants which partially replace coke; pulverized coal and natural gas are typical injectants.) Assuming the fuels to contain 90% carbon on average, and since nearly all of this carbon exits the steel plant as CO_2, a fuel rate of 450 kg is equivalent to a CO_2 intensity of 1485 kg/THM – by far the largest part of the total CO_2 intensity of steelmaking.

Blast furnace ironmaking is expected to remain the main source of new iron units; this paper gives some estimates of the extents to which further reductions in carbon intensity from blast furnace ironmaking is possible. A recent European assessment suggested that blast furnace process intensification options such as increased pulverized-coal injection or natural gas injection would yield only modest energy savings [2]. However, if the focus is on CO_2 emissions rather than on total energy use, there would be an advantage to using fuels – such as natural gas – which have lower carbon contents relative to energy content.

Here, evaluations of four approaches to lowering the carbon intensity of blast furnace ironmaking are summarized. Two of these involve increased use of natural gas, one would use the blast furnace as a coal gasifier (employing a gas turbine to generate electricity using the resulting high calorific value blast furnace gas), and one would strip carbon dioxide from the blast furnace off-gas, recycling carbon monoxide and hydrogen to the blast furnace.

Calculations

The quite different approaches (to decreasing the carbon intensity of blast furnace ironmaking) were quantified by using a mass and energy balance similar to that first described by Rist and Meysson [3]. Conditions at the furnace exit (hot metal temperature and composition, slag basicity and temperature) were calculation inputs, as was the temperature in the wüstite reserve zone. The gas phase was assumed to reach thermal and chemical equilibrium with metallic iron and wüstite in the wüstite reserve zone. Additional mass and energy balances were used to calculate the top gas temperature and composition, and the raceway adiabatic flame temperature (the latter was calculated by assuming that the combustion products in the raceway are CO, H_2 and N_2). Changes in blast furnace productivity were estimated by assuming the production rate to be limited by the bosh gas flow rate [4].

Operating window

Any change in the fuel input of the blast furnace would affect not only the overall energy requirements of the process, but also the local energy balance at the tuyères and the upper region of the blast furnace shaft. The overall energy requirements determine the coke rate; the local energy balance at the tuyères determines the flame temperature, and the average top-gas temperature is controlled by the energy balance of the upper shaft. The flame temperature needs to be high enough – typically in the range 1900 to 2000°C – to ensure adequate melting and tapping of the hot metal and slag. The top-gas temperature also needs to be sufficiently high to avoid condensation of water at the top of the furnace.

Tuyère injection of cold fuels – such as natural gas or pulverized coal – tends to decrease the flame temperature; this quenching effect is counteracted by increasing oxygen enrichment of the blast air, or a higher blast temperature, or both. Increased oxygen enrichment lowers the top-gas temperature, because less nitrogen is available to carry heat to the upper shaft. Levels of oxygen enrichment that meet both the flame-temperature requirement and the top-gas requirement are conveniently shown in "operating windows" [5], of which an example is given in Figure 1.

Figure 1 demonstrates that, for natural-gas injection, oxygen enrichment is essential to maintain the flame temperature. Actual blast furnaces (for which data points are shown in the figure) operate close to the minimum required oxygen enrichment. The figure also shows that the flame-temperature and top-gas temperature constraints impose a maximum limit on possible levels of oxygen enrichment; this maximum limit is around 150 kg natural gas per THM. Beyond this maximum limit, there is no oxygen enrichment level which would meet both the flame-temperature limit and the top-gas temperature limit.

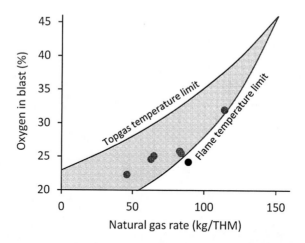

Figure 1. Calculated allowable operating window (shaded region), showing combinations of natural-gas injection rates and oxygen enrichment levels which would satisfy both the flame-temperature and top-gas temperature constraints. Calculated for a blast temperature of 1100°C, 1900°C minimum flame temperature, and 110°C minimum top-gas temperature. Data points are operating conditions of North American blast furnaces injecting only natural gas [6].

Calculated carbon intensity of blast furnace variants

In the rest of the paper, results on the calculated carbon intensity of four blast furnace variants are presented. In each case, the assumptions and base-line conditions are stated, or references to the originally reported calculations are provided, or both. The variants are tuyère injection of natural gas (requiring increased oxygen enrichment; see Figure 1), partial prereduction of the blast furnace burden in a separate direct-reduction unit, the top-gas recycling oxygen blast furnace (as studied in the ULCOS program) and the "Blast Furnace plus" approach. The main elements of last of these are a *lowered* blast temperature and maximal oxygen enrichment, to ensure that the blast furnace top gas has a sufficiently high calorific value to be used as fuel in a combined-cycled gas-turbine electric power plant. Given the constraints on natural-gas injection (Figure 1) this approach would only be suited to pulverized-coal injection; pulverized coal has a much smaller quenching effect on the flame temperature than does natural gas, allowing a combination of lower blast temperature and high fuel injection rate to be used.

The examples show that incremental decreases in the carbon intensity are possible with all of these approaches; only sequestration of CO_2 (which would be an option with the oxygen blast furnace) would allow for decreases in CO_2 intensity of more 15%. Preliminary estimates of costs indicate that the approaches that maximize natural-gas use (increase tuyère injection, and feed prereduction) have the potential to decrease hot-metal cost *and* decrease CO_2 intensity; this would not be the case for the oxygen blast furnace.

Tuyère injection of natural gas

Main differences (compared with conventional blast furnace):
Partial replacement of coke with tuyère injection of natural gas;
higher oxygen percentage in blast air required to maintain flame temperature

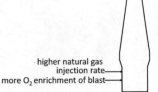

higher natural gas injection rate
more O_2 enrichment of blast

Current status: Employed at many blast furnaces in North America, at rates up to 125 kg natural gas per THM [6].
Typical calculated figures: (compared with base case of 100 kg/THM pulverized coal injection; energy requirement for oxygen 0.3 kWh/Nm^3; CO_2 intensity of electricity 0.53 kg CO_2/kWh; values are for 100 kg natural gas per THM)
CO_2 intensity: net 69 kg CO_2 per THM reduction (after allowing 28 kg CO_2 per THM for oxygen production)
Extra energy use relative to base case: 53 kWh/THM electricity (for oxygen)
Coke rate: approximately 30 kg/THM lower; Productivity: approximately 18% higher

Prereduced feed

Main differences (compared with conventional blast furnace):
Partial replacement of iron ore (pellets) with metallic iron
Current status: Employed at AK Steel Middletown [6]; MIDREX plant for pellet reduction (2 million tons per year capacity) to be constructed for voestalpine in Corpus Christi, TX [7].

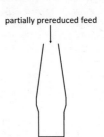

partially prereduced feed

Typical calculated figures: (compared with base case of 100 kg/THM pulverized coal injection and no prereduction; energy requirement for oxygen 0.3 kWh/Nm^3; CO_2 intensity of electricity 0.53 kg CO_2/kWh; 200 kg natural gas per tonne of direct-reduced iron [DRI]; values are for 100 kg DRI per THM and tuyère injection of 100 kg natural gas per THM)
CO_2 intensity: net 130 kg CO_2 per THM reduction (after allowing 28 kg CO_2 per THM for oxygen production)
Extra energy use relative to base case: 53 kWh/THM electricity (for oxygen)
Coke rate: approximately 66 kg/THM lower; Productivity: approximately 29% higher

Top-gas recycling (TGR), or oxygen blast furnace (OBF) [8]

Main differences (compared with conventional blast furnace):

O_2 instead of blast air;
CO_2 removed from top gas and CO recycled to furnace;
loss of export gas; extra fuel required to preheat recycled CO and O_2.
Large energy requirement for CO_2 capture.
Large increase in cost.

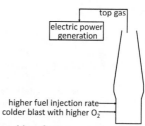

Current status: "Plans are currently being developed to test this principle on a commercial scale blast furnace. This will take place in the next phase of the ULCOS project, ULCOS II requiring an initial R&D investment of several hundred million Euros." [8]

Typical calculated figures [4,9]:
CO_2 intensity: Decrease of 116 kg CO_2 per tonne hot-rolled coil compared with conventional blast furnace base case; decrease of 976 kg CO_2 per tonne hot-rolled coil if CO_2 were sequestered

Cost: 10% increase in production cost; CO_2 avoidance cost $56 (US) per tonne CO_2 (excluding sequestration).

Extra energy use relative to base case: 100 kWh/THM electricity (mainly for oxygen plant; excluding CO_2 compression); 2 GJ/THM (560 kWh/THM) steam (for CO_2 capture)

Coke rate: approximately 150 kg/THM lower; Productivity: approximately 25% higher

"BF plus" [10]

Main differences (compared with conventional blast furnace):
Blast furnace used as a coal gasifier.
Lower blast temperature;
increased pulverized coal injection rate;
higher oxygen enrichment;
rich top-gas used for power generation
(combined-cycle gas turbine)

Current status: Similar projects to generate electricity from blast furnace gas recently started at ArcelorMittal Indiana Harbor works [11], and JFE Chiba works [12]

Typical calculated figures: 300 kg pulverized coal injection per THM; 36% O_2 blast; blast temperature 900°C; efficiencies: 55% for electrical power generation; 60% for stoves)

CO_2 intensity: Small increase in carbon from fuel (50 kg CO_2 per THM) because of colder blast; 300 kg CO_2 per THM credit for electricity (assuming 0.56 kg CO_2/kWh), hence net reduction in CO_2 intensity of 250 kg CO_2 per THM

Cost: Dominated by coke saving, cost of pulverized coal, and income from electricity sales

Net electrical power generated: 0.52 MWh/THM

Coke rate: approximately 150 kg/THM lower; Productivity: approximately 25% higher

Conclusions

After more than two centuries of blast-furnace ironmaking, the process is already highly optimized. Of the variants considered here, those using natural gas are the most promising for current operations. The recent availability of lower-cost natural gas in North America offers an opportunity to lower ironmaking cost and carbon intensity, without a substantial increase in process complexity.

Acknowledgment

Support of this project by the industrial members of the Center for Iron and Steelmaking Research is gratefully acknowledged.

References

1. worldsteel association, "Steel's contribution to a low carbon future," worldsteel position paper, March 2013. Available at www.worldsteel.org.
2. N. Pardo, J.A. Moya and K. Vatopoulos, *Prospective Scenarios on Energy Efficiency and CO_2 Emissions in the EU Iron & Steel Industry*, Policy Report, European Commission Joint Research Centre, Institute for Energy and Transport, 2012.
3. A. Rist and N. Meysson, "A dual graphic representation of the blast-furnace mass and heat balances," *Journal of Metals*, 12(4) (1967), 50-59.
4. P.C. Pistorius, "Technical and economic evaluation of top gas recycling blast furnace ironmaking," Proceedings of the Fray International Symposium, 27nd Nov 1st Dec. 2011, Cancun.
5. M. Geerdes, R. van Laar, and R. Vandershteyn, "Low-cost hot metal: the future of blast furnace ironmaking," *Iron & Steel Technology*, 8(3) (2011), 51-56.
6. "AIST 2013 North American Blast Furnace Roundup," *Iron & Steel Technology* 10(3) (2013), pp. 218-221.
7. "voestalpine signs contract with Siemens and Midrex for DR Plant," http://www.midrex.com/press-detail.cfm?news_id=1305&cat_id=5. Last visited October 4, 2013.
8. ULCOS website, http://www.ulcos.org/en/research/blast_furnace.php. Last visited October 4, 2013.
9. L. Hooey, A. Tobiesen, J. Johns and S. Santos, "Techno-economic study of an integrated steelworks equipped with oxygen blast furnace and CO_2 capture," *Energy Procedia* 37 (2013), 7139-7151
10. M.D. Lanyi, J. Cao and J.A. Terrible, "A Sensible Route to Energy Efficiency Improvement and CO_2 Management in the Steel Industry." *AISTech 2012 Conference Proceedings*, Atlanta, GA, May 7-10, 2012, pp. 605-613.
11. "Energy Department, ArcelorMittal Partnership Boosts Efficiency of Major Steel Manufacturing Plant," http://energy.gov/articles/energy-department-arcelormittal-partnership-boosts-efficiency-major-steel-manufacturing, last visited October 4, 2013
12. "GE to Supply Japan's JFE Steel Co. with Multi-Fuel, On-Site Power Solution at Chiba Steel Mill," http://www.genewscenter.com/Press-Releases/GE-to-Supply-Japan-s-JFE-Steel-Co-with-Multi-Fuel-On-Site-Power-Solution-at-Chiba-Steel-Mill-392a.aspx, last visited October 4, 2013.

Energy Technology 2014: Carbon Dioxide Management and Other Technologies
*Edited by: Cong Wang, Jan de Bakker, Cynthia K. Belt, Animesh Jha, Neale R. Neelameggham,
Soobhankar Pati, Leon H. Prentice, Gabriella Tranell, and Kyle S. Brinkman*
TMS (The Minerals, Metals & Materials Society), 2014

GREEN SLAG SYSTEM DESIGN DURING CONTINUOUS CASTING

Juan Wei[1], Wanlin Wang[1], Boxun Lu[1], Lejun Zhou[1] and Kun Chen[1]

[1]School of Metallurgy and Environment, Central South University, Changsha, Hunan, 410083,
China

Key words: F-free Mold Flux, Viscosity, Crystallization.

Abstract

Slag system is an indispensable part during the continuous casting process, and playing various important roles like absorbing inclusions, modifying heat transfer rate, preventing oxidation of molten steel, etc. to ensure the smooth process and excellent slab quality. However most slags contains more than 8% fluorides which tend to introduce a series of environmental and health problems. This study involves the research of the removal of fluorine through the chemical composition optimization. Through the studies of CCT, DHTT, XRD and SEM, it has been successfully founded that the crystalline phase of $Ca_{11}Si_4B_2O_{22}$ discovered in this study shows very similar crystallization property as Cuspidine ($Ca_4Si_2O_7F_2$), which is the most important crystalline phase in the current fluorine bearing slags. Moreover, the optimized free/low fluorine slag shows the opportunity to replace conventional high-fluorine mold slag for casting medium carbon steels that is regarded as the bottleneck of green slag development for continuous casting.

Introduction

Mold flux has been widely used in continuous casting to control heat transfer, lubricate the strand, absorb inclusions, provide insulation and prevent oxidation of the molten steel[1-3]. It usually contains fluorides, like CaF_2, NaF etc., which are used to control the viscosity and solidification temperature of mold fluxes [4-5]. However, the presence of fluoride leads to corrosion of equipment, serious environmental pollution (toxic volatiles and water pollution), and high processing cost of cooling water[6-7]. Therefore, the study of low or free fluorine mold flux is of interest and attracts worldwide attention.

S. Y. Choi et al. [4] used B_2O_3 as a substitute for cuspidine in steel B2PB2) however, B_2O_3 content was extremely high (up to 30 w%). Other studies have investigated F-free mold flux by adding or optimizing components as MgO, Li_2O, B_2O_3 et al[8-11]. However, most of these focused on the melting and viscosity issues or for low or ultra-low carbon steels, where crystallization tendency is not significantly strong. Very few studies regarding to the development of low/free Fluorine mold flux for casting medium carbon (MC) steels have been conducted, due to the difficulty of the crystallization and heat transfer control without the formation of cuspidine.

When casting medium carbon (MC) steels, longitudinal cracking occurs frequently, which arise from thermal stresses within the solidified shell due to the volumetric shrinkage during the δ-γ phase transformation. In order to minimize the stresses, the in-mold heat transfer for casting MC steels needs to be reduced with the aim to keep a thin and uniform solidified shell, which could be achieved maintaining a thick solid crystalline layer of slag film, as the crystallization of mold flux tends to block radiative heat transfer and increase the interfacial thermal resistance by the formation of mold/slag gap. As a result, the formation of cuspidine during the mold flux crystallization is critical in controlling the forming of cracks, as cuspidine that with a strong crystallization tendency and relatively low crystallization temperature would create a thicker crystallized film and still keep a thin liquid flux layer, such that it could reduce heat transfer, resist the ferrostatic pressure in the mold and keep lubrication at the same time. Also, cuspidine has a relatively lower thermal conductivity compared with other crystalline mold flux phase like calcium silicate ($CaO\cdot SiO_2$) and gehlenite ($2CaO\cdot Al_2O_3\cdot SiO_2$), et. al.[12-13]. It would be a technical obstacle if fluorine was removed and cuspidine ($Ca_4Si_2O_7F_2$) couldn't be formed.

Therefore, a successful F-free powders should exhibit similar (i) viscosity, and (ii) crystallization behaviors, as those in the F-bearing powder. Thus, the aim of this study is to investigate above behaviors for a proposed F-free mold slags through the design of chemical composition to achieve a green slag system for continuous casting.

1. Experiment

1.1 Apparatus and Procedure

A. Rotating Viscometer and Experiment Procedure. The high temperature viscosity of liquid mold flux was measured by using rotating viscometer. This instrument measures the torque of spindle rotated at fixed speed in a crucible filled with the liquid slag. The crucible was heated from room temperature to 1723K in an electric furnace. After holding at 1723K for 10min, the furnace temperature decreased at a rate of $5Kmin^{-1}$ and the viscosity was measured simultaneously.

B. Single and Double Hot Thermocouple Technique. SHTT and DHTT were adopted in the experiments. During SHTT tests, a slag sample was firstly placed on a B-type thermocouple and heated directly. The CCD camera was used to record the melting and crystallization [14-15]. While for DHTT tests, a temperature gradient is set between two thermocouples [16]. The detailed information about the experimental heating profile was shown in **Figure 1**.

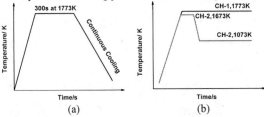

(a)   (b)

Figure 1. Thermal profiles of (a) SHTT tests and (b) DHTT tests

1.2 Preparations of Mold Flux Powder

The major chemical composition of mold slags are shown in **Table I**, among which Sample A is a typical industrial mold flux for casting medium carbon steel, Sample B-E are designed F-free slags. Pure chemical reagent $CaCO_3$, SiO_2, Al_2O_3, MgO, Na_2CO_3, CaF_2 and Li_2CO_3 were used as raw materials.

The raw materials were decarburized in a programmable furnace prior to the fusion process. Then, the decarburized mold flux was melted at 1773K (1500°C) in a graphite crucible placed into a silicon-molybdenum electric furnace. After holding 300s to homogenize its chemical composition and eliminate bubbles, the mold flux was quenched from its molten state into cold water to achieve a fully glass phase. Next, the glass sample was subjected to crushing and grinding to powders for the following SHTT and DHTT tests.

Table I The Chemical Compositions of Slags （w%）

Sample	R	CaO	SiO_2	MgO	Al_2O_3	Na_2O	B_2O_3	Li_2O	F
A	1.25	43.06	34.44	2	4	7.5	0	1	8
B	1.15	41.72	36.28	2	4	8	6	2	0
C	1.25	39.44	31.56	2	4	11	10	2	0
D	1.15	39.58	34.42	2	4	10	8	2	0
E	1.25	38.33	30.67	2	4	12	10	3	0

2. Results and Discussion

2.1 Viscosity

Table II shows the viscosity at 1573K and the break temperature of samples listed in Table I. According to **Table II,** for Sample A, the viscosity was 0.136 Pa•s, and the break temperature was 1526K (1253°C). While for the F-free mold slags, viscosity ranged from 0.113 Pa•s to 0.256 Pa•s, most of these F-free slags had higher viscosity than Sample A except Sample E, which exhibited minimum value of 0.113 Pa•s. Besides, all F-free slags had lower break temperature compared to that of Sample A, with a range from 1443K (1170°C) to 1493K (1253°C).

The relationship between viscosity and temperature was shown in **Figure 2**. According to Fig.2, obviously, Sample B was more similar to that of Sample A than the other samples; besides, the break temperature of Sample B (1493K) was the closest to that of Sample A, though the viscosity value at 1573K was a little higher. The reason was that, in Sample B, F was removed and the basicity decreased from 1.25 to 1.15. As both of F and basicity acted as network breakers, removing or decreasing of them would directly lead to complexity of slag network. Still, B_2O_3 was added to this system with an amount of 6% (in w%). And B_2O_3 was known as a network former; and it could also lower the melting temperature remarkably.

Table II Viscosity (at 1573K) and Break Temperature of Mold Slags

Sample	A	B	C	D	E
viscosity/Pa•s	0.136	0.256	0.155	0.209	0.113
$T_{br}/°C$	1253	1220	1188	1180	1170

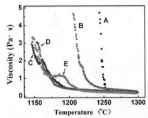

Figure 2. The relationship between viscosity and temperature of all samples

2.2 SHTT Experiments

A. CCT tests. CCT diagrams of mold flux samples were shown in **Figure3 (a)–(e)**. The CCT diagrams indicated that all mold slags used in the experiments were with great crystallization tendency with critical cooling rates all over 30K/s, which was the upper limit of the facility.

Still, **Table III** summarized the crystallization temperature of slags when the cooling rate was fixed at certain values. According to Table III, all F-free slags crystallization temperatures were lower than Sample A, which inferred the F-free slags' crystallization tendency were not as strong as that of Sample A; but anyway, when cooling rate was fixed at 1K/s, 5K/s, 10K/s or 30K/s, Sample C was the most similar to Sample A. While at 20K/s, Sample B had the nearest crystallization temperature to Sample A, and next to it was Sample C. Compared to Sample A, in Sample C, F was removed, 10% (in w%) B_2O_3 was added, at the same time, Na_2O and Li_2O content increased by 3.5% and 1% (in w%) respectively. Previous studies [17-19] on the development of low F mold fluxes for casting medium carbon steels suggested Na_2O is effective in improving the crystallization behavior, and B_2O_3 tends to inhibit crystallization. Thus, after F was removed, increased B_2O_3 and Na_2O content, the interaction of the two opposite effects influenced the overall crystallization behavior, which prevented it from changing considerably.

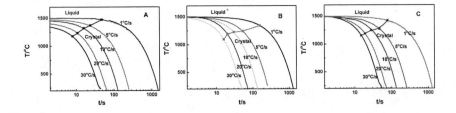

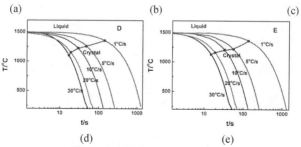

(a)   (b)   (c)

(d)   (e)

Figure 3. CCT diagrams of slags

Table III Crystallization Parameters of CCT Diagrams of the Tested Mold Fluxes (°C)

Sample	A	B	C	D	E
1°C/s	1473.1	1341.7	1419.1	1355.9	1360.6
5°C/s	1369.8	1240.7	1269.9	1257.1	1219.1
10°C/s	1320.3	1226.1	1246.5	1226.5	1208.0
20°C/s	1234.4	1193.2	1186.9	1154.9	1164.4
30°C/s	1179.0	1096.2	1151.1	1102.5	1131.3

2.3 DHTT Experiments

Figure 4 was crystallization images standing for each stage of Sample A. While **Figure 4 (a)** referred to position I (194 seconds), where CH-1 was at 1773K (1500°C), while CH-2 was 1673K (1400°C) due to a slightly adhesive force need to stretch the melting mold flux. Owing to strong crystallization tendency, when temperature of CH-2 dropped, crystals precipitated very fast, as was clearly at position II (196 seconds, **Figure 4(b)**). And at position III (**Figure 4(c)**) and position IV (**Figure 4(d)**), more crystals formed and grew up. **Figure 4(e)** was corresponded to position V (237 seconds), which showed the moment that CH-2 was just decreased to 1073K (800°C) and became dark, and crystallization nearly ended. **Figure 4(f)** was the moment the structure of mold flux was in a relatively steady state - distribution of mold flux layers was apparent. Owing to strong crystallization tendency, at the end of crystallization process, only crystalline layer (95.37%) and a thin liquid layer (only 4.63%) were formed.

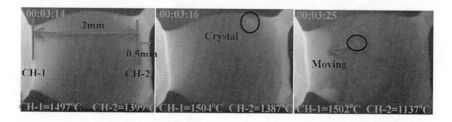

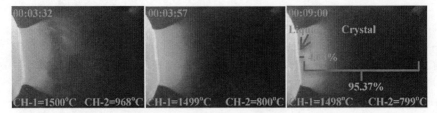

Figure 4. Crystallization process of Sample A under simulated thermal gradient

By adopting the same method, crystallization processes of F-free slags (except Sample C and E) under same thermal gradient were obtained, and steady states of slags were shown in **Figure 5**. From Figure 5, crystalline layer were 70.14% and 66.52% for Sample B and D respectively. While for Sample C, it crystallized completely during melting slag stretching between two thermocouples, which agreed with CCT tests that Sample C was the easiest to crystallize; and Sample E failed to be stretched owing to its low viscosity (Table II, 0.113Pa•s).

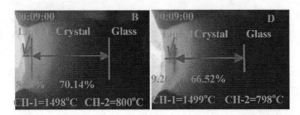

Figure 5. The steady states of several F-free slags

2.4 XRD and SEM Analysis

In order to identify the phase composition of crystals, XRD, SEM and EDS were adopted, as shown in **Figure 6.** According to Figure 6(a), cuspidine was formed in F bearing Sample A; while in F-free mold slags Sample B and Sample D (Figure 6 (b) and (c)), two phases were found - $Ca_{11}Si_4B_2O_{22}$ as the main crystalline phase, together with some $Ca_{14}Mg_2(SiO_4)_8$.

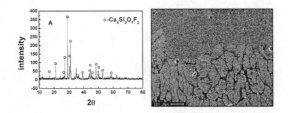

Element	Ca	Si	Al	Mg	Na	O	F
Wt%	59.67	26.32	2.02	0.95	2.40	5.49	3.15

(a)

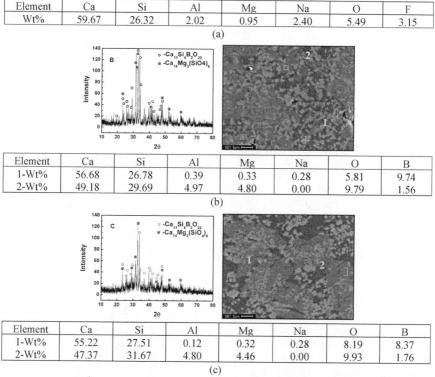

Element	Ca	Si	Al	Mg	Na	O	B
1-Wt%	56.68	26.78	0.39	0.33	0.28	5.81	9.74
2-Wt%	49.18	29.69	4.97	4.80	0.00	9.79	1.56

(b)

Element	Ca	Si	Al	Mg	Na	O	B
1-Wt%	55.22	27.51	0.12	0.32	0.28	8.19	8.37
2-Wt%	47.37	31.67	4.80	4.46	0.00	9.93	1.76

(c)

Figure 6. XRD and SEM results of tested slags.

3. Conclusions

In this article, SHTT and DHTT were adopted to investigate the crystallization behavior of mold fluxes, Rotating Viscometer was used to study the viscosity, and XRD, SEM and EDS were also adopted to identify the crystalline phases. The main conclusions are summarized as follows:

(1) The viscosity experiments showed that the viscosity of Sample A at 1300°C is within the range to that of the four F-free slags, and the break temperature of F-free slags are all lower than that of Sample A. However, comparing the viscosity and temperature relationship curve of all five slags, Sample B is the best one, as it shows the closest T_{br} and viscosity - temperature curve to that of Sample A.

(2) SHTT tests indicate that at certain cooling rates, Sample C and Sample B had the closest crystallization behavior compared to that of Sample A, which means that the crystallization temperatures of above two F-free slags varies similarly as that of reference Sample A, when

temperature is decreased at different cooling rates.

(3) DHTT tests results show that Sample B and Sample D, the two successful F-free slags in DHTT tests, have steady states of three layers - liquid layer, crystalline layer and glass layer. The results agree with SHTT and viscosity results, which inferred that appropriate crystallization tendency and viscosity are two important factors affecting the DHTT tests.

(4) XRD and SEM results suggest that cuspidine is the major crystalline phase precipitated in F-bearing mold flux Sample A. As for F-free slags Sample B and Sample D, a mixture of $Ca_{11}Si_4B_2O_{22}$ and $Ca_{14}Mg_2(SiO_4)_8$ is formed. Thus, the crystalline phase $Ca_{11}Si_4B_2O_{22}$ (together with $Ca_{14}Mg_2(SiO_4)_8$) found in F-free slags is very likely to act as a substitute for cuspidine.

(5) Considering all of the results, Sample B is best of all F-free samples, which is very likely to act as a substitute for F-bearing slags used in medium carbon steels.

Reference:

[1] A.Yamauchi, K. Sorimachi, T. Sakuraya and T. Fujii: *ISIJ Int*, 33(1993),140.

[2] K.C. Mills, A.B. Fox, Z. Li, and R.P. Thackray: *Ironmaking Steelmaking*, 32(2005), 26.

[3] W.L. Wang and A.W. Cramb: *ISIJ Int.*, 45(2005), 1864.

[4] S.Y. Choi, D.H. Lee, D.W. Shin, S.Y. Choi, J.W. Cho, and J.M. Park: *J.Non Cryst. Solids*, 345&346(2004), 157-160.

[5] H. Nakada and K. Nagata: *ISIJ Int.*,46(2006), 441.

[6] A.I. Zaitsev, A.V. Leites, A.D. Litvina, and B.M. Mogutnov: *Steel Research*, 65(1994), 368.

[7] G.H. Wen, S. Sridhar, P. Tang, X. Qi and Y.Q. Liu: *ISIJ Int.*, 47(2007), 1117-1125.

[8] N. Hideko and N. Kazuhiro:*ISIJ Int.*, 2006, vol. 46(3), pp. 441-449.

[9] M. Hayashi, T. Watanabe, H. Nakada and K. Nakata: *ISIJ Int.*, 46(2006), 1805-1809.

[10] Q. Wang, Y. Lu, S. He, K.C. Mills, and S. Li: *Ironmaking Steelmaking*,38(2011) 4:297-301.

[11] H. Nakada, H. Fukuyama and K. Nagata: *ISIJ Int.*, 46(2006), 1660-1667.

[12] R. Taylor.Physical properties of casting powders: Part 3 thermal conductivities of casting powders[J]. *Ironmaking Steelmaking*, 15(1998)4, 187-194.

[13] V. Riboud. Lubrication and heat transfer in a continuous casting mold[C].*Steelmaking Conference Proceedings*. ISS, Warrendale, 1979, 78-92.

[14] Y. Kashiwaya, K. Ishii: ISSTech 2003 Conference Proc., *Iron and Steel Society*, Indianapolis, IN, U.S.A, 1021.

[15] Y. Kashiwaya, C.E. Cicutti, A.W. Cramb, and K. Ishii: *ISIJ Int.*, 38(1998), 348-356.

[16] K. Shimizu, T. Suzuki, I. Jimbo and A. W. Cramb: "An Investigation on the Vaporization of Fluorides from Slag Melts" *Iron and Steelmaking Conference*, 79(1996),723-729.

[17] B. Lu, W.Wang, J. Li, H. Zhao, D. Y. Huang: *Met Mat. Transc. B*, 44(2013)2, 365-377.

[18] J. Li, W. Wang, J. Wei, D. Huang and H. Matsuura: *ISIJ Int.*, 50(2012), 2220-2225.

[19] L. Zhou, W. Wang, J. Wei and B. Lu: *ISIJ Int.*, 54(2013), 664-671.

Energy Technology 2014: Carbon Dioxide Management and Other Technologies
Edited by: Cong Wang, Jan de Bakker, Cynthia K. Belt, Animesh Jha, Neale R. Neelameggham,
Soobhankar Pati, Leon H. Prentice, Gabriella Tranell, and Kyle S. Brinkman
TMS (The Minerals, Metals & Materials Society), 2014

INNOVATIVE HYDROGEN PRODUCTION PROCESS UTILIZING THERMAL AND CHEMICAL ENERGIES OF STEELMAKING SLAG

Hiroyuki Matsuura, Fumitaka Tsukihashi

The University of Tokyo; 5-1-5 Kashiwanoha, Kashiwa, Chiba 277-8561 Japan

Keywords: Steelmaking, Slag, Waste Energy Recovery, Hydrogen

Abstract

Emission of CO_2 gas from steelmaking industry covers more than 12 % of total CO_2 emission in Japan. Though many efforts have been made to curtail energy consumption and high energy efficiency of Japanese steel production process has been achieved so far, further reduction of CO_2 emission is strongly requested to develop sustainable steel production process and produce "green material". Recently, authors have proposed the innovative hydrogen production process, in which water vapor is reacted with discharged steelmaking slag to produce hydrogen gas. This reaction can be achieved without any external energy source and operated by huge thermal energy associated with steelmaking slag and FeO or unrecovered Fe droplets that are relatively rich in the slag. In this paper, thermodynamic calculation results with various initial conditions in terms of hydrogen productivity were presented. Laboratory-scale experiments were also conducted to clarify the validity of thermodynamic calculation and the reaction rate in realistic conditions.

Introduction

Ironmaking and steelmaking process from iron ore through blast furnace and BOF is an essential steel production route for Japanese steel industries to supply high quality steel products in large quantity to society stably. Blast furnace – BOF process requires huge amount of fossil fuels such as mainly coal, resulting the emission of considerable amount of CO_2 gas. Since the emission of CO_2 gas from steel industries, about 147 Mt (FY2011), accounts for approximately 12 % of domestic CO_2 gas emission in Japan (FY2011),[1] it is an urgent issue for steel industries to develop environmental-friendly ironmaking and steelmaking process and reduce CO_2 emission.

Various kinds of gases and by-products generated from ironmaking and steelmaking processes have been recycled as much as possible at present. Enormous thermal energy is also generated from processes and released to surrounding atmosphere as various forms such as sensible heats of gases, molten steels or by-products, and as heat losses. However, these energies are not utilized before final dissipation to atmosphere. Although various technologies to utilize thermal energy from ironmaking and steelmaking processes have been attempted to develop, practical application of these technologies has not been achieved because the thermal energy is not always useful as an energy source.

In steelmaking process by BOF, fluxing agent such as CaO is added during reefing and thus about 100 kg/t-steel of steelmaking slag are generated. Since the steelmaking slag is discharged at temperature range between 1873 and 1923 K after discharge of refined molten steel, it also has large thermal energy. Amount of generated BOF slag is 11.0 Mt/y (FY2012) in Japan.[2] Assuming the amount of steelmaking slag is 10 Mt/y and its average heat capacity from room temperature to 1923 K is 1 kJ/kg·K,[3] unutilized heat is estimated to be 16 PJ/y, which is equal to annual energy consumption of 420 thousand families in Japan.[4]

Steelmaking slag contains relatively large concentration of FeO in the range from 20 to 30 mass%. Therefore, reduction of H_2O gas by FeO in slag described as reaction (1) is expected,

$$2\ FeO\ (l) + H_2O\ (g) = Fe_2O_3\ (s) + H_2\ (g) \quad \Delta G° = -82130 + 102.7T\ \text{J/mol[5]} \quad (1)$$

From Gibbs free energy change of reaction (1), this reaction does not proceed forward spontaneously in standard states of reactants and products, and progress of the reaction to equilibrium is expected. In addition, no heat or only small heat supply may be necessary because reaction (1) is exothermic, and thus it is considered that thermal energy required for stable operation could be supplied sufficiently by sensible heat of steelmaking slag. From above reasons, the development of energy-saving H_2 production process without further energy consumption would be possible. Tata Steel has also researched the fundamental technology on the H_2 production process and a pilot scale has clarified the production of 2.1 m^3 of H_2 gas from 12 t of BOF slag.[6]

However, FeO and Fe_2O_3 contents in the slag decrease and increase, respectively, with progress of reaction (1). This compositional change may result in the increment in solid fraction or viscosity of slag affecting the physical properties of slag such as fluidity and thus operational condition of H_2 gas production process. Therefore, sufficient understanding of the physical chemistry of reaction between slag containing FeO and steam gas is important for precise operation of the process.

In this paper, development of H_2 production process by using BOF steelmaking slag discharged at steelmaking temperature range was considered. Thermodynamic calculations of the reaction between FeO–CaO–SiO_2 slag system and H_2O–Ar gas was conducted to examine the effects of temperatures, compositions of gas and slag on the behavior of H_2 gas production or the physicochemical properties of slag.[7] The measurement of the generation behavior of H_2 gas by reaction between FeO-containing slag and H_2O–Ar gas was also conducted at 1723 K to clarify the effect of slag compositions on the behavior of H_2 gas production.[8]

Thermodynamic Calculation[7]

Calculation Conditions

Thermodynamic calculation software FactSage 6.2 was used for estimation of equilibrium state between FeO-containing slag and H_2O-containing gas, and H_2 gas generation behavior. The detail of the calculation conditions is given elsewhere.[7] Firstly, the initial condensed phases including molten slag and solid oxides were prepared by inputting oxide constituents with prescribed compositions and equilibrating at initial slag temperature. Subsequently the prepared condensed phases were equilibrated with H_2O–Ar gas. In the case of the preparation of initial condensed phases, FeO, CaO and SiO_2 were input to be 1 kg as a total, and then pure compounds and solutions were considered as final stable candidates at predetermined temperature. In this stage, the formation of gas phase was not allowed. Phases other than molten slag such as metallic or $2CaO·SiO_2$ phase were formed in some calculations and these phases were also taken into account for slag-gas equilibrium calculation. In the slag-gas equilibrium calculation, pure compounds and solutions were considered in the equilibrium calculation as final stable candidates. For the estimation of change in composition of condensed phases and system temperature with proceeding of reaction between slag and gas, 10 L of H_2O–Ar gas at predetermined gas temperature were introduced in the system and equilibrated with condensed phases at the condition of no enthalpy change of the system. Condensed phases such as molten slag or various solids after equilibrium calculation were input as initial condensed phases in the next calculation, and 10 L of new H_2O–Ar gas were introduced and equilibrated again. The above equilibrium calculation between slag and gas was repeated by decreasing temperature less than 1722 K. Change in slag compositions and system temperature with process of reaction between slag and gas were estimated.

Calculation Results

Effect of Slag Compositions and Gas Temperature   Figure 1 shows the change in the amount of produced H_2 gas with volume of introduced H_2O gas in the case of initial slag compositions with mass%FeO = 20 or 40, and mass%CaO/mass%SiO$_2$ = 2 or 3. The amount of produced H_2 gas is the accumulated amount of H_2 gas produced by repetitive slag-gas equilibrium calculation. Increase in introduced gas temperature increased the amount of produced H_2 gas in all calculation conditions, which is because temperature drop by equilibrium between gas and slag became smaller with increasing gas temperature and thus more H_2O gas could react with FeO contained in slag. Increase in initial FeO content in slag increased the amount of produced H_2 gas. Figure 2 shows the change in solid fraction of slag phase with introduced gas volume. Note that the fraction of solid phase was nearly zero in the case of (b). Most of solid phase was $2CaO \cdot SiO_2$. With larger initial FeO content, solid fraction decreased because the slag composition became closer to the liquidus, which is favorable to maintain the fluidity of slag for the stable operation. As expected, the increase in CaO/SiO$_2$ ratio of slag increased the solid fraction. Reaction between slag and gas decreased and increased FeO and Fe_2O_3 contents, respectively, resulting in the increase in the solid fraction. Figure 3 shows the change in production ratio of H_2 gas with introduced H_2O gas volume, where production ratio of H_2 gas is defined as the molar ratio of the amount of generated H_2 gas to that of introduced 10 L H_2O–Ar gas in one equilibrium calculation. The effect of gas temperature or initial FeO content on the production ratio of H_2 gas was small. The production ratio of H_2 gas was the largest at the beginning of reaction, from 60 to 85 %, and decreased drastically with increasing introduced gas volume. Figure 4 shows the change in slag temperature with introduced gas volume. Although reaction (1) is exothermic, system temperature decreased with increasing introduced gas volume because temperature of the introduced gas was lower than that of slag and thus thermal energies of slag and generated heat by reaction (1) were flown out as sensible heat of exhaust gas. Slag temperature increased once after decreased at the initial stage, and then decreased again with increasing introduced gas volume. This transient temperature increase was due to the increase of solid fraction as shown in Fig. 2, because FeO in slag was oxidized by H_2O gas to Fe_2O_3 and liquidus temperature changed. In the present calculation conditions, temperature increased transiently because solid fraction drastically increased during slag-gas reaction and heat was generated more than that required for compensation of slag temperature decrease. However, the drastic increase in the amount of produced H_2 gas due to drastic increase of slag temperature was not observed.

Effect of Initial Slag Temperature   The effect of initial slag temperature on the generation behavior of H_2 gas by the reaction of slag and gas was also examined by changing slag temperature from 1873 K to 1973 K. Increase in the initial slag temperature almost linearly increased the equilibrium temperature or decreased the solid fraction of slag, while the production ratio of H_2 gas did not change, because the extra thermal energy provided by increasing initial slag temperature compensated only small portion of energy required for the slag-gas system.

Effect of H_2O Partial Pressure of Introduced Gas   Figure 5 shows the change of total produced amount and production ratio of H_2 gas with partial pressure of H_2O of introduced H_2O–Ar gas in the case of 20 mass% of initial FeO content, 2 of mass%CaO/mass%SiO$_2$ ratio, 1273 K of initial gas temperature and 1933 K of initial slag temperature. Increase in H$_2$O partial pressure increased the produced H$_2$ amount and maximum was about 0.75 mol-H$_2$/kg-slag, while the production ratio of H$_2$ gas against introduced H$_2$O gas decreased to about 7 %. Since inert gas species in introduced gas such as Ar gas in this case also requires thermal energy to increase gas temperature but its energy is released without efficient utilization, the increase of H$_2$O partial pressure increased the produced H$_2$ amount.

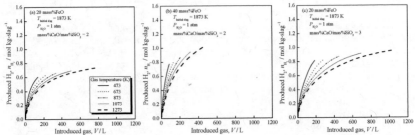

Figure 1. Change of the amount of H$_2$ gas produced by reaction between H$_2$O–Ar gas and various kinds of slag with introduced gas volume.

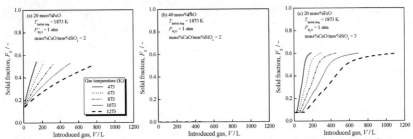

Figure 2. Change of the solid fraction of various kinds of slag after equilibrium with H$_2$O–Ar gas with introduced gas volume.

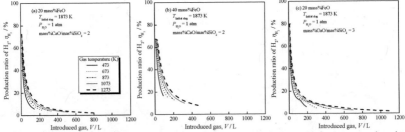

Figure 3. Change of the production ratio of H$_2$ gas by reaction between H$_2$O–Ar gas and various kinds of slag with introduced gas volume.

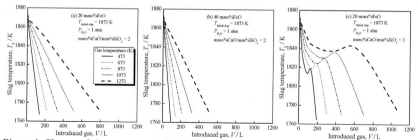

Figure 4. Change of slag temperature by reaction between H_2O–Ar gas and various kinds of slag with introduced gas volume.

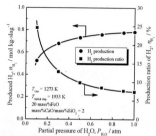

Figure 5. Relationship between the total produced amount or total production ratio of H_2 gas and the partial pressure of H_2O in the introduced gas.

Measurement of H_2 Gas Generation Behavior[8]

<u>Experimental</u>

Figure 6 shows the experimental apparatus, which consists of a gas circuit for preparation of H_2O–Ar gas with constant H_2O partial pressure and an electric furnace with a mullite reaction tube (O.D. 50mm, I.D. 42 mm, length 1000 mm).[8] Three Erlenmeyer flasks with distilled water were connected in series and put inside a thermostat bath kept at 333 K. Argon gas was continuously supplied into flasks with flow rate of 300 cm³/min and 0.2 atm H_2O–Ar gas was obtained. Further Ar gas was mixed with flow rate of 500 cm³/min to increase the gas flow rate and finally 0.086 atm H_2O–Ar gas was blown to the surface of molten slags through a mullite tube (O.D. 6 mm, I.D. 4 mm) which tip was kept at the position of 25 mm above the slag surface. A Pt crucible (upper diameter 36 mm, bottom diameter 22 mm, height 40 mm, volume 30 cm³) was employed to keep 10 g of molten slag at Ar atmosphere. After the slag sample was melted, H_2O–Ar gas was introduced onto the sample. Count of reaction time started when the gas switched to H_2O–Ar gas. After prescribed reaction time passed, H_2O–Ar gas was changed to Ar gas and the reaction tube was purged, and then the Pt crucible was quickly taken out of the reaction tube and quenched by blowing Ar gas. Weights of samples with the Pt crucible before and after each experiment were measured, and chemical compositions of slags before and after experiments were analyzed.

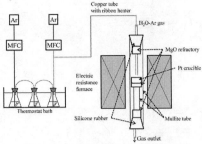

Figure 6. Schematic diagram of the experimental apparatus.

Results

Four kinds of slags were prepared with various FeO contents and mass%CaO/mass%SiO$_2$ ratios, while Al$_2$O$_3$, MgO and P$_2$O$_5$ contents were almost maintained constant, hereafter shown as slags A (47 mass%FeO, mass%CaO/mass%SiO$_2$ = 0.81), B (48 %FeO, %CaO/%SiO$_2$ = 1.28), C (36 %FeO, %CaO/%SiO$_2$ = 0.92) and D (28 %FeO, %CaO/%SiO$_2$ = 0.81), respectively. Approximately 10 g of above slags reacted with 0.086 atm H$_2$O–Ar gas with flow rate of 875 cm^3/min for 15 to 120 min at 1723 K.

Figure 7 shows the relationship between weight increase and reaction time for four slag samples. Weights of all slags increased monotonously with time. The increment at the initial 15 min was the largest and similar for all slags. On the contrary, the increasing rates of slag weights after 15 min were different between slags, which is due to the difference in FeO content of slags. When all FeO in slags are oxidized to Fe$_2$O$_3$, the estimated increase in slag weights are 0.520, 0.538, 0.406 and 0.312 g for slags A, B, C and D, respectively. Therefore, the reaction ratios calculated from weight increase were 43.0, 50.5, 32.2 and 32.2 % for slags A, B, C and D, respectively. The reaction ratio increased with increasing initial FeO content in slag. This difference would be due to the change of conditions of molten slag surface, such as partial solidification of slag surface because of the oxidation of FeO to Fe$_2$O$_3$.

FeO and Fe$_2$O$_3$ contents decreased and increased respectively while contents of other oxides did not change clearly, which means FeO was oxidized to Fe$_2$O$_3$. Comparing slags A and B, FeO content of slag B decreased more than that of slag A. Initial FeO contents of these slags were almost the same, however mass ratio of CaO to SiO$_2$ for slag B was larger than that for slag A. Activity of FeO in slags relative to pure liquid FeO estimated for the FeO–CaO–SiO$_2$ system[9] at 1723 K are 0.56 and 0.63 for slags A and B, respectively. Considering reaction (1), larger FeO activity increases the forward reaction rate more. Therefore, FeO in slag B was oxidized faster than that in slag A.

The inlet and outlet gases were analyzed by a quadrupole mass spectrometer and significant increase in the ion current corresponding to H$_2$ gas of outlet gas was observed, indicating the production of H$_2$ gas by reaction between FeO-containing slag and H$_2$O-containing gas.

From above consideration, the amount of produced H$_2$ gas was calculated by Eq. (2) from the change of FeO content of each slag,

$$n_{H_2} = \frac{-\Delta C_{FeO}}{100} \times W_{slag} \times \frac{1}{71.85} \times \frac{1}{2} \qquad (2)$$

where n_{H_2} is the amount of produced H_2 gas in mole, ΔC_{FeO} is the change of FeO content of slag in mass%, and W_{slag} is the initial weight of slag in gram. The calculated amount of H_2 gas is shown in Fig. 8.

Present experiments were conducted for slags with mass%CaO/mass%SiO$_2$ ratio from 0.8 to 1.3 and temperature at 1723 K because of the regulation of experimental setup. Therefore, the direct estimation regarding H_2 gas generation behavior from BOF steelmaking slag is difficult. However, for example, in the case of slag D which initial FeO content was 28 mass%, 0.59 mol-H$_2$/kg-slag were generated after reaction with 0.086 atm H$_2$O–Ar gas for 120 min. Assuming amount of converter slag generated after one time operation with 300 t capacity BOF is 30 t and the same efficiency for H$_2$ production by slag–gas reaction is expected, 17700 mol or 396 Nm^3 of H$_2$ gas would be produced from this slag. Considering that practical BOF slag contains fine steel particles, the amount of produced H$_2$ gas is expected to increase further.

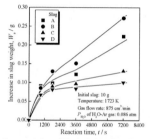

Figure 7. Increase in slag weight with time by reaction between H$_2$O–Ar gas.

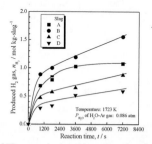

Figure 8. Amount of produced H$_2$ gas calculated based on the change in FeO content of slags.

Conclusions

Development of innovative H_2 production process with BOF steelmaking slag was considered to utilize thermal and chemical energies of discharged slag from steelmaking process. Firstly thermodynamic calculations of reaction between steelmaking slag and H_2O–Ar gas were conducted. Increase in initial FeO content increased the amount of generated H_2 gas and increase in gas temperature was effective to produce more H_2 gas by minimizing temperature drop. Solid fraction would become an important property to be controlled. Secondly, measurement of generation behavior of H_2 gas by reaction between FeO-containing slag and H_2O–Ar gas was done at 1723 K with four kinds of synthesized FeO–CaO–SiO_2–Al_2O_3–MgO–P_2O_5 slag. From 32.2 to 50.5 % of FeO contained in slags was oxidized to Fe_2O_3 by 0.086 atm H_2O–Ar gas for 120 min. Maximal amount of generated H_2 gas estimated from composition change was 0.59 mol-H_2/kg-slag for 120 min at present experimental conditions, namely 17700 mol or 396 Nm^3-H_2 from 30 t of BOF slag discharged from 300 t capacity BOF.

References

1. Greenhouse Gas Inventory Office of Japan, "National Greenhouse Gas Inventory Report of JAPAN" (Report, Ministry of the Environment of Japan, 2013), 2-5.

2. Nippon Slag Association, "Statistics on Iron and Steel Slag (FY2012)" (Report, Nippon Slag Association, 2013), 8.

3. K. Ogino, A. Nishiwaki, Y. Habara and T. Nishino, "Measurement of Heat Content of Slag," *Tetsu-to-Hagané*, 65 (1979), S179.

4. Agency for Natural Resources and Energy, "2010 Annual Report on Energy (Japan's Energy White Paper 2010)" (Report, Ministry of Economy, Trade and Industry of Japan, 2011), 85.

5. E. T. Turkdogan, *Physical Chemistry of High Temperature Technology* (New York, NY: Academic Press, 1980), 11-12.

6. Tata Steel, Media report on *"Tata Steel develops hydrogen production tech, granted PCT"*, (October 08, 2008), http://www.tata.com/media/reports/inside.aspx?artid=C/GtiSlbk+A=.

7. H. Matsuura and F. Tsukihashi, "Thermodynamic Calculation of Generation of H_2 Gas by Reaction between FeO in Steelmaking Slag and Water Vapor," *ISIJ International*, 52 (2012), 1503-1512.

8. M. Sato, H. Matsuura and F. Tsukihashi, "Generation Behavior of H_2 Gas by Reaction between FeO-Containing Slag and H_2O–Ar Gas," *ISIJ International*, 52 (2012), 1500-1502.

9. Verein Deutscher Eisenhüttenleute ed., *SLAG ATLAS 2nd Edition*, (D-Düsseldorf: Verlag Stahleisen GmbH, 1995), 243.

Energy Technology 2014: Carbon Dioxide Management and Other Technologies
Edited by: Cong Wang, Jan de Bakker, Cynthia K. Belt, Animesh Jha, Neale R. Neelameggham,
Soobhankar Pati, Leon H. Prentice, Gabriella Tranell, and Kyle S. Brinkman
TMS (The Minerals, Metals & Materials Society), 2014

A LABORATORY EXPERIMENT STUDY ON IRON PHASE FORMATION DURING HYDROGEN REDUCTION OF IRON OXIDES IN THE MOLTEN SLAG

Chuanjie Cai, Jing Chen, Zemin Zhuang, Xuebin Hao, Shaobo Zheng*

Shanghai Key Laboratory of Modern Metallurgy & Materials Processing, Shanghai University
Shanghai 200072, China

Keywords: Hydrogen metallurgy, Smelting reduction, Iron particles

Abstract

Carbon which was used as heating and reducing agent in traditional iron and steel making process at present brought us large amounts of greenhouse gas harmful gases and dust seriously leading to the fog and haze condition. As one of ideal global new energies, Hydrogen has good features such as rich resource, high heat efficiency, cleanness, renewable and transportation. With the commercialization of hydrogen energy, the world energy structure will be changed greatly. Although many scholars have done a lot of researches and long-term exploration on hydrogen metallurgy, researches on the formation of metal iron phase in the process of hydrogen reduction of iron oxides in slag was rarely reported, which had important theoretical significance on the development of hydrogen smelting reduction process. Based on the thermodynamics and kinetics calculation, A laboratory experiment has been designed to research the iron phase formation during hydrogen reduction of iron oxides in the molten slag in this paper. SEM and EDS techniques were also used to observe and analyze the iron morphology. According to detection results, the influence of temperature, reaction time and surface tension on the morphology of the iron phase reduced were showed.

Introduction

With the increasingly severe global warming greenhouse effect, governments are paying more attention to no-carbon and low-carbon energy development and utilization. Hydrogen has received widespread attention around the world, which has rich resources, high thermal efficiency, large energy density, use clean, transportation, storage, renewable etc [1].The development of the hydrogen economy is new competition areas of the world economy for the 21st century. China should assess the situation, and achieve the important strategy of peaceful development as the realization of new industrialization, by establishing a "hydrogen economy" industry revolution which replaces fossil energy. The current hydrogen metallurgy process includes hydrogen direct reduction, smelting reduction and direct steelmaking process. Hydrogen plasma hydrogen metallurgy is the use of hydrogen as a reducing agent to replace carbon reductant, to reduce carbon dioxide emissions, and to ensure the sustainable development of ironmaking industry.

Production scale of iron and steel is so huge, but there is a problem where a large amount of hydrogen can be obtained. At present the mature technology of hydrogen production includes transformation of petroleum fuel pyrolysis and oxidation method [2], hydrogen and transformation of coal gasification. As used in high carbon energy, this technology still can't avoid carbon dioxide emissions, and also is involved the efficiency problem, therefore not suitable for hydrogen metallurgy. Another problem is water electrolysis hydrogen [3] production. Due to the current electricity is given priority to coal, problems also exist in carbon dioxide emissions, and is also not suitable for hydrogen metallurgy. Future of hydrogen production will be microbial hydrogen production technology, solar hydrogen production [4] and the nuclear waste heat hydrogen production, There is no carbon dioxide emissions [5], to achieve these hydrogen production technologies will impossible, even though the current large-scale hydrogen production [6], but for the development of hydrogen metallurgy points the rest assured.

For the hydrogen reduction of iron oxides in most research on kinetics and reaction mechanism of the hydrogen reduction of nucleation, only for research on solid floating austenite reduction [7], while there is no related reports for the molten state reduction nucleation involving reports of reducing agent used in carbon, hydrogen reduction of molten iron oxide nucleation. In this paper, by hydrogen reduction of iron oxides in slag, observation and analysis of the process in iron nucleation theory [8] and the morphology of iron [9] were done. In the metallurgical reaction process, product of the new generation often must pass through the nucleation and growth of new phase nucleus, their rate under certain conditions, may also become part of the reaction rate. So the nucleation and growth of iron reaction of hydrogen reduction of iron oxides in slag is of great significance. Before experiment, a thermodynamic analysis of hydrogen reduction of iron oxide in the slag were done, the critical nucleation radius reduction temperature of different slag iron interface tension and the critical nucleation energy [9] were calculated. According to the thermodynamic analysis of results of experiments on different reduction temperature and reduce the surface tension of iron, using SEM&EDS to observe and analyze the reduction reaction of the specimen surface morphology and iron slag iron oxide in the experiments under different conditions and the hydrogen mechanism.

Experiment and Procedures

This experiment materials dosage was such few that in order to ensure the accurate of the data. Each component adopts analytical pure chemical reagents. Iron oxide of analytically pure reagents isolates air heating, and the decomposition of ferrous oxalate was prepared. In order to study the hydrogen reduction of the molten slag iron oxide, the slags for experiment should be in accordance with the smelting reduction furnace slag [10] composition preparation. CaO, SiO_2, Al_2O_3 and MgO style in slags sample analysis were made of pure chemical reagent preparation, composition of which lists in Table 1. Afterwards, proportional weighing amount of CaO, SiO_2, Al_2O_3 and MgO four oxide were blend into the graphite crucible, and to improve the accuracy of the experiment, the graphite crucible was put into induction furnace, heated to molten mixture and taken out after fully mixing, cooling and crushing, the process of which is called premelting slags.

Table 1. The compositions of the preparation slags

oxide	FeO	CaO	SiO2	Al2O3	MgO
wt%	5-10	32.5-35	32.5-35	15	10

In order to ensure the appearance of reduced iron core cannot change in the cooling process, the high purity quartz tube was used as furnace tube to make it draw from the chamber of furnce easily to rapid cooling. Experimental apparatus is shown in figure 1. Quartz tube length is 1000 mm, inner diameter 80 mm, thickness of 6 mm, one end of the block, quartz purity 99.999%.

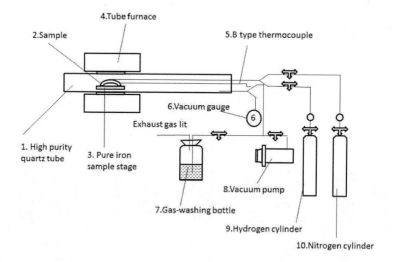

Fig.1 Experimental apparatus

Then the experiment works as following steps:

(1) Discharging air: We mixed quaternary slag samples and FeO in proportion as 9:1, a total of 0.5 g, used tableting machine to press , placed it in a pure iron in the center of quartz tube, closed the valve, put quartz tube in tube furnace hearth, adjusted the sample position at the center of the high temperature area. And then opened the N_2 valve, aerated N_2, We used vacuum pump to extract air to the limit vacuum - 0.1 MPa, and closed vacuum pump. After that, we opened the N_2 valve, and aerated nitrogen until nitrogen filled with quartz tube, we repeated 5 times to remove the air in the quartz tube as far as possible. After the above process, we began to heat up and opened the vent valve and N_2 valve, aerated N_2 as gas protection.

(2) Heating: We set up the heating program of tube type resistance furnace, started to heat up according to the program.

(3) Reduction and fast cooling: when the temperature reached the set temperature, aerate hydrogen after sample melts into a homogeneous slag phase, at a gas flow rate of 3L/min. light the exhaust after the water filtered it. In order to ensure the morphology of iron nuclear in reduction cannot change after cooling process, taken quartz tube out of the furnace immediately after reduction was completed, so that the sample rapidly cooled and solidified. After cooling, we opened the valve and removed the sample, and analyzed it by SEM&EDS.

The reduction time is 4, 6, 8, 10s, temperature is 1400C; reduction time of different temperatures is 6s; the reduction time of adding FeS slag is 6S, and temperature is 1400 ℃; reduction time of adding niobium concentrate residue is 6S, temperature is 1400 ℃.

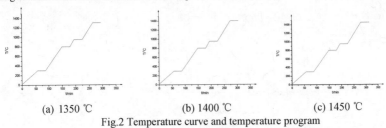

(a) 1350 ℃   (b) 1400 ℃   (c) 1450 ℃

Fig.2 Temperature curve and temperature program

Experiments Results and Results Analysis

(1)The morphology observation pictures by SEM at different temperatures are shown in Fig.3.

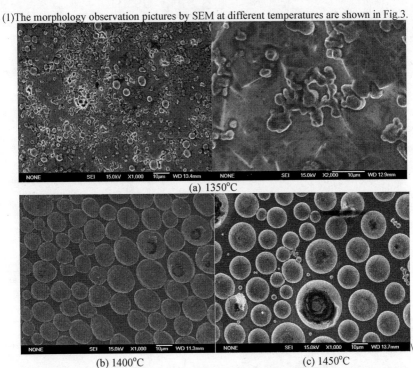

(a) 1350°C

(b) 1400°C   (c) 1450°C

Fig.3 Morphology observation at different temperature

Twenty SEM images of morphology observation of samples at different positions obtained at 1400 °C and 1450 °C. They were respectively dealt with Image-Pro Plus software and statistic of the diameter distribution of iron particles was gathered. According to the statistical results, the diameter of iron particles obtained at 1400 °C mostly ranges from 4μm to 15μm, accounting for 79 percent of the whole amount, and iron particle of which the diameter surpasses 30μm didn't appear; Whereas the diameter of iron particles obtained at 1450 °C mostly ranges from 4μm to 15μm, accounting for 59 percent of the whole amount, and the percentage of iron particles of which the diameter ranges from 20μm to 65μm reaches 22.4 percent. It can be indicated that more iron particles in larger bulk appeared with the increasing reduction time, for higher temperature enhances the rate of chemical reaction and initial rate of component in slag, favoring the generation of iron particles in larger bulk.

(2)Figure 4 restores the morphology of iron generated at different times at 1400°C. The reduction time in (a) was 4s, and iron particles didn't have enough time to grow up once they formed; the reduction time in (b) was 6s, it could be seen that the iron particles had grown into hemisphere; The reduction time in（c）and（d）were 8s and 10s respectively, and many sheets of iron was generated at this moment.

(a)4s　　　　　　　　　　　　　　(b)6s

(c)8s

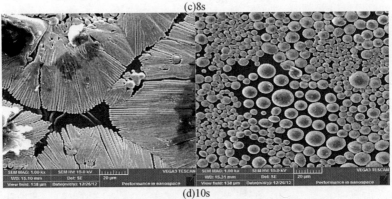

(d)10s

Fig.4 Morphology observation at different time at 1400 °C

(3) Scanned by cold field emission scanning electron microscopy at high resolution, surface morphology of iron particles obtained at different reaction temperatures was observed at different magnifications in Figure 5. At the experimental temperature, the iron generated was solid-state, and formed a crossing crack region, and the crossing strip region existed at the top of some iron particles, which is shown in Fig.5 (b)&(c) respectively, this is because the hydrogen dissolved in iron diffused to the reducing slag at the interface of slags and irons, and reacted with FeO in it, resulting in the generation of bubbles. With the reaction process, the atmospheric pressure inside the bubbles became larger, leading to implosion of iron particles, which is shown in Fig.4. At this experimental temperature, the iron obtained from the reduction reaction with hydrogen is pure iron and solid-state, and the crystal structure of the iron is body-centered cubic, once nucleation process finishes, the iron nucleus begins to grow up radically from center to around in the form of two-dimensional crystal nucleus, which is shown in Fig.5.

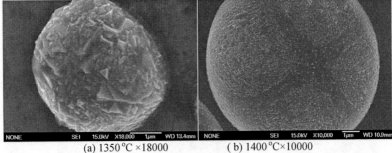

(a) 1350 °C ×18000   (b) 1400 °C×10000

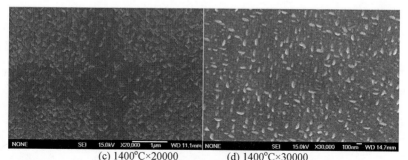

(c) 1400°C×20000   (d) 1400°C×30000
Fig.5 High resolution surface morphology of iron particles

Conclusion

Irregular spherical iron particles are generated at 1350 ℃ under reduction, because the iron grows a certain direction due to the slow diffusion. With the increase in the rate of diffusion of iron, iron particles are spherical cap at 1450℃ and 1400℃, and many iron particles with larger radius are observed compared with those in 1350℃. The iron particles dissolved after the formation of hydrogen diffusion to the interface of slag iron reduction in iron oxide in the slag, the formation of air bubbles, along with the reaction pressure increase of iron particles in bubble burst open, hydrogen will burst after a molten slag at the bottom of the hole bursted open, and direct reduces iron particles.

References

[1] J. Edelson, " Method for Reducing Particulate Iron Ore to Molten Iron With Hydrogen as Reductant" *United States Patent: America*, 5, 464, 464 .1995-11-07.

[2] H. Hiebler and J.F Plaul, "Hydrogen Plasma Smelting Reduction An Option For Steelmaking on the Future," *Metalurgija*, 43(2004)3: 155-162.

[3] T. Nagasaka, M. Hino and S. Ban-ya, " Interfacial Kinetics of Hydrogen with Liquid Slag Containing Iron Oxide," *Metallurgical and Materials Transactions B*, Volume 31, Issue 5 : 945-955.

[4] M.E. Choi and H.Y. Sohn, " Development of Green Suspension Ironmaking Technology Based on Hydrogen Reduction of Iron Oxide Concentrate: Rate Measurements," *Ironmaking and Steelmaking*, 37（2）: 81-88.

[5] J. Bessieres, A. Bessp and J.J. Heizman, "Iron Oxide Reduction Kinetics by Hydrogen," *Int. J. Hydrogen Energy*,5: 585-595.

[6] S. Hayashi and Y. Iguchi, "Influence of Gangue Species on Hydrogen Reduction Rate of Liquid Wustite in Gas-Conveyed Systems," *ISIJ International*, Vol. 35 (1995), No. 3, pp. 242-249.

[7] S. Ban-ya, Y. Iguchi and T. Nagasaka, " Rate of Reduction of Liquid Wustite with Hydrogen" *Tetsu to Hagane*, 1984;70:1689-96.

[8] S. Hayashi and Y. Iguchi, "Surface Segregation of Calcium Oxide in Wustite and Its Effects on the Reduction," *Iron Steel Inst. Jpn. Int*. 1994;34:555-61.

[9] S.L. Teasdale and P.C. Hyes, "Kinetics of Reduction of FeO from Slag by Graphite and Coal Chars," *ISIJ International*, Vol.45(2005),No.5,pp642-650.

[10] C. G. Poper, " Nucleation and Growth Theory in Zeolite Synthesis," *Microporous and mesporous materials*, 1998,21: 333-336.

Energy Technology 2014
Carbon Dioxide Management and Other Technologies

SYMPOSIUM: ENERGY TECHNOLOGIES AND
CARBON DIOXIDE MANAGEMENT

Carbon Dioxide Management

Session Chairs:

Leon H. Prentice
Soobhankar Pati

Energy Technology 2014: Carbon Dioxide Management and Other Technologies
Edited by: Cong Wang, Jan de Bakker, Cynthia K. Belt, Animesh Jha, Neale R. Neelameggham,
Soobhankar Pati, Leon H. Prentice, Gabriella Tranell, and Kyle S. Brinkman
TMS (The Minerals, Metals & Materials Society), 2014

CO₂ EMISSION REDUCTION THROUGH INNOVATIVE MOLTEN SALT ELECTROLYSIS TECHNOLOGIES USING INERT ANODES

Diyong Tang, Huayi Yin, Wei Xiao, Xuhui Mao, Dihua Wang*

School of Resource and Environmental Sciences, Wuhan University, Wuhan 430072 China
wangdh@whu.edu.cn

Keywords: CO_2 emission reduction, CO_2 capture and utilization, molten salt electrolysis, inert anode

Abstract

The world production of crude steel reached 1.55 billion tons in 2012, which generated ~2.5 billion tons of CO_2. Electrochemical metallurgy especially through high temperature molten salt electrolysis with renewable electricity stands for a great opportunity for producing "green" metals. In recent years, electrolytic production of iron without emission of CO_2 in different molten salt systems has been reported. A cost-affordable inert anode is the key to produce acceptable eco-friendly iron by the novel molten salt electrolysis processes. Based on the authors' research, this paper will discuss iron production by molten salt electrolysis using a nickel alloy inert anode. On the other hand, the paper also covers a novel molten salt CO_2 capture and electrochemical transformation (MSCC-ET) process, in which CO_2 is effectively converted to value-added carbon and oxygen in a molten bath using a cost-affordable non-consumable SnO_2 anode.

1. Introduction

Energy and materials are cornerstones to support the development of human society. The traditional primary metal production, especially iron and steel, and electricity production are mainly based on the reactions between carbon (coal) and oxygen (represented by equations (1) and (2) respectively), which produce huge amount of CO_2.

$$M_xO_y + 0.5y\ C = x\ M + 0.5y\ CO_2\ (g) + Q_1 \qquad (1)$$

$$C + O_2\ (g) = CO_2\ (g) + Q_2 \qquad (2)$$

In 2012, the world production of crude steel reached 1.55 billion tons, of which 46.3 %, ~0.7165 billion tons were produced in China. Based on the average level of 1.7t CO_2 emission per ton steel, more than 1 billion tons of CO_2 generated from steel industry in China in the year. Besides the iron and steel industry, thermal power plant leads the CO_2 emission in both USA and China. The total consumed electricity in China increased to 4959 billion kWh in 2012, ~70% of it was from coal fired power plant. The average CO_2 emission for electricity production in the typical thermal power plant in China is around 0.78 kg kWh^{-1}. It accounts for ~ 3.868 billion tons of CO_2 emission from thermal power plant in 2012 in China.

Although the energy efficiency of current metallurgical industry and power plant is being improved, using cleaner or renewable reductants / fuels is crucial to reduce CO_2 emissions. With the rapid development of renewable energy such as solar, wind and hydropower in recent years and its growing expectation in future, novel electrochemical technologies, which bridge the gap between chemicals and energy, shall play a unique role to a low-carbon industry and society.

Production of iron by electrolysis of iron compounds using electricity from non-carbon sources will result in lower CO2 emissions than conventional iron ore reduction with carbon. Early efforts for producing iron by electrolysis go back more than 100 years [1]. Due to the increasing concern of CO_2 emission in the new century, the research on the electrolysis production of iron and steel is drawing intensive attention [2-11]. Electrolysis was attempted in the temperature range from room temperature to 1600°C, in the medium of aqueous solution to high temperature molten salts. Generally speaking, the current density of an electrolysis cell will determine its productivity. The current density in aqueous solution is always in the order of tens of mA cm^{-2}. It is industrially acceptable for production of Pb, Zn, Mn, Cu, Co, Ni, but it is too low to meet the market requirement of iron and steel. The reduction kinetics can be accelerated greatly by simply increasing the temperature, for example, the current density for electrolysis production of Al is one order higher than that of Cu, Ni. In recent years, the evaluated molten salt systems for iron production include molten hydroxide [4,5], molten chloride [6,7], molten carbonate [8-11] and molten oxide [12,13]. The current efficiency in molten NaOH [4,5], CaCl$_2$ [6], Na$_2$CO$_3$-K$_2$CO$_3$ [10,11] at medium temperature (500-900°C) was reported to be around or more than 90%. However, a successful low-carbon iron electro-metallurgical process relies on a cost-affordable inert anode, which is still a great challenge in molten chloride [14] and high temperature molten oxide [13,15].

Besides reducing the generation of CO_2 in metallurgy and power industry, CO_2 capture, utilization and storage (CCUS) is another important pathway to keep an acceptable atmospheric CO_2 level. Electrochemical reduction of CO_2 was investigated in aqueous solution, room temperature ionic liquids (RTILs) and high temperature molten salts. Carbon film was electrochemically deposited on a steel electrode in LiCl-KCl-K$_2$CO$_3$ melts using CO_2 as feedstock by Ito et al.[16] Licht reported a solar thermal carbon capture process in molten Li$_2$CO$_3$ to electrochemically reduce CO_2 to CO and C at 750-950 °C [17]. Using CO_2 as feedstock in molten carbonate salts, mild steel and titanium were recently electro-carburized to form an effective case hardened surface [18,19]. Kaplan et al. reported conversion of CO_2 to CO by electrolysis in molten Li$_2$CO$_3$ at 900°C using a cell comprising a titanium cathode and a graphite anode [20]. Although the reported high thermodynamic efficiency was more than 85%, a graphite anode in the melt was consumable and generated carbon oxide. More recently, Yin et al. successfully transformed CO_2 to value-added carbon and oxygen in molten ternary carbonates using a SnO$_2$ inert anode [21,22], suggesting molten salt CO_2 capture and electrochemical transformation (MSCC-ET) process a potentially practical pathway of CCUS.

In this talk, the authors' recent works on electro-producing iron without CO_2 emission by electroreduction of solid iron oxide in molten binary carbonates and preparing value-added carbon and oxygen by splitting CO_2 in molten ternary carbonates using affordable inert anodes are briefly introduced.

2. Experimental

2.1 Electrolysis of solid Fe$_2$O$_3$ in molten Na$_2$CO$_3$-K$_2$CO$_3$

Fe$_2$O$_3$ (analytical purity, Sinopharm Chemical Regent Co. Ltd.) powder was pressed into pellets (~20 mm in diameter and ~2 mm in thickness) through die-pressing at a pressure of 6 MPa and then sintered at 800 °C in air for 2 hours to ensure a reasonable mechanical strength. The pellets were served as cathode in electrolysis. A Ni10Cu11Fe alloy rod (20mm in diameter) was used as anode. An alumina crucible containing 500 g anhydrous Na$_2$CO$_3$-K$_2$CO$_3$ (59:41 in molar ratio, analytical purity, Sinopharm Chemical Regent Co. Ltd.) was the cell, which was dried in air at 250 °C for 24 h and then the temperature of the cell was raised to 750 °C in Ar atmosphere to allow the melting of the salt. The cell is schematically shown in Figure 1a. Pre-electrolysis was performed at 1.8 V between an anode and a foamed nickel film cathode for about 2 h to further remove residual water and, if any, other impurities from the molten salts. Cyclic voltammetry measurements were performed using an Fe_2O_3 coated iron wire working electrode (FCE), a Ni10Cu11Fe alloy rod counter electrode and a Ag^+/Ag reference electrode (a silver wire dipped into a mixture of Ag_2SO_4 (0.1 mol kg^{-1}) and eutectic carbonate melt ($Li_2CO_3:Na_2CO_3:K_2CO_3$ = 43.5:31.5:25 mol %) in a close-one-end mullite tube). The cyclic voltammetry (CV) measurements were conducted on a CHI1140a electrochemical workstation (Shanghai Chenhua Instrument Co. Ltd.). Electrolysis was controlled by a DC power system (Shenzhen Neware Electronic Ltd.) between the Fe_2O_3 pellets cathode and the alloy anode. More experimental details can refer to literature [10,11].

2.2 Electrochemical reduction of CO_2 in molten Li_2CO_3-Na_2CO_3-K_2CO_3

Anhydrous Li_2CO_3, Na_2CO_3, K_2CO_3 with analytical purity were purchased from Sinopharm Chemical Regent Co., Ltd. SnO_2 anodes (STANNEX E) were obtained from Dyson Thermal Technologies. The electrolysis cell consisted of an alumina crucible filled with 500g of ternary carbonate ($Li_2CO_3:Na_2CO_3:K_2CO_3$ = 43.5:31.5:25 mol%). The crucible was sealed in a steel reactor heated by a tube furnace. In order to remove impurities and residual water, pre-electrolysis was conducted using a SnO_2 anode and a nickel sheet cathode under constant cell voltage of 1.5V for 2 hours. CO_2 was continuously bubbled into the melt by an alumina tube during electrochemical tests and electrolysis experiments. CV measurements were performed on an electrochemical workstation (CHI1140a) using a three-electrode configuration with a silver wire quasi-reference electrode. Constant cell voltage electrolysis was carried out in the melt using a computer-controlled DC power system (Shenzhen Neware Electronic Co. Ltd.) equipped with a Ni sheet cathode and a SnO_2 rod anode. The cell is schematically shown in Figure 2a. The experimental temperature was between 450°C to 650°C. More experimental details can refer to literature [21,22].

3. Results and Discussion

3.1 Electrolytic production of iron using a Ni10Cu11Fe inert anode in binary carbonates

Figure 1b presents a typical CV of a solid Fe_2O_3 powder electrode in molten Na_2CO_3-K_2CO_3 at 750°C. It can be seen that there are two reduction peaks (c1 and c2) before the deposition of alkaline metals (c3). Iron product was obtained by potentiostatic electrolysis at -1.75V as evidenced by XRD test (Figure 1d), demonstrating that c2 is related to the generation of metallic iron. This is consistent with the thermodynamic of the system as shown in Table 1. Detailed investigation by potentiostatic electrolysis and XRD tests [11] reveals that the pathway of Fe_2O_3 reduction in molten Na_2CO_3-K_2CO_3 includes three steps (3-5). And minor carbon may deposit on the reduced iron to form Fe-C through reaction (1-1) in Table 1.The carbon content was found to be in the range of 0.035% to 0.760%, depending on the applied electrolysis cell voltage [10,11].

c1: $Fe_2O_3 + Na^+ + e = NaFe_2O_3$ (3)

c1: $2NaFe_2O_3 + Na^+ + e = 3NaFeO_2 + Fe$ (4)

c2: $NaFeO_2 + 3e = Fe + Na^+ + 2O^{2-}$ (5)

I-t curve during constant voltage electrolysis is shown Figure 1c. It behaves a typical kinetics of electroreduction of solid pellet cathode in molten salts [4-6]. Based on the i-t curves, current efficiency for the electrolysis of solid Fe_2O_3 in the binary carbonates was calculated [10,11]. The current efficiency can be as high as 95%, higher than that obtained in molten alkaline [4,5] and molten oxide [12,13]. The energy consumption of electrolysis was 2.87 kWh/kg-Fe[10], which can be further optimized by reducing the cell voltage. Oxygen evolution on the anode was detected by GC and the Ni10Cu11Fe alloy anode did not consumed and covered by a dense and conductive thin oxide layer [10]. Therefore, an energy efficiency molten salt electrolysis process with zero carbon emission, consisted of a solid iron oxide cathode, a NiFeCu inert anode and Na_2CO_3-K_2CO_3 electrolyte was demonstrated.

We would further highlight the merits of using molten Na_2CO_3-K_2CO_3 as electrolyte. The first, it provides enough wide potential window for iron oxide reduction; the second, unlike the high solubility of Fe_2O_3 in molten Li_2CO_3, the solubility of iron oxide, nickel oxide in the melt is quite low so that a solid-state reduction can take place, which avoids the dendritic deposit; the third, comparing with molten chloride and super-high temperature molten oxide, a cheap nickel alloy anode works well; the fourth, high quality iron product with controllable carbon content can be obtained. Iron prepared by electrolytic reducing haematite pellets in molten sodium hydroxide at 530°C contained oxygen between 3 and 7 wt% which is needed to re-melt to separate iron from the slag [4,5]. Electrodeposited iron from molten Li_2CO_3 was dendrite and contained lots of Li_2CO_3 [8,9]. Removing Li_2CO_3 from the product will produce extra-cost and is not easy because the solubility of Li_2CO_3 in water is quite low and decomposition of Li_2CO_3 will take place at the melting temperature of the product.

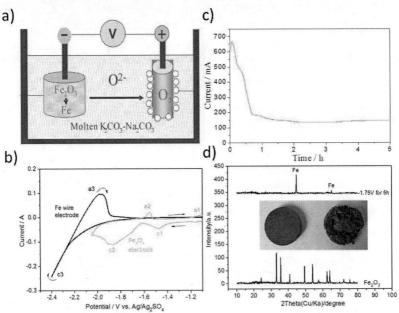

92

Figure 1. (a) Schematic of production of iron without CO_2 emission by electrochemical reduction of solid Fe_2O_3 in molten Na_2CO_3-K_2CO_3 using a NiFeCu rod inert anode on which oxygen evolution takes place; (b) CVs of iron and solid Fe_2O_3 electrode in the melt; (c) Typical i-t curve during electrolysis of Fe_2O_3 pellets cathode equipped with a NiFeCu alloy anode; (d) Typical electrolytic iron product and its XRD patterns.

Table 1. Gibbs free energy and decomposition voltages of Na_2CO_3, K_2CO_3, $FeCO_3$ and typical oxides at 750 °C (data from HSC 5.0).

	Reactions	ΔG (kJ)	ΔE (V)
1-1	$Na_2CO_3 = Na_2O + C + O_2$ (g)	569.85	1.48
1-2	$2Na_2CO_3 = 4Na$ (l)$+O_2$ (g)$+2CO_2$ (g)	810.71	2.10
1-3	$2FeCO_3 = 2Fe + O_2$ (g)$+2CO_2$ (g)	196.87	0.51
1-4	$K_2CO_3 = K_2O + C + O_2$ (g)	638.68	1.65
1-5	$2K_2CO_3 = 4K$ (l)$+ O_2$ (g)$+2CO_2$ (g)	926.60	2.40
1-6	$Fe_2O_3 = 2Fe + 1.5 O_2$ (g)	553.58	0.96
1-7	$6Fe_2O_3 = 4Fe_3O_4+O_2$ (g)	178.57	0.46
1-8	$2Fe_2O_3 = 4FeO+O_2$ (g)	316.89	0.82
1-9	$2FeO = 2Fe + O_2$ (g)	395.13	1.02
1-10	$2NaFeO_2 = 2Fe + Na_2O + 1.5O_2$ (g)	706.76	1.22

3.2 Electrochemically splitting CO_2 to carbon and oxygen using a SnO_2 inert anode in ternary carbonates

Due to the narrow potential window (determined by the decomposition voltage of Na_2O and K_2O), Na_2CO_3-K_2CO_3 binary electrolyte is not suitable for electrochemical transformation of CO_2. From the thermodynamic data as shown in Table 2, Na_2O/K_2O will spontaneously transformed into Na_2CO_3/K_2CO_3 and Li_2O in the Li_2CO_3-Na_2CO_3-K_2CO_3 ternary system (reactions 2-9, 2-10). The reactions (2-1) and (2-7) are the most favorable carbon generation reactions. The decomposition voltage of Li_2O at 500°C is 2.57V, providing plenty of reduction window of carbonate and CO_2. Furthermore, the Gibbs energy change for the reaction between Li_2O and CO_2 (2-11) is -103.34 kJ mol^{-1}, suggesting CO_2 is readily captured by Li_2O if any Li_2O is presented in the melt. In fact, Li_2O will be generated by reaction (2-1) during the electrolysis. Figure 2b presents the CO_2 equilibrium pressure at different Li_2O activity. It can be seen that the CO_2 pressure can be as low as 10^{-4} even the activity of Li_2O in the melt is just in the order of 10^{-3}. The reduction product of CO_2 / carbonate in molten carbonates can be C or CO, depended on the temperature and applied potential as shown in Figure 2c. At the temperature below 900°C, production of C is more thermodynamically favorable. Considering the potential variation on the cathode, a potential window above 200-300mV is necessary for eliminating the side-reaction of producing CO. Therefore, electrolysis in the temperature range of 450-650°C was performed using a nickel cathode and a SnO_2 anode with a cell voltage between 3.0V and 5.0V. Amorphous carbon was obtained (Figure 2d). The current efficiency can reach 90% [21]. Furthermore, The obtained carbon exhibits a BET surface area excess of 414 m^2 g^{-1}, being advantageous for its use for capacitive and adsorbent materials in energy storage and pollutant removing. NiFeCu alloy was found unstable in the ternary system due to the high solubility of oxide in the melt, a dense oxide layer is not able to form. Fortunately, SnO_2 rod behaves effective in the melt and keeps stable for more than 500 hours in the laboratory. Oxygen gas was detected in the outlet gas, demonstrating that CO_2 is successfully split into carbon and oxygen.

Comparing with previously reported CCS processes, the MMCS-ET process has some unique characteristics favoring its large-scale application: working at the lower temperature and intrinsic ability to capture for CO_2; capable of producing value-added carbon materials and oxygen; high flux transformation of CO_2 at reasonable current efficiency using affordable electrodes; amenable to combine with a renewable energy source like solar power.

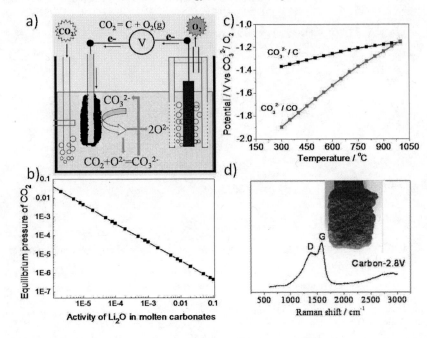

Figure 2. (a) Schematic of molten salt CO_2 capture and electrochemical splitting to carbon and oxygen in molten Li_2CO_3-Na_2CO_3-K_2CO_3 using a SnO_2 rod inert anode on which oxygen evolution takes place; (b) Equilibrium pressure of CO_2 with different activity of Li_2O in the ternary carbonates at 500 °C; (c) Deposition potential of CO_3^{2-}/C and CO_3^{2-}/CO versus temperature; (d) Raman spectra of the typical electrolytic carbon.

Table 2 Thermodynamics data and dissociation potential of typical reactions (data from HSC 5.0)

	Reactions (500°C)	ΔG(kJ / mol)	ΔE(V)
2-1	$Li_2CO_3 = Li_2O + C + O_2(g)$	498.86	1.29
2-2	$Na_2CO_3 = Na_2O + C + O_2(g)$	602.06	1.56
2-3	$K_2CO_3 = K_2O + C + O_2(g)$	674.66	1.75
2-4	$Li_2CO_3 = 2Li(l) + C + 1.5O_2(g)$	995.62	1.72
2-5	$Na_2CO_3 = 2Na(l) + C + 1.5O_2(g)$	881.66	1.52
2-6	$K_2CO_3 = 2K(l) + C + 1.5O_2(g)$	927.40	1.60
2-7	$CO_2(g) = C + O_2(g)$	395.5	1.02
2-8	$Li_2O = 2Li(l) + O_2(g)$	993.50	2.57
2-9	$Na_2O + Li_2CO_3 = Na_2CO_3 + Li_2O$	-103.2	

| 2-10 | $K_2O + Li_2CO_3 = K_2CO_3 + Li_2O$ | -175.8 |
| 2-11 | $Li_2O + CO_2 \ (g) = Li_2CO_3$ | -103.34 |

4. Perspective

Along with the rapid development of renewable electricity and increasing concern of global climate change, novel molten salt electrolysis technologies can play important roles in the reduction of CO_2 level in atmosphere, either through zero-emission iron and steel metallurgy or capture and utilization of CO_2. Binary and ternary molten carbonates have been demonstrated here effective media for electrolysis using cost-affordable inert anodes. And the demonstrated molten carbonate electrolysis processes sound effective in energy efficiency and product quality, although further optimization is still needed.

Acknowledgements

The authors thank the NSFC (Grant Nos. 20873093, 51071112), MOE of China (NCET-08-0416) and the Fundamental Research Funds for Central Universities of China for financial support.

References

1. R. H. Aiken, *U.S. Pat.* 816142 (1906).
2. A. Allanore, H. Lavelaine, G. Valentin, J.P. Birat and F. Lapicque, *J. Electrochem.Soc.*, 155, 125 (2008)
3. A. Allanore, J. Feng, H. Lavelaine and K. Ogle, *J. Electrochem.Soc.*, 157, 24 (2010)
4. A. Cox, D.J. Fray, *J. Appl. Electrochem.*, 38, 1401 (2008).
5. A. Cox, D.J. Fray, *Ironmak. Steelmak.*, 35, 561 (2008).
6. G.M. Li, D.H. Wang, Z. Chen, *J. Mater. Sci. Technol.*, 25, 767 (2009).
7. S.L.Wang, G.M. Haarberg, E. Kvalheim, *J. Iron Steel Res. Int.*, 15, 48 (2008).
8. S. Licht, B.H. Wang, *Chem. Commun.*, 46,7004 (2010).
9. S. Licht, H.J. Wu, Z.H. Zhang, H. Ayub, *Chem. Commun.*,47, 3081 (2011).
10. H.Y. Yin, D.Y. Tang, H. Zhu, Y. Zhang, D.H. Wang, *Electrochem. Commun.*, 13, 1521 (2011).
11. D.Y. Tang, H.Y. Yin, W. Xiao, H. Zhu, X.H. Mao, D.H. Wang, *J. Electroanal.Chem.*, 689, 109 (2013).
12. D.H.Wang, A.J. Gmitter, D.R. Sadoway, *J. Electrochem.Soc.*, 158, E51 (2011).
13. A. Allanore, L. Yin, D. R. Sadoway, *Nature*, 497, 353–356 (2013).
14. H.Y. Yin, L.L. Gao, H. Zhu, X.H. Mao, F.X. Gan, D.H. Wang, *Electrochim. Acta*, 56, 3296 (2011).
15. H. Kim, J. Paramore, A. Allanore, and D.R. Sadoway, *J. Electrochem. So*c., 158, E101 (2011).
16. H. Kawamura, Y. Ito, *Journal of Applied Electrochemistry*, 30, 571 (2000)
17. S. Licht, B.H. Wang, H.J. Wu, *Journal of Physical Chemistry C*, 115, 11803(2011)
18.N.J. Siambun, H. Mohamed, D. Hu, D. Jewell, Y.K. Beng, G.Z. Chen, *Journal of the Electrochemical Society*, 158, H1117 (2011).
19. V. Kaplan, E. Wachtel, I. Lubomirsky, *Journal of the Electrochemical Society* ,159, E159 (2012).

20. V. Kaplan, E. Wachtel, K. Gartsman, Y. Feldman, I. Lubomirsky, *Journal of the Electrochemical Society*, 157, B552 (2010).
21. H. Yin, X. Mao, D. Tang, W. Xiao, L. Xing, H. Zhu, D.H. Wang, D.R. Sadoway, *Energy & Environmental Science*, 6, 1538 (2013).
22. D.Y. Tang, H.Y. Yin, X.H. Mao, X.H. Mao, W. Xiao, D.H. Wang, *Electrochimica Acta*, (accepted)

Energy Technology 2014: Carbon Dioxide Management and Other Technologies
Edited by: Cong Wang, Jan de Bakker, Cynthia K. Belt, Animesh Jha, Neale R. Neelameggham,
Soobhankar Pati, Leon H. Prentice, Gabriella Tranell, and Kyle S. Brinkman
TMS (The Minerals, Metals & Materials Society), 2014

STUDY ON UTILIZATION OF CYCLIC HEAT STEWED STEEL SLAG WASHING WATER TO MINERALIZE CO_2

DOU Zhihe[1], ZHANG Zimu[1], LIU Yan[1], Lv Guozhi[1], ZHANG Ting'an[1], JIANG Xiaoli[1]

DOU Zhihe, Lecturer, E-mail: douzh@smm.neu.edu.cn, Shenyang, Liaoning 110819, China
ZHANG Zimu, PhD Candidate, Shenyang, Liaoning 110819, China
LIU Yan, Professor, E-mail: liuy@smm.neu.edu.cn, Shenyang, Liaoning 110819, China
ZHANG Ting'an, Professor, E-mail: zta2000@163.net, Shenyang, Liaoning 110819, China
JIANG Xiaoli, Engineer, E-mail: jiangxl@smm.neu.edu.cn, Shenyang, Liaoning 110819, China
([1]Key Laboratory of Ecological Utilization of Multi-metal Intergrown Ores of Ministry of Education, School of Materials and Metallurgy, Northeastern University, Shenyang, Liaoning 110819 , China)

Keywords: steel slag; heat stewing technology; mineralization of CO_2; calcium carbonate

Abstract

Cyclic heat stewed slag washing water carries high content of alkali and calcium and may cause serious tube scaling and impose an adverse impact on normal production. This study proposes a solution to utilize washing water by absorbing and mineralizing CO_2, potentially restraining scaling and being able to create calcium carbonate powders. Effects of water flux on water quality, utilization ratio of CO_2 and yield of calcium carbonate powder have been investigated; meanwhile, features of calcium carbonate powder are also represented. On the other hand, effects of factors such as as injection flux, pressure of CO_2 on utilization ratio of CO_2, morphology and purity of calcium carbonate are also investigated. The results show unambiguously that after soften injection of CO_2, calcium ions were removed effectively via precipitation, alleviating the scaling problem. High purity yet fine carbonate powder could be achieved.

Introduction

Nowadays, 0.34 tons of blast furnace slags will be produced when producing 1 ton of pig iron; ensuing that, 0.12 tons of steel slag will be generated with 1 ton steel production while 1.57 billion tons of steel output all over the world in the year of 2012. It is therefore a critical issue to realize zero discharge of steel slags to achieve sustainable development and cyclic economy of iron and steel industry. In recent years, research institute of MCC and China JINGYE engineering company developed a heat stewed steel slag self-decomposition technology GENERATION III to solve the problem of instability of steel slags and realize the zero discharge. However, the problem of circulation line scaling which may cause the clogging, even exploding, need to be solved. The principles of the technology are: directly spraying water to the high temperature steel slag when it is poured into the hot closed pot, realizing the self-decomposition via utilizing the waste heat of steel slag. The washing water is recycled to achieve a cyclic utilization. The process is shown in Figure 1.

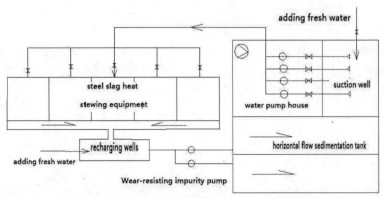

Figure 1. Flow chart of the steel slag heat stewed technology

In this process, high hardness, high alkalinity and high pH value of cyclic water will lead to an unstable state of water cyclic system and serious scaling issue. The main deposition is $CaCO_3$ which can only be removed by labor intensive work. Aiming on this problem, ZHANG Ting' an etc. created a technology of injecting CO_2 to soften the heat stewed steel slag washing water to prevent scaling, successfully solving the scaling problem in the heat stewed process and receiving the calcium carbonate powder as a secondary product, which realized the high valued mineralization of CO_2 off gas.

Experimental

<u>Detections and analysis</u>

Due to the fact that the cyclic washing water is of a high alkalinity and calcium content without impurities such as magnesium, silicon and iron, during the experiment, it is the pH value that was detected in the soften process (there is a one-to-one correspondence between the pH value and the calcium content). The phase composition of calcium carbonate powder was detected by X-ray diffraction analyzer（Rigaku D/max Ⅲ B）; The microstructures and particle size distribution were detected by S8010 scanning electron microscope and laser particle size instrument.

<u>Experiment principles and procedures</u>

The step of XRD analysis was 0.02. The main phases of steel slags are CaO, SiO_2, FeO, Fe_2O_3, MnO and Al_2O_3 etc. Slags and iron are separated after the heat stewed process; meanwhile, cyclic washing water is found to carry high hardness and alkalinity due to the dissolution of CaO and MgO, which causes scaling in the tube during the cooling process. According to Figure 2, the main deposition is $CaCO_3$. The pH value of the water is as high as 12-14, with a calcium ion content of 0.45~0.60g/L. However, impurities such as magnesium, silicon and iron were not found. It can be deduced that the high content of calcium and high value of alkalinity are the main reason of scaling in the cyclic water.

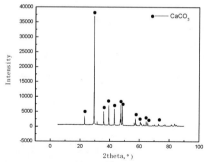

Figure 2 XRD pattern of the deposition in the tube

The results in this paper are all from industrial experiments in Environmental Resources Development co. of MCC in XINYU in 1.1 million tons/a steel slag heat stewing system.

Results and Discussion

Principles of the soften process

According to the characters of the deposition and the washing water, this study proposes a method that injecting CO_2 off gas to the washing water directly to soften the calcium deposition, then returning the filtered liquid to the cyclic system and obtaining calcium carbonate powder as a secondary product. The principle of the soften process is:

$$Ca^{2+}+CO_2+2OH^-\rightarrow CaCO_3\downarrow+ H_2O$$

Water quality analysis

Figure 3 shows the effects of water influx on the softening results under the condition of a CO_2 injection flux of $12m^3/h$ and a pressure of 0.5Mpa,which can keep stable gas flow rate and easily investigate the influence of different water influx.. At a constant injection flux and pressure of CO_2, the pH value of the washing water increased with the increasing water influx. The higher water influx is, and the faster pH value increases. When the injection flux of CO_2 was at $12m^3/h$, the water influx must be higher than 120 m^3/h, otherwise the pH value of the softened water was so high that the calcium ion cannot precipitate fully, and the scaling would still exist. But the washing water influx cannot be lower than $60m^3/h$ as well, otherwise the amount of CO_2 would be excessive, and the generated calcium carbonate would redissolve in the water, leading a high calcium content.

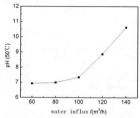

Figure 3 pH of cyclic water with different water influx

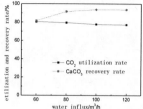

Figure 4 CO_2 utilization and $CaCO_3$ recovery rate with different water influx

CO$_2$ utilization and CaCO$_3$ recovery rates

According to Figure 4, under the condition of a CO$_2$ injection flux of 12m^3/h and a pressure of 0.5MPa, the higher washing water influx would lead to a constant CO$_2$ utilization rate of 75%, by controlling the pH value of the treated water,. But a relative low water influx would lead to a low CaCO$_3$ recovery rate; it is because of the redissolution of calcium carbonate deposition, which is adverse to prevent scaling. Under that condition, the CaCO$_3$ recovery rate was 80%. When the water influx increased, the recovery rate could increase to 90% or higher. Therefore, to guarantee a good washing water softening effect, it is necessary to control an appropriate flux of water and gas to recovery the CaCO$_3$ powder.

Representation of CaCO$_3$ powder

CaCO$_3$ in the washing water was collect by a pocket filter and the size of filter bag was 2 to 5um. After filter pressing with air, the moisture rate was less than 50%. Then the CaCO$_3$ with about half water was taken out of the filter bag, baked in the drying oven and ground to be powdery.

Figure 5 shows the XRD patterns of the CaCO$_3$ powder obtained under the condition of a CO$_2$ injection flux of 12m^3/h and a pressure of 0.5MPa, all the products are only CaCO$_3$. The ICP analysis of Ca, Mg, C, O, Fe, Al and etc. shows that the purity of CaCO$_3$ powder is higher than 96%. When the washing water influx is too low, the purity of CaCO$_3$ powder would significantly decrease. Because at a constant injection flux of CO$_2$, a low washing water influx would lead to the redissolution of CaCO$_3$, followed by a decrease of whiteness and purity. According to Figure 6, controlling the pH value of treated water at 9, a washing water influx of 80m^3/h, the purity of CaCO$_3$ can reach 98%. But when the washing influx was at 60m^3/h, the purity was just 95%. Careful analysis dictates the impurity to be CaO, whose content was lower than 5%.

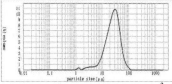

Figure 5 XRD patterns of the CaCO$_3$ powder   Figure 6 Purity of CaCO$_3$ powder

Figure 7 shows the particle size distributions of CaCO$_3$ powders under different washing water influx. According to the figures, the mean grain sizes of CaCO$_3$ were between 20~30μm, the minimum size appeared at the water influx of 80m^3/h.

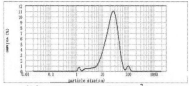

（a）water influx of 120m^3/h   （b）water influx of 100 m^3/h

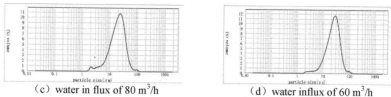

(c) water in flux of 80 m³/h   (d) water influx of 60 m³/h

Figure 7 Contrast of particle sizes in different water influx

Figure 8 shows the SEM photos of calcium carbonate obtained at different gas and liquid ratio. It can be seen that the CaCO₃ powders demonstrate cube morphology at higher water influx. However, virgate crystals populate substantially when the washing water influx decreases to 100m³/h. So it is the case for 80 and 60 influxes.

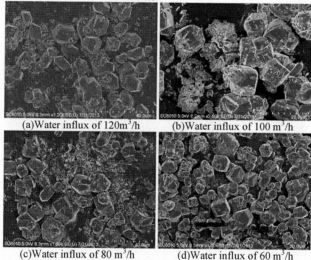

(a)Water influx of 120m³/h   (b)Water influx of 100 m³/h

(c)Water influx of 80 m³/h   (d)Water influx of 60 m³/h

Figure 8 SEM photos of calcium carbonate obtained in different water influx

Figure 9 shows the whiteness variation of calcium carbonate as a function of water influx. It can be seen that washing water influx had negligible effect on the whiteness of calcium carbonate, which is all between 86-88%.

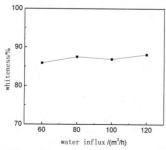

Figure 9 Whiteness of the calcium carbonate particles with different water influx

Conclusion

1. After softening treatment on steel slag in heat stewed washing water, calcium ions have been removed effectively, and the tube scaling problem has been alleviated.
2. Injection of CO_2 enables mineralization of CO_2, of which the absorbing rate is as high as 75%. Meanwhile, high value-added $CaCO_3$ is obtained as a secondary product with a whiteness of 86-88%, particle sizes of 20-30μm and purity of 96%.

Acknowledgement

This study is supported by the National Natural Science Foundation of China (No. U1202274, 51004033, 50934005, 50974035, 51204040, 51374064 and 51074047), the National HighTechnology Research and Development Program ("863" Program) of China (2010AA03A405 and 2012AA062303), Operation Expenses for Universities' Basic Scientific Research (N100302005), and The Fund of Doctoral Program of Higher Education (20050145029).

References

1. ZHENG Xinye. "Global carbon dioxide emissions situation and strategy", *International Observe*, 2010, 6: 55-57.
2. JIANG Huaiyou. "Policies for CO_2 emission reduction and prospects for CO_2 geological storage underground", *Sino-global Energy*, 2007, 12 (5):7-13.
3. ZHANG Ting'an, et al. Anti-scale and anti-corrosion method of water circulation line base on the Heat atewed steel slag technology, Chinese patent, 2009, 200810229145.3.
4. YANG Guilan. Bury the carbon diaoxide. *Digest of Science and Technology*, 2001, 11: 84-85.
5. MA Qianqian. "Research survey of CO_2 emission reduction technology", *Liaoning Chemical Industry*, 38(3) (2009): 176-179.
6. LI Xianyong. "General stiation and development of discharge-reduction and sealing-up-utilization technology of CO_2", *Electrial Equipment*, 2008, 9(5): 710.
7. Herzogh, Eliassonb, Kaarstado. "Capturing greenhouse gases", *Scientific American*, 2000, 2: 54-61.
8. GENG Xiaojie, et al. "Choosing Sites to bury CO_2 in abandoned oil reservoirs", *Resources & Industries*, 2009, 11(6): 69-73.
9. Eloneva S, et al. "Preliminary assessment of a method utilizing carbon dioxide and steelmaking slags to produce precipitated calcium carbonate", *Applied Energy*, 2011, 90(1-SI): 329-334.
10. Florin N，Fennell P，Synthetic. "CaO-based Sorbent for CO2 Capture"，The 10th International Conference on Greenhouse Gas Control Technalogies, 2011, 4: 830-838.

Recent Advances in Carbon Dioxide Mineralization to Nano-size Calcium Carbonate Utilizing Wastewater

ZHANG Ting-an[1], ZHAO Hongliang[1], LIU Yan[1], DOU Zhihe[1], LV Guozhi[1],ZHAO Qiuyue[1], LI Yan[1]

ZHANG Ting-an, Professor, E-mail: zta2000@163.net,Shenyang, Liaoning 110819, China
ZHAO Hongliang, Doctor, E-mail: jayjayzhl@126.com, Shenyang, Liaoning 110819, China
LIU Yan, Professor, E-mail: liuy@smm.neu.edu.cn, Shenyang, Liaoning 110819, China
([1]Key Laboratory of Ecological Utilization of Multi-metal Intergrown Ores of Ministry of Education, School of Materials and Metallurgy, Northeastern University, Shenyang, Liaoning 110004 , China)

Keywords: carbon dioxide; high alkaline wastewater; mineralization; nano-size calcium carbonate

Abstract

Abstract: How to reduce carbon dioxide emissions has become a vexing problem, meanwhile, there exist high energy consumption and low efficiency problems in the existing CO_2 capture and storage technologies. For CO_2 comprehensive Utilization, this paper researches a new technology of absorbing the CO_2 emission with high alkaline calcium or high alkaline wastewater sludge mineralization of calcium and invent a core device for multi-level jet carbonation. The industrial test has been carried out in China Xinyu slag Engineering Co., Ltd. The wastewater flow is $150 m^3/h$ in the pilot scale. The prepared calcium carbonate has a whiteness greater than 90%, particle size less than 150nm. The mineralization of aluminum exhaust with waste carbide indicates that the CO_2 absorption rate is greater than 99.0%, CO_2 concentration in the exhaust gas is lower than 0.08%, the purity of mineralized calcium carbonate is greater than 99.0% and particle size is less than 100 nm.

Instruction

According to the statistic data from IEA (2009), the carbon dioxide discharge from fossil fuel consumption in China has exceeded it in the US by the year of 2007, China, has become the largest emitter of carbon dioxide all over the world. Meanwhile, the economic scale is still on an increasing trend in the next 50-100 years, leading to a continuous increment in carbon dioxide discharge. According to the international energy agency forecasts, the energy-relative CO_2 discharge in China will rise to 11,615 million tons, 91.3% higher than it in 2007. In the year of 2012, the CO_2 gross discharge has reached 10,000 million tons, obviously higher than the predicted value. It is a tremendous task to reduce the emission. Non-ferrous metals industry is one of the largest emitters of carbon dioxide among the industrial circle, the whole industry might discharge a huge amount of carbon dioxide (over 600 million tons), in which the electrolytic aluminum industry is the largest energy consuming industry and CO_2 emitter. At present, the CO_2 discharge has focused by the whole world due to the environmental problems it causes; meanwhile, supplement of new carbon source is needed due to the exhausting oil and coal resources, so that various countries in the world are attaching importance of the recycle and purification of CO_2. Many countries are studying on the cyclic utilization of carbon dioxide. Gas CO_2 is mainly applied in chemical, pharmaceutical raw materials, beverage filling agent, refrigerant, inert medium, solvents, pressure source, the welding protective gas, antioxidant and fire extinguishing agent, etc. Solid CO_2 is mainly applied in penicillin production, fish, cream, ice cream and other food storage and transportation at low temperature, etc. nowadays, such methods as absorption method, membrane separation, adsorption, etc. are widely used to recycle CO_2, with such problems as high investment cost and technical constraints etc.

Steel industry, as one of China's national economic development pillar industries, is raising water consumption and pollution with its rapid development. In the year of 2009, the amount of waste water with high alkali content has been higher than 3,000 million tons in steel industry, accounting for 12% of the total industrial discharge. The heat stewed steel slag self-decomposition technology GENERATION III to solve the problem of instability of steel slags, realizing the zero discharge; meanwhile made huge amount (1,000 million tons of waste water discharge) of waste water with high calcium and alkali content, and terrible scale in the tube, strongly impacting the normal producing process in a negative way. Moreover, PVC industry discharges more than 40 million tons of carbide slag slurry waste with high alkali and calcium content. Therefore, high alkali and calcium content wastewater comprehensive treatment has become the bottleneck of restricting the sustainable development of the whole industry.

On one hand, aiming at treat the huge amount of CO_2 gas discharge, and dealing with enormous amount of wastes with high calcium content, ZHANG Ting' an etc. created "a method to prevent scale and corrosion based on heat stew steel slag cyclic washing water" and "an equipment to trap and mineralize CO_2 in exhaust gas from electrolytic aluminum", "a method to trap and mineralize CO_2 in exhaust gas from electrolytic aluminum" etc. and raised several theories such as "efficient use CO_2 emissions from metallurgical industry(such as steel, aluminum electrolytic)", "solving the scaling problem caused by high alkali and calcium waste water system in steel industry" and "mineralizing CO_2 from steel and electrolytic aluminum industry by carbide slag", reducing emissions of carbon dioxide in metallurgy industries, also enhancing the technology to deal with the waste water with high alkali and calcium content, realizing the comprehensive utilization of secondary resources.

Technique principles

Technology to inject CO_2 to the high alkali and calcium waste water to prepare $CaCO_3$

Focusing on the characters of heat stewed slag cyclic washing water (high calcium content & high pH value), created "a method to mineralize high alkali and calcium waste water by injecting CO_2 to prepare calcium carbonate and equipment". By details, this technology is injecting CO_2 to the heat stewed slag cyclic washing water under a mechanical stirring condition to realize soften modification, meanwhile, producing calcium carbonate as a secondary products. This technology can easily integrate with heat stewed steel slag self-decomposition technology GENERATION III to realize a full utilization of steel slag.

Technical principles: pumping the high calcium content waste water into the gas-liquid reaction system, then injecting CO_2 off gas under the efficient mechanical stirring condition, realizing the washing water soften and CO_2 mineralization at the same time, after filtering, the clear liquid reenters into the washing system. The reaction formula is as:

$$Ca^{2+}+ CO_2+2OH^-\rightarrow CaCO_3\downarrow+H_2O$$

A method to absorb the aluminium electrolytic exhaust gas via waste carbide slag slurry

Aiming at the characters of waste carbide slag, created a method to absorb the aluminum electrolytic exhaust gas via waste carbide slag slurry, and designed muti-level jet mineralization absorption equipment, shown in fig 1 and fig 2.

Technical principles: pumping the calcium contained water to the jet reactor and mixing with the CO_2+CO gas to Complete mineralizing absorption reaction, obtaining slurry containing calcium carbonate powder. The unabsorbed CO off gas escapes from the top of the jet reactor, then pressurized liquefied to CO fuel gas. Then transporting the slurry to the solid-liquid separation system via airtight tube, obtains calcium carbonate powder and clear liquid after filtering. clear liquid returns to calcium ion generator to realize recycling.

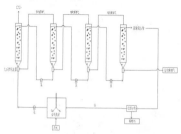

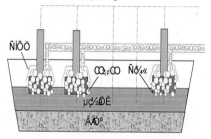

| Fig 1 flow chart of the aluminum electrolytic exhaust gas absorption technique | fig 2 sketch diagram of aluminum electrolytic anode gas trapping equipment |

Research progress

Research of bubble ultra- micronization model in progress of CO_2 injection

The process of absorbing CO_2 mineralization involves three phase reaction system about the gas liquid solid. According to the process and purpose of CO2's absorbing with teel slag circulating water used to dealing with steel slag and the waste carbide slag , The author studied the mixing effect on the bubble behavior and absorption characteristics in the system of gas-liquid mixing process .(Having formed three invention patents, patent number：ZL200810011863.8, application number：201310380610.4, 201310380583.3）。

Analyzing various axial and radial flow impeller to bubbles' influence degree of refinement within the system. Select the type of impeller blade. As shown in the fig, examining the mixing mode's influence of bubble mineralization and the dispersion characteristics in elaboration Mineralization absorption system. Eccentric stirring is better than Center mixing pattern in the effect on the bubble ultra-micronization. It can The increase of gas distribution effectively, make gas distribution relatively uniform and greatly improve the efficiency of air bubble.

(a) with the diverting tube (b) without diverting tube   (a) center   (b) eccentric center

Fig 4 the diversion tube's influence on the   Fig 3 the bubble dispersion behavior with

bubble distribution characteristics   eccentric center stirring and center stirring

Adopting the way of add diversion tube in absorption equipment improves the retention time of absorption and absorption efficiency. The results show that, as shown in fig4, adding diversion tube in absorption equipment can improve the gas holdup of system. Diversion tube limits and strengthened the liquid circulation, focuses on the bubble, reduces the mixing time, speeds up the reaction, so the volumetric mass transfer coefficient and gas utilization rises with the increase of diversion tube. To improve the efficiency of reaction, put forward that increasing venture jet reaction device in the place of high alkali waste water into the mineralized absorption equipment. By absorbing CO_2 gas through the reactor's vacuum characteristics, it can improve the mixing characteristics of two phases (As shown in fig5) and realize high mineralization of CO_2 absorption by multistage absorption way (As shown in fig1).

Ug=3.54 m·s^{-1} Ug=10.61 m·s^{-1} Ug=17.68 m·s^{-1} Ug=24.76m·s^{-1}

Fig 5 the fig of Bubble ultra-micronization in jet reactor

The research of steel slag hot braised washing water absorbing CO_2 mineralization

As shown in fig6, after sequential processing, the water wash scale of 1 million t/a steel slag hot braised circulating wash water system, which is at China metallurgical environmental resources development co., LTD, is completely removed.

Fig7 is, in different conditions, the wash water treatment termination of pH and TDS diagram in the solution. The figshows , when PH between 9 and 10 , the TDS of washing water is smaller . Now it below added new water (About 110 mg/L). It Instructs the dissolved solids in water is less if pH control within the scope and high recovery rate of calcium carbonate can get. If PH is high, calcium is not fully precipitate out wash water. If PH is low, calcium carbonate is in the form of calcium bicarbonate to return back to the wash water and affect the yield.

A before treatment   B after treatment

Fig 6 water quality situation before and after processing

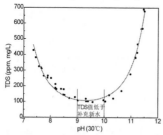

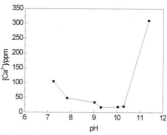

Fig 7 the wash water treatment termination of pH and TDS diagram in the solution in different conditions

Fig 8 The wash water treatment termination of pH and calcium ion concentration diagram in the solution in different conditions

As shown in fig 8,it is the calcium ion concentration for washing water under different processing end pH. Looking at the initial from calcium content, the wash water's temperature into the processing plant is at about 55 degrees, and its' PH is 11.3~11.4. At this point the high concentration of calcium ions in the water for washing is 270 ~ 310 parts per million, far higher than calcium ion concentration in the Supplementary water (40ppm). After processing, controlling the pH 9 ~ 10 range, the calcium ion concentration in the water can be dropped to the lowest range. Its content is lower than the added new water. So controlling the process termination pH between 9 and 10, washing water pH is still high, but because the washing water calcium content is very low, also it won't produce cause scarring in the loop. In addition, the magnesium ion and iron ion content wash water is very low. Magnesium ions content is below 0.5 PPM, and iron ion below 0.2 PPM.

The calcium carbonate sizes are all between 20-30μm prepared in different pH values, when reacting pH at 8-9, reunion is common; the average particle size is about 30μm. The particle size shows an increasing trend with the increasing water influx, intake pressure has little effect on the particle sizes. Under the stirring condition, the reunion appears much less, and the particle sizes are smaller as well. Moreover, under the condition of the secondary filter, calcium carbonate with small particle sizes is also obtained, which is with an average grain size of 20μm.in addition, there are a few nano sized particles among the calcium carbonate powder, which grain size is between 100nm-500nm, as shown in Fig 9. The whiteness of calcium carbonate powder is 95.3%.

Fig9. Nano-sized calcium carbonate particles

Research based on the absorption of aluminum electrolytic exhaust mineralization of waste carbide slag

As shown in figure1, level 3 countercurrent mineralization absorption equipment is adopted, which absorbs carbon dioxide from the exhaust gas from electrolytic aluminum with waste carbide slag slurry as mineralizer .The results are shown in fig10 and fig11. The figshows that when pH between 8 ~ 10, soluble ion content of carbide slag slurry is lowest. Main component in the solution is calcium hydroxide. Calcium ion concentration in the initial effluent is 340 ~ 370 parts per million. Access to the low concentration of carbon dioxide, pH dropped to 10, and calcium ion concentration in the solution is to minimize range, about 90 PPM. The results agree with steel slag water treatment process. When pH is between 9 ~ 10, it is helpful for better reaction of calcium ion in the solution. And after the treatment, magnesium ion and iron ion in total amount of carbide slag slurry is less than 1 PPM.

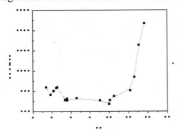

Fig10 the wash water treatment termination of pH and TDS diagram in the solution in different conditions

Fig11 the wash water treatment termination of pH and calcium ion concentration in the solution in different conditions

The fig12 shows that the average particle size of calcium carbonate particles under different pH is between 10 ~ 15 μm. When pH is between 9 and 10, the average particle size of calcium carbonate particles obtained from the youngest, around 10 μm. The fig13 shows that with the increase of processing termination of pH, the purity of calcium carbonate particles increases after the first decreases, and when pH ranges between 9 and 10, purity of calcium carbonate in more than 95, up to more than 99%.The fig14 shows that calcium carbonate powder is polyhedron block structure. When the pH is 9 ~ 10, the white degree of calcium carbonate powder is higher, up to 85%.

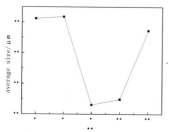

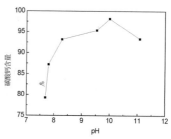

Fig12 the average particle size of calcium carbonate particles under different pH

Fig13 the wash water treatment termination of pH and the purity of calcium carbonate diagram in the solution in different conditions

Fig14 A single particle morphology of calcium carbonate produced by carbide slag slurry

Prospect and expectation

This project has been completed. The results show that after processing steel slag thermal cycle of washing water's pH value become from 11 ~ 14 to below 8.5 and it thoroughly solved the pipeline scaling phenomenon. Processing one ton of circulating wash water can get 1.2 kg of calcium carbonate. And calcium carbonate of the whiteness is up to 95%, the purity over 98.0%, and particle size is between 50 nm to 10 μm. According to corresponding national standards，the technical index is better than that of papermaking grade and food grade calcium carbonate.

The experimental results showed that after CO_2 emissions in aluminum electrolysis is absorbed by reactors , the residual gas in the CO_2 concentration below 0.08%.And CO_2 mineralization absorption rate can reach more than 95%. The purity of Calcarea carbonic is 99.0%, whiteness is more than 85.0%, and particle size less than 20 μm. the technical index is better than that of papermaking grade and food grade calcium carbonate. It realizes the comprehensive utilization of waste carbide slag and high mineralization using CO2 emission.

This project has organized an independent technology and equipment to treat high calcium alkali waste water by CO_2 injection and mineralization to prepare high valued products, solved the scaling problem in the heat stew process; also, realized the mineralization of CO_2 via Waste carbide slag to achieve the idea of "Using waste treat waste". Assuming that the steel slag production in China is more than 120 million tons per year, 30 million t /a CO_2

discharge and 40 million t /a waste carbide slag discharge, 1,600 million RMB Yuan could obtain as direct economic benefit by treating heat stew slag washing water; meanwhile, 25 million tons of calcium carbonate could be obtained by treating CO_2 off gas and waste carbide slag with a benefit over 20 billion RMB Yuan,CO_2 emission reduction of 700 million RMB Yuan.

Acknowledgement

This research was supported by the National Natural Science Foundation of China (Nos. U1202274, 51004033, 50934005, 50974035, 51204040, 51374064 and 51074047), the National HighTechnology Research and Development Program ("863"Program) of China (2010AA03A405 and 2012AA062303), Operation Expenses for Universities' Basic Scientific Research (N100302005), and The Doctoral Fund of EDU gov (20050145029).

References

1. ZHENG Xinye. Global carbon dioxide emissions situation and strategy[J].*International Observe*, 2010.(6): 55-57.
2. JIANG Huaiyou. Policies for CO_2 emission reduction and prospects for CO_2 geological storage underground. *Sino- Global Energy*, 2007.12 (5): 7-13.
3. YANG Guilan. Bury the carbon dioxide [J]. *Digest of Science and Technology*, 2001.(11) : 84-85.
4. MA Qianqian. Research survey of CO2 emission reduction technology [J]. *Liaoning Chemical Industry*, 2009.38(3): 176-179
5. LI Xianyong. General stiation and development of discharge-reduction and sealing-up-utilization technology of CO2 [J]. *Electrical Equipment*, 2008.9(5): 710.
6. Herzogh, Eliassonb, Kaarstado. Capturing greenhouse gases [J].*Scientific American*, 2000.(2):54-61.
7. GENG Xiaojie, WANG Hongliang, ZHANG Li, etc.Choosing Sites to bury CO_2 in abandoned oil reservoirs [J]. *Resources & Industries*, 2009.11(6):69-73
8. Eloneva S, Said A, Fogelholm CJ, etal. Preliminary assessment of a method utilizing carbon dioxide and steelmaking slags to produce precipitated calcium carbonate[J].*Applied Energy*, 2011, 90(1- SI): 329-334
9. Florin N，Fennell P，Synthetic CaO-based Sorbent for CO_2 Capture[C],*10TH International Conference on Greenhouse Gas Control Technologies*[C], 2011, 4:830-838.
10. Wang Yuanbo Shen Yuesong Zhu Shemin. Reclmation and Recycle of Carbide Slag [J]. *Environmental Engineering*, 2008, 26: 256-258.
11. LIU Ming-yan, YANG Yang, XUE Juan-ping, HU Zong-ding. Measuring Techniques for Gas−Liquid−Solid Three-phase Fluidized Bed Reactors [J]. *The Chinese Journal of Process Engineering*, 2005.5(2):217-222.
12. BoyerC.,Duquenne A.M.,Wild G. Measuring techniques in gas--liquid and gas-liquid-solid reactors[J].*Chem.Eng.Sei.*,2002.57:3185-3215.

DEVELOPMENT OF MATERIALS-BY-DESIGN FOR CO_2 CAPTURE APPLICATIONS

Izaak Williamson and Lan Li

Department of Materials Science and Engineering, Boise State University, Boise, ID 83725-2090

Keywords: Carbon capture, porous solid, density functional theory, van der Waals-DFT.

Abstract

The efficient separation and storage of CO_2 from power plant flue gases can reduce the amount of CO_2 released into the atmosphere and mitigate global warming. Potential candidates for industrial applications are solid sorbent materials. Crucial factors to control the efficiency of porous sorbent materials include the framework and pore structure, and the chemical and physical reactivity of CO_2 within the pores. Computational modeling approaches, based on density functional theory (DFT) and van der Waals-DFT, have been applied to nanoporous solid – manganese dioxide α-MnO_2. We found that the types and charges of cations as dopants in the α-MnO_2 and the concentration of CO_2 influence the structural features of α-MnO_2, which control its CO_2 selectivity performance within the flue gases.

Introduction

Carbon capture and storage is a technique to separate large quantities of CO_2 from power plant flue gases before being released into the atmosphere. Porous solids are a potential candidate as sorbents to "trap" CO_2 through chemical or physical adsorption on their porous surfaces[1]. Such separation efficiency is determined by the size and shape of its pores and by the interaction between the porous surface and the gas adsorbates. The incorporation of cations into the pore can act as a means of structural support and alter the pore diameter to be comparable to the kinetic diameter of the gas adsorbate, such as CO_2.

A porous solid of particular interest for CCS is manganese dioxide α-MnO_2, an octahedral molecular sieve (OMS) composed of edge-sharing MnO_6 octahedra that are linked into a nanoporous framework[2]. The framework consists of (1×1) and (2×2) tunnels, referred to as OMS-2. Due to the larger tunnel size, cations and gas adsorbates prefer to reside along the (2×2) tunnel axis of the OMS-2. As seen in Figure 1, a potassium cation K^+ is located in the (2×2) tunnel, known as cryptomelane-type MnO_2. Its pore diameter of ~ 4.6 Å gives it a good CO_2 selectivity performance, because the diameter is comparable to the kinetic diameter of 3.3 Å for CO_2[3]. Alternatively, replacing potassium K^+ with other cations, e.g. barium Ba^{2+}, sodium Na^+ or lithium Li^+, in the OMS-2 tunnel is of great research interest, due to the effects of their concentrations, types and charges on the CO_2 selectivity efficiency and sorption hysteresis phenomenon[2,4].

Our recent studies[4] have shown the structural effects of adding K^+ and Ba^{2+} cations to α-MnO_2 and reveal the stronger interaction of the porous surface with Ba^{2+} than with K^+. We also revealed that cations act as a "gate keeper", weakening the adsorption of CO_2 and blocking the diffusion of CO_2 along the OMS-2 tunnel[4,5]. The proposed mechanism indicated that adsorbed CO_2 molecules diffuse along the tunnel axis until they encounter a cation, e.g. K^+. In order to bypass K^+, they force the cation to shift off its original position to "give way". Nudging the cation aside forms an unstable transition state for CO_2 in the OMS-2. We estimated that the energy barrier (or activation energy) for CO_2 diffusion is 5.44 eV in the α-MnO_2 with K^+ dopant and 6.87 eV in the α-MnO_2 with Ba^{2+}.

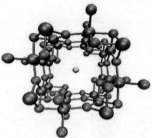

Figure 1. Atomic structure of α-MnO_2 doped with cation K^+, known as OMS-2. The purple, red, and green spheres represent Mn, O, and K atoms, respectively.

Based on the previous studies, we further discuss the structural effects of adding CO_2 at varying concentrations in this paper. The existence of a hysteresis loop in the CO_2 sorption isotherms at pressures greater than 7 bar implied that the concentration of CO_2 plays a crucial role in CO_2 sorption properties[2]. We also compare this to the structural effects of K^+ and Ba^{2+} dopants.

Computational Methods

Structural calculations were performed using the Vienna *ab-initio* Software Package (VASP) within density-functional theory (DFT) [6]. The generalized gradient approximation (GGA) was used with the Perdew Burke Ernzerhoff solid (PBEsol) formalism to approximate the exchange correlation potential[7]. Projector-augmented wave (PAW) pseudopotentials were used to expand the plane wave basis sets to a cutoff energy of 400eV[8,9]. The Brillouin zone integration was performed with a 2x2x2 Monkhorst-Pack k-mesh. To treat the magnetism of Mn, the GGA+U scheme by Liechtenstein et al was used to induce an effective on-site Coulomb potential (U) of 2.8eV and an effective on-site exchange potential (J) of 1.2eV[10]. We also applied DFT calculations with the inclusion of van der Waals interactions using the empirical underpinnings of the DFTD2 approach to the system, in comparison with the non-vdW DFT calculations, to determine the effect of the vdW interactions on cell parameters and CO_2 sorption behavior in the OMS-2[11]. Structural relaxation and total energy minimization were well converged until residual forces were reduced to 0.01eV/Å or less while the total energy to 10^{-6} eV.

Results and Discussion

In Table 1, the cell parameters of undoped α-MnO_2 are a = 9.702 Å, b = 8.568 Å, c = 9.685 Å, and β = 90°. When K^+ is added to form $KMn_{24}O_{48}$, the cell parameters become a = 9.702 Å, b = 8.554 Å, c = 9.731 Å, and β = 89.960°. K^+ lengthens the cell in the *c* direction but shrinks it in the *b* direction, leading to a slight increase in the cell volume of 0.30%. Likewise, the addition of Ba^{2+} alters the cell to a = 9.713 Å, b = 8.542 Å, c = 9.735 Å, β = 89.950°, and a cell volume increasing by 0.32%. These computed results suggest that adding cations, whatever type and charge they are, has a minor effect on the α-MnO_2 structure. By adsorbing one CO_2 molecule in the cell of $K/BaMn_{24}O_{48}$ (i.e. $(CO_2)_{1/24}K_{1/24}MnO_2$ or $(CO_2)_{1/24}Ba_{1/24}MnO_2$), the cell volume is further increased by 1.00% and 0.97%, respectively. This structural deformation becomes worse as the concentration of CO_2 increases. The maximum amount of CO_2 stored in the cell of $K/BaMn_{24}O_{48}$ is expected to be 8 CO_2 molecules, i.e. $(CO_2)_{1/3}K_{1/24}MnO_2$ or $(CO_2)_{1/3}Ba_{1/24}MnO_2$.

Table 1. Structure parameters of α-MnO_2 with K^+ and Ba^{2+} dopants including cell parameters (Å), unit cell volume V (Å^3), cell angle β (°), and % change in cell volume. The "opt." represents systems where CO_2 is located in an "optimum" position of the OMS-2 tunnel. The "trans." refers to CO_2 located in the "transition" position of the OMS-2 tunnel.

	CO_2 position	Cell Parameters, a b c (Å)			V (Å^3)	Cell Angle β	Changing %
$Mn_{24}O_{48}$		9.702	8.568	9.685	805.04	90.0	
$KMn_{24}O_{48}$		9.702	8.554	9.731	807.48	89.960	0.30
$BaMn_{24}O_{48}$		9.713	8.542	9.735	807.64	89.950	0.32
$KMn_{24}O_{48}$ + 1CO_2	opt.	9.797	8.545	9.712	813.06	90.370	1.00
	trans.	9.697	8.563	10.084	837.31	89.240	4.01
$BaMn_{24}O_{48}$ + 1CO_2	opt.	9.747	8.536	9.769	812.83	89.939	0.97
	trans.	9.777	8.570	10.136	849.27	89.785	5.49
$KMn_{24}O_{48}$ + 2CO_2	opt.	9.994	8.570	9.826	841.55	90.090	4.53
	opt. & trans.	9.778	8.592	10.068	845.82	89.601	5.07
$BaMn_{24}O_{48}$ + 2CO_2	opt.	10.004	8.563	9.812	840.31	89.573	4.38
	opt. & trans.	9.911	8.570	10.094	857.35	89.855	6.50

The K^+ cation has a greater effect on the structure of OMS-2 than Ba^{2+}. This data, also shown in Table 1, is taken with CO_2 in the "optimum" position which is in front of the cation along the tunnel axis (Figure 2a). The structure experiences its largest distortion when CO_2 is wedged against the cation, offset from the tunnel axis, forming the "transition" state (Figure 2b). With one CO_2 in the transition state, this leads to volume increases of 4.01% and 5.49% for K^+ and Ba^{2+}, respectively, which is more than four times the increase of having CO_2 in the optimum position. This pattern is not consistent, however, when a second CO_2 is added. The volume increase for two adsorbed CO_2, where one molecule is in the optimum position and the other is in the transition position (Figure 2c), is 5.07% and 6.50% for K^+ and Ba^{2+}, respectively. K^+ induces the larger volume deformation than Ba^{2+} does in the optimum state. In contrast, when CO_2 is in a transition state, Ba^{2+} leads to the larger volume increases, implying a higher energy cost required to offset Ba^{2+} if CO_2 bypasses Ba^{2+}. This prediction agrees well with the estimation of binding energy between cation and porous surface. The binding energy of 8.34 eV/Ba^{2+} cation evidences a stronger Ba - α-MnO_2 interaction than the K - α-MnO_2 interaction (4.48 eV/K^+ cation).

(a)   **(b)**   **(c)**

Figure 2 Atomic structures for CO_2 adsorbed in the cell of $KMn_{24}O_{48}$, including one CO_2 molecule located in (a) the optimum and (b) transition positions; and two CO_2 molecules, where one is in the optimum position and the other in the transition position.

For an accurate depiction of structural and energetic effects in porous solid systems, it is usually important to account for nonlocal correlation energy due to van der Waals (vdW) forces[12]. At the small concentration of CO_2 in the OMS-2, the diffusion barrier is energetically high, obtained from both DFT and vdW-DFT calculations (Figure 3). As the concentration of CO_2 increases, this barrier dramatically decreases to 0.13 eV/CO_2 for DFT and 0.82 eV/CO_2 for vdW-DFT. Our results suggest that CO_2 can bypass the cation to further diffuse in the OMS-2 at the high concentrations (e.g. two or more CO_2 molecules per $KMn_{24}O_{48}$).

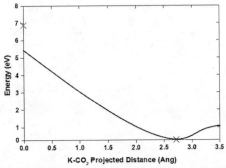

Figure 3. Potential energy vs. K^+-CO_2 projected distance (defined as the difference in the K^+ position and one of the O positions in CO_2 when projected onto the tunnel axis). The results plotted with the black curve are obtained from DFT calculations while the vdW-DFT results are in red crosses. Electrostatic forces dominate the K^+-CO_2 interaction when CO_2 diffuses between the optimum (zero potential energy) and transition (largest potential energy) positions. At the K^+-CO_2 distance beyond the optimum position, van der Waals forces dominate the interaction[13].

Conclusions

We have used DFT and vdW-DFT to investigate the structural effects of incorporating K^+/Ba^{2+} dopants and adsorbing CO_2 into the α-MnO_2 system. We found that for conditions where CO_2 is in the energetically optimum position within the big tunnel, the type of dopant has relatively small effect on the volume deformation of the structure. Increasing the CO_2 concentration leads to the larger volume deformation. If beyond the saturated amount of CO_2, the porous surface would be damaged, and α-MnO_2 would lose its octahedral molecular sieve framework with (1×1) and (2×2) tunnels. Our computational studies provide detailed insights into the atomic structure and carbon capture properties of porous solids and facilitate the design and scale-up production of one-dimensional solids for efficient carbon capture applications.

Acknowledgement

The authors are grateful for Eric Cockayne, Laura Espinal, and Winnie Wong-Ng at the Institute of Standards and Technology for helpful discussion on the DFT calculations and providing experimental data and analysis, as well as Anais E. Espinal and Steven L. Suib at the University of Connecticut for providing OMS-2 materials and SEM images for experiments. All the calculations were performed at NIST and HPC at Boise State.

114

References

1   Y. Cui, H. Kita, K. Okamoto, Chemical Communications (2003) 2154.

2   L. Espinal, W. Wong-Ng, J.A. Kaduk, A.J. Allen, C.R. Snyder, C. Chiu, D.W. Siderius, L. Li, E. Cockayne, A.E. Espinal, S.L. Suib, Journal of the American Chemical Society 134 (2012) 8.

3   R.N. Deguzman, Y.F. Shen, E.J. Neth, S.L. Suib, C.L. O'young, S. Levine, J.M. Newsam, Chemistry of Materials 6 (1994) 815.

4   L. Li, E. Cockayne, I. Williamson, L. Espinal, W. Wong-Ng, Chemical Physics Letters 580 (2013) 6.

5   E. Cockayne, L. Li, Chemical Physics Letters 544 (2012) 6.

6   G. Kresse, J. Furthmuller, Physical Review B 54 (1996) 11169.

7   J.P. Perdew, A. Ruzsinszky, G.I. Csonka, O.A. Vydrov, G.E. Scuseria, L.A. Constantin, X.L. Zhou, K. Burke, Physical Review Letters 100 (2008).

8   P.E. Blochl, Physical Review B 50 (1994) 17953.

9   G. Kresse, D. Joubert, Physical Review B 59 (1999) 1758.

10   A.I. Liechtenstein, V.I. Anisimov, J. Zaanen, Physical Review B 52 (1995) R5467.

11   S.J. Grimme, J. Comput. Chem. 27 (2006) 1787.

12   K.G. Ray, D. Olmsted, N. He, Y. Houndonougbo, B.B. Laird, M. Asta, Physical Review B 85 (2012).

13   J. Klimes, D.R. Bowler, A. Michaelides, Physical Review B 83 (2011).

Energy Technology 2014: Carbon Dioxide Management and Other Technologies
Edited by: Cong Wang, Jan de Bakker, Cynthia K. Belt, Animesh Jha, Neale R. Neelameggham,
Soobhankar Pati, Leon H. Prentice, Gabriella Tranell, and Kyle S. Brinkman
TMS (The Minerals, Metals & Materials Society), 2014

The GHG Emissions List Analysis of Aluminum Industry in China

Yuanyuan Wang, Hao Bai*, Guangwei Du, Yuhao Ding, Kang Zhou

School of Metallurgical and Ecological Engineering,
University of Science and Technology Beijing, Beijing, 100083, China
*Corresponding author: baihao@metall.ustb.edu.cn

Abstract

In this paper, the list analysis of GHG emissions of aluminum industry was conducted based on the data from typical electrolytic aluminum plants of a China's northwest province. The results show that under the conditions of average anode effect coefficient as 0.16 times/(cell• day) and anode duration as 5 minutes per day, equivalent emissions of CO_2 reached 1.23 ton per ton of aluminum, in which PFC emissions accounted for about 71%. Compared with foreign advanced level, which is 0.87 ton per ton of aluminum under the conditions of average anode effect coefficient as 0.01 times/(cell•day) and anode duration as 0.46 minutes per day, there is much room for China's aluminum industry to reduce GHG emissions. Since anode effect plays an important role on PFCs generation in electrolytic aluminum process, how to control anode effect strictly becomes the key factor for remarkable reduction of GHG emissions in China's aluminum industry.

Keywords: aluminum electrolysis; GHG; PFCs; anode effect; emission reduction

Introduction

Technology of aluminum electrolysis is a mature and efficient process of aluminum production. Due to the huge demand on aluminum worldwide, so resource consumption, energy consumption and pollutant emissions in the process of aluminum production are still large. The global output of aluminum reached 45193 kt in 2012. A large amount of CO_2 and PFCs are produced in the process of aluminum production, and the greenhouse effect and environmental pollution caused by emissions of them is a concern. PFCs belong to a powerful greenhouse gas, thus emission reduction of PFCs is very important for aluminum electrolysis industry. At present, China's aluminum production accounts for about 33% of the world's aluminum production, and controlling PFCs emission has become the key point of pollution control of aluminum electrolysis industry in China [1].

Taking the case of a typical electrolytic aluminum plant of a China's northwest province, this paper studies the list analysis of GHG emissions of aluminum industry in China. GHG emissions of a China's northwest province were calculated through investigating existing situation of aluminum electrolysis industry in the province, knowing capacity, output, main production technology and equipment level of aluminum electrolysis industry, and using the calculation framework and methods of GHG emissions of aluminum electrolysis industry, which were put forward by the intergovernmental panel on climate change (IPCC), the international aluminum association and other organizations at home and abroad. The list analysis of GHG emissions of aluminum electrolysis industry will offer useful information in promoting emission reduction of GHG in aluminum electrolysis industry, improve the efficiency of energy utilization.

1. Generation Mechanism of GHG of Aluminum Electrolysis Industry

GHG in aluminum electrolysis industry includes direct GHG, which mainly includes CO_2 generated by anode consumption and PFCs generated by anode effect, and indirect GHG (CO_2 from alumina production process, anode production process and electricity production process). Generation of direct GHG has a close relationship with the anode effect coefficient and the anode duration. PFCs as powerful greenhouse gas, can't decompose easily in the atmosphere, and exist for a long time. The lifetime and global warming potential (GWP) of PFC_s were shown in table Ⅰ.

Table I. The lifetime and global warming potential (GWP) of PFCs [4]

Gas Molecular Formula	Number	Lifetime/year	Global Warming Potential(GWP)		
			100 year	100 year	500 year
			From the second assessment report (AR2) by IPCC	From the third assessment report (AR3) by IPCC	From the forth assessment report (AR4) by IPCC
CF_4	PFC-14	50000	6500	7390	11200
C_2F_6	PFC-116	10000	9200	12200	18200
C_3F_8	PFC-218	2600	7000	8830	12500

According to Table I, lifetime of CF_4 in the atmosphere is at least 50000 years, and its greenhouse effect is as at least 6500 times as that of CO_2.

Electrode reaction of normal aluminum electrolysis is shown as follows:

$$2Al_2O_3+3C{=\!=}4Al+3\ CO_2 \tag{1}$$

PFCs are formed as intermittent by-products during operational disturbance in the process of aluminum electrolysis. When the concentration of alumina in the cells gets too low, reactions between the carbon anode and the cryolite bath can occur generating the PFCs CF_4 and C_2F_6 .These gases have global warming potential of 6500 and 9200 respectively. Reaction equations are shown as follows:

$$NaAlF_6+3/4C{=\!=}3/4CF_4+Al+3NaF \tag{2}$$

$$NaAlF_6+C==1/2C_2F_6+Al+3NaF \tag{3}$$

The temperature of electrolytic cell will increase when the side reaction of anode effect happens; high temperature damages the conditions of production process of aluminum electrolysis, resulting in the addition of energy consumption.

2. Methods of GHG Emissions List Analysis of Aluminum Electrolysis Industry

2.1 Estimation Methods of CO2 Emission

This paper estimates CO_2 emission only generated by anode consumption. All possible electrode reactions of normal aluminum electrolysis are shown as follows:

$$2Al_2O_3+3C=4Al+3CO_2 \tag{4}$$

$$2Al_2O_3+4C=4Al+2CO_2+2CO \tag{5}$$

$$2Al_2O_3+6C=4Al+6CO \tag{6}$$

Reaction (4) takes place when current efficiency is 100%,and in this reaction,1mol of aluminum produced generates 3/4mol of CO_2, that is to say, to produce 1t of aluminum will generate 1122kg of CO_2, during which, 333kg of carbon anode will be consumed.
Reaction (5) takes place when current efficiency is 75%, in this reaction, 1t of aluminum produced consumes 444kg of carbon anode.
Reaction (6) takes place when current efficiency is 50%, in this reaction, 1t of aluminum produced consumes 666kg of carbon anode.
According to reaction (4), theoretical consumption of carbon anode is 333kg/t Al. However, the reaction shown as equation (7) of aluminum in electrolyte and CO_2 causes aluminum loss, which makes current efficiency decline. So consumption of carbon anode is 333/η kg/t Al (η refers to the current efficiency).

$$2Al+3CO_2=Al_2O_3+3CO \tag{7}$$

2.2 Estimation Methods of PFCs Emissions

PFCs emissions of aluminum electrolysis process are calculated with the unified formulas recommended by IPCC. The unified formulas are shown as follows:

$$M_1(CF_4) = S \times AEF \times AED \tag{8}$$

Where, S refers to the coefficient related to technical parameters, process technology of electrolytic cell being good or bad and duration of anode effect being long or short both can decide the value of S, the range of S for prebaked cell as $0.14 \sim 0.16$ and for soderberg cell as $0.07 \sim 0.11$; M_1 (CF_4) refers to the CF_4 emission of 1t of aluminum, kg/t Al; AEF refers to the anode effect coefficient, times/ (cell•day); AED refers to the anode duration, min.

$$M_2(C_2F_6) = M_1(CF_4) \times [\frac{F(C_2F_6)}{CF_4}] \tag{9}$$

Where, M_2 (C_2F_6) refers to the C_2F_6 emission of 1t of aluminum, kg/t Al; F (C_2F_6) / CF_4 refers to the weight ratio of C_2F_6 and CF_4, and its value changes with different technology. For prebaked cell, the value is 0.121.

Emissions of C_2F_6 and CF_4 can be calculated by formulas (8) and (9). And Emissions of C_2F_6 and CF_4 should be converted into equivalent emission of CO_2. Formulas of conversion can be shown as follows:

$$Eeq = 6500 \times M_1(CF_4) + 9200 \times M_2(C_2F_6) \ [5] \tag{10}$$

Where, Eeq refers to the equivalent emission of CO_2 converted, kg/t Al.

3. Estimation of GHG Emissions of Aluminum Electrolysis Industry in a China's Northwest Province

3.1 Estimation of GHG Emissions of an Aluminum Electrolysis Plant in a China's Northwest Province

The production equipment and facilities of an aluminum electrolysis plant in a China's northwest province are advanced, and its annual output is about 35×10^4t. The plant uses 240KA electrolytic cell. The production level of the plant can represent production level of a China's Northwest Province.

The technical parameters of each electrolytic cell at runtime of an aluminum electrolysis plant in a China's northwest province are shown as follows in table II, which were the average value of the data from four months of electrolytic cell running in 2012.

Table II. The technical parameters of 240KA electrolytic cells

Cell No.	Daily output (t)	Current efficiency (%)	Anode effect coefficient (times/(cell•day))	Anode duration (min)
1	1.83	94.41	0.16	5
2	1.81	93.87	0.16	5
3	1.83	94.49	0.17	5

GHG Emissions of No.1, No.2 and No.3 electrolytic cells were calculated according to data shown in Table II. The calculation results are shown as follows in table III.

Table III. The calculation results of GHG Emissions of No.1, No.2 and No.3 electrolytic cells

Cell	PFCs		CO$_2$			
	CF_4 (10^{-1}kg/t Al)	C_2F_6 (10^{-2}kg/t Al)	Consumption of carbon anode (kg/t Al)	CO_2 (kg/t Al)	CO_2 converted (kg/t Al)	The total (kg/t Al)
1	1.20	1.45	352.71	352.45	872	1224.45
2	1.13	1.38	354.86	354.58	872	1226.58
3	1.28	1.55	354.86	352.13	872	1224.13

GHG emissions of that aluminum electrolysis plant can be estimated as about 1225.05kg/t Al by the above results shown in table III, in which PFCs emissions accounted for about 71%. This value can be used to estimate GHG emissions of all aluminum electrolysis plants in a China's northwest province.

3.2 Estimation of GHG Emissions of Aluminum Electrolysis Industry in a China's Northwest Province

The production level of each aluminum electrolysis plant in a China's northwest province is close, thus GHG emissions of each aluminum electrolysis plant can be estimated by 1225.05kg/t Al.
The technical parameters and output of all aluminum electrolysis plants in a China's northwest province are shown as follows in table IV.

Table IV. The technical parameters and output of all aluminum electrolysis plants

Plant	Capacity (10 kt)	Annual output (10 kt)	Cell type	Current intensity (KA)	The number of cells (set)
1	20	20	Prebaked cell	180	260
2	20	20	Prebaked cell	200	268
3	70	35	Prebaked cell	400	660
4	60	30	Prebaked cell	350	320
5	10	5	Prebaked cell	240	106
6	50	50	Prebaked cell	350	552
7	35	35	Prebaked cell	240	370

GHG emissions of all above aluminum electrolysis plants can be calculated by the value of GHG emissions, which was 1225.05kg/t Al. The calculation results are shown as follows in table V.

Table V. GHG emissions of each aluminum electrolysis plant

Plant	Annual output (10 kt)	The number of cells (set)	Annual GHG emissions (t)
1	20	260	245010
2	20	268	245010
3	35	660	428768
4	30	320	367515
5	5	106	61252
6	50	552	612520
7	35	370	428768
The total	195	2536	2388848

According to table V, annual GHG emissions of aluminum electrolysis industry in a China's northwest province can be estimated as about 2388.848 kt.

4. The GHG Emissions Estimation of Aluminum Industry in China

Six aluminum electrolysis plants were selected from different provinces in China, and GHG emissions of them were calculated in 2012, shown as follows in figure 1. Cell type, technology used and technical parameters of six aluminum electrolysis plants are shown as follows in table VI.

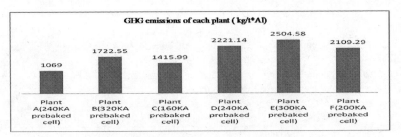

Fig.1. GHG emissions of each plant (kg/t Al)

Table VI. Cell type, technology used and technical parameters of each plant

Plant	Cell type	Technology used	Average anode effect coefficient (times/(cell•day))	Average current efficiency (%)
A	240KA prebaked cell	Cell made by graphite	0.062	92.43
B	320KA prebaked cell	Full graphite cathode	0.120	94.50
C	160KA prebaked cell	Anode with slot	0.114	94.06
D	240KA prebaked cell	Control temperature of electrolyte	0.163	94.27
E	300KA prebaked cell	"Five low" technology	0.188	93.62
F	200KA prebaked cell	Strengthen electric current	0.153	92.88

According to the data of figure 1, plant with the minimal GHG emissions among six aluminum electrolysis plants was plant A, being 1069kg/t Al. Plant with the maximum GHG emissions was plant E, being 2504.58kg/t Al. The maximum value was more than two times than the minimum value. It demonstrates that GHG emissions of aluminum electrolysis pants in China vary widely and electrolysis technology of some pants is still backward. The average value of GHG emissions of six aluminum electrolysis plants was 1823.19 kg/t Al, and GHG emissions of aluminum electrolysis industry in China can be estimated as 1.8 t/t Al, Compared with foreign advanced level, which is 0.87 t/t Al, there is much room for China's aluminum industry to reduce GHG emissions. Through analysis of technical parameters and GHG emissions of the six aluminum electrolysis plants, it can be clearly concluded that GHG emissions have a great relationship with anode effect coefficient.

Through the experiments in the six aluminum electrolysis plants, some issues can be discussed and analyzed.

(1) CO_2 emission of aluminum electrolysis with cell made by full graphite was more than that of aluminum electrolysis with cell made by partial graphite, but PFCs emissions of them were almost the same. However, PFCs belong to powerful greenhouse gas and PFCs emissions accounted for the most of equivalent emissions of CO_2 after PFCs were converted into equivalent emission of CO_2. Electrolytic cell made by full graphite or partial graphite had no effect on GHG emissions.

(2) GHG emissions of aluminum electrolysis with full graphite cathode were much less than that of aluminum electrolysis with other type graphite cathodes.

(3) Anode with slot can cut down cell pressure, make operation of electrolytic cell be stable and improve current efficiency by 1%. So anode with slot can lower GHG emissions of aluminum electrolysis.

(4) "Five low" technology, refers to low temperature of electrolyte, low ratio of molecule, low concentration of alumina, low anode effect coefficient and low voltage. GHG emissions of aluminum electrolysis can be greatly reduced through application of "Five low" technology.

(5) Strengthening electric current had no direct relationship with emission reduction of GHG.

GHG emissions of aluminum electrolysis industry in China can be estimated as 1.8 t/t Al, and this value can be used to estimate GHG emissions of aluminum industry in China from 2008 to 2012. The calculation results are shown as follows in figure 2.

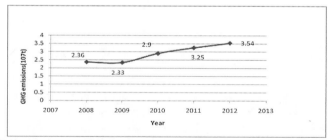

Fig.2. GHG emissions estimation of aluminum industry in China from 2008 to 2012(10^7t)

According to figure 2, the average value of annual GHG emissions of aluminum industry in China from 2008 to 2012 was 2.88×10^7t. GHG emissions of aluminum industry in China from 2008 to 2012 increased quickly year by year, the total growth amounted to 1.18×10^7t, and the total growth rate reached 5%. Increase of GHG emissions of aluminum industry in China can't be overlooked, and emission reduction countermeasures of GHG of aluminum industry in China should be taken seriously.

Three aspects as follows should be considered to reduce GHG emissions of aluminum industry.

(1)Controlling the anode effect coefficient and the anode duration could reduce GHG emissions. It can be seen that the anode effect coefficient and the anode duration are two key process parameters, which are directly related to GHG emissions according to the calculation formula of PFCs emission. The methods of controlling the anode effect coefficient and the anode duration mainly include anode effect terminating and anode effect prediction.

(2)Feedstock of alumina should be optimized by using control system of point type feedstock. By this way, the occurrence of the anode effect caused by reduction of concentration of alumina can be reduced.

(3) Property of electrolytic cell should be optimized by improving automatic control system of aluminum electrolysis process, and balance control of material and energy can be solved.

5. Conclusions

(1)In this paper, the results show that under the conditions of average anode effect coefficient as 0.16 times/(cell•day) and anode duration as 5 minutes per day, GHG emissions of aluminum electrolysis industry in a China's northwest province reached 1.23 ton per ton of aluminum. Compared with foreign advanced level, which is 0.87t per ton of aluminum under the conditions of average anode effect coefficient as 0.01 times/(cell•day) and anode duration as 0.46 minutes per day, there is much room for China's aluminum industry to reduce GHG emissions.

(2)PFCs emission accounted for about 71% of GHG emissions of aluminum electrolysis industry in a China's northwest province.

(3) GHG emissions of aluminum electrolysis industry in China were much larger than the foreign advanced level, and there is still a large room in reducing GHG emissions of aluminum electrolysis industry and there is still a long way to go to reduce GHG emissions.

(4) GHG emissions of aluminum electrolysis industry can be reduced through controlling the anode effect coefficient and the anode duration.

References

[1]Z. Xiang. "Research on the approaches to emission reduction of PFCs of aluminum electrolysis industry," Environmental Science and Technology, 24 (5) (2011), 60-66.

[2]Aluminum for Future Generation[R].International Aluminum Institute, 2009.

[3]Result of 2009 Anode Effects Survey[R]. International Aluminum Institute, 2010.

[4]W. X. Li. "Analysis on GHG emission and reduction of aluminum electrolysis industry," Low Carbon Development of Non-ferrous Metals Industry.

[5]K. J. Sun and H.B.Hu. "Analysis and expectation of PFCs reduction of aluminum electrolysis," Light Metals, (6) (2009), 42-45.

[6]Y. Jin, Y. L. Zhu, and C. Y. Shen. "Analysis and expectation of GHG reduction of aluminum electrolysis," Yunnan Metallurgy, 39 (2010), 170-173.

[7]C. F. Zhao, S. Z. Zhang, and X. Huang. "Research of PFCs reduction of aluminum electrolysis industry," Light Metals, (10) (2008), 26-29.

Energy Technology 2014
Carbon Dioxide Management and Other Technologies

SYMPOSIUM: ENERGY TECHNOLOGIES AND
CARBON DIOXIDE MANAGEMENT

Novel Technologies and Life Cycle Assessment

Session Chairs:

Neale R. Neelameggham
Jan de Bakker

Energy Technology 2014: Carbon Dioxide Management and Other Technologies
Edited by: Cong Wang, Jan de Bakker, Cynthia K. Belt, Animesh Jha, Neale R. Neelameggham,
Soobhankar Pati, Leon H. Prentice, Gabriella Tranell, and Kyle S. Brinkman
TMS (The Minerals, Metals & Materials Society), 2014

Pure Oxygen Anodes™ for Low- or Zero-Carbon Energy Efficient Metal Oxide Reduction

Adam Powell[1], Matthew Earlam[1], Salvador Barriga[1]

[1]INFINIUM Inc., 3 Huron Drive, Natick, MA 01760-1314, USA

Keywords: Molten salt electrolysis, Magnesium, Aluminum, Rare Earths, Neodymium, Lanthanum, Zirconia, Primary Production, Pure Oxygen Anodes

Abstract

The inert anode is a key enabling technology for dramatic energy and emissions reduction in extractive metallurgy. In molten salts, the materials challenge is nearly insurmountable: the anode must conduct electrons at high temperature in a pure oxygen and molten salt environment. A solid electrolyte, *e.g.* zirconia, between the salt and anode removes the molten salt stability constraint, and can act as a container for a liquid metal anode. This can be liquid silver for oxygen-generating inert anodes, or several metals for natural gas-fueled anodes; it is possible to switch between these modes. The solid electrolyte brings several other benefits: it eliminates molten salt contamination of the anode gas, creates a reducing environment enabling low-cost steel vessels, increases current efficiency, and eliminates carbon contamination of the product. New developments presented here, including current collector and low-cost anode designs, amplify these benefits for producing aluminum, magnesium, rare-earths, and other metals.

Introduction

Reactive metals such as magnesium, aluminum and rare-earths play increasing roles in efficient transportation and clean energy. In response to new fuel economy standards, automotive OEMs are planning large increases in use of aluminum and magnesium to reduce vehicle weight. Rare-earth permanent magnet motors and generators exhibit better power-to-weight and efficiency than induction motors and other designs, making this today's primary motor technology for hybrid and electric vehicles, and for wind turbines. Nickel metal hydride (NiMH) batteries with lanthanum are dominant in hybrids. In 2011, transportation comprised 28.6 out of 103.1 EJ of US energy consumption [1] and 1,845 out of 5,471 million metric tonnes (MMT = 10^9 kg) CO_2e of US greenhouse gas (GHG) emissions [1], creating big targets for economy-wide reductions.

Unfortunately processes for making reactive metals for these applications are very inefficient and heavily-polluting. In primary aluminum production, Hall-Héroult cell electricity use [2] is more than double the reaction enthalpy due to heat losses through frozen cryolite side-walls and anode gases, graphite anode plants add significant energy use and GHG emissions [2] (see Table 1). Magnesium is even worse: silicothermic reduction by the Pidgeon process in China produces about 85% of the world's magnesium using coal or coke gas [3]. Chloride electrolysis begins with high-purity MgO and requires carbochlorination or hydrochlorination and costly chloride dehydration [4]—rendered unnecessary for aluminum by the Hall-Héroult cell. Production of neodymium and lanthanum with graphite anodes leads to inconsistent carbon content: small cells make 1 kg batches for analysis, sorting and sale by carbon content, with very high labor and analysis costs. And vertical electrodes give no incentive to reduce anode effect: some reports indicate that China's rare-earth metal industry emits more CF_4 and C_2F_6—with global warming potential (GWP) 6500 and 9200 times CO_2 respectively—than China's aluminum industry [5].

Metal (ΔH, ΔG at 1000°C)	Aluminum	Magnesium	Neodymium	Lanthanum
Oxide ΔH, ΔG, kWh/kg metal	8.70, 6.22 [6]	6.95, 5.27 [6]	1.73, 1.40 [7]	1.80, 1.43 [7]
C reduction ΔH, ΔG, kWh/kg	5.73, 3.32	4.75, 2.88	1.18, 0.79	1.22, 0.81
CH_4 reduction ΔH, ΔG	5.61, 3.56	4.66, 3.05	1.16, 0.82	1.20, 0.83
Current industrial practice energy usage, kWh/kg	15.6 electric 5.7 anode 21.3 total [2]	100 Pidgeon/ 40 MgCl$_2$ e⁻, 91 avg [3]	No data	No data
Direct GHG, kg CO₂e/kg	1.83	26/6, 23 avg	No data	No data
Electricity GHG 0.35 kg/kWh	5.42 [2]	1/12, 2.5 avg	No data	No data
Pure Oxygen Anodes est. electrical energy, kWh/kg	10.5-13 O₂, 8-10 CH₄	11-14 O₂ 8-10 CH₄	No model	No model
POA direct GHG, O₂/CH₄	0/0.45	0/0.5	0-0.1	0-0.1

Table 1: Metal production energy and GHG emissions, changes using Pure Oxygen Anodes.

Pure Oxygen Anodes™

A new technology could dramatically reduce reactive metals production costs, energy use and emissions by placing a zirconia solid electrolyte between the molten salt and anode, effectively creating a Pure Oxygen Anode. Figure 1 shows a schematic aluminum oxide electrolysis cell using the Hall-Héroult process and this technology. Both involve feeding the oxide into a molten salt solvent, a cathode where cations reduce to metal, and an anode taking in oxygen ions. But unlike graphite anodes which make CO_2 on the surface, Pure Oxygen Anodes use closed-end zirconia tubes to selectively withdraw oxygen ions from the molten salt to the anode inside, often a liquid metal. This leads to several direct benefits for production of all metals:

- There is **no graphite anode plant,** and **no carbon contamination** of product metal.
- With no salt-anode contact, there is **no anode effect or perfluorocarbon formation.**
- One can inject **natural gas fuel** into the tubes, reducing electricity use and cutting direct electrolysis CO_2 emissions in half *vs.* graphite.
- Due to zirconia's selectivity, there is **no HF or other volatile contamination of anode gases**, just high-purity O₂, or fuel combustion products.
- Anode gas purity enables **anode gas energy recovery** impossible with HF contamination, and **eliminates many permitting requirements.**
- Separating out anode gases creates a reducing environment in the molten salt and metal, enabling the use of a **low-cost steel vessel,** with a lining for rare-earth metal production.
- Separating anode gases also **enables production of metal vapor** such as magnesium and zinc.
- Insulating the steel vessel **dramatically reduces heat losses** *cf.* frozen side-wall in Hall-Héroult.
- With uniform temperature in the crucible, there is no salt crust, and **simpler oxide feeding.**
- Vertical electrodes increase electrode area and **reduce cell footprint by 50-80%.**

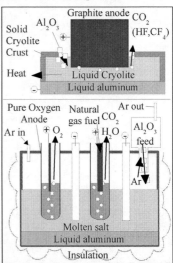

Figure 1: Conventional Hall-Héroult (above) and Pure Oxygen Anode (below) Al₂O₃ reduction cells.

The above benefits can potentially reduce energy losses for aluminum electrowinning by 60% or more, and dramatically reduce costs as well. The ability to produce metal vapor **enables direct magnesium oxide electrolysis** with in-line distillation, skipping chlorination and chloride dehydration and producing high-purity Mg from lower-purity MgO raw material. And because carbon contamination constrains batch size, this technology **enables neodymium batch sizes 10-100 times larger,** dramatically reducing per-unit labor, analytical, and capital costs.

Using a solid electrolyte is not a new concept. In the late 1960s, Marincek [8] and Schmidt-Hatting and Huwyler [9] at Swiss Aluminum investigated the use of a zirconia solid electrolyte for aluminum production, as did Ginsberg and Wilkening [10], and later Minck [11]. More recently, Rapp [12] and LaCamera and Ray [13] added new configurations and electrolytes respectively. Finally, Uday Pal's work on new molten salt compositions [14-17] increased solubilities of many oxides *e.g.* CaF_2-MgF_2 dissolves 10 wt% MgO, and reduced zirconia corrosion rate to as low as 0.2 μm/hr, making this commercially feasible.

This paper will briefly discuss transportation benefits of reactive metals, then focus on recent enabling innovations for Pure Oxygen Anodes™ in more detail, and conclude with overall energy balances for aluminum and magnesium production using this technology.

Transportation Benefits of Reactive Metals

Transportation benefits fall into two categories: reduced weight using the light metals aluminum and magnesium, and lower fuel use by hybrid vehicles or lower emissions by electrification.

Motor Vehicle Weight Reduction

Although the relationship between fuel use and vehicle weight depends heavily on the duty cycle, with weight affecting "city" driving more than "highway" driving, Koffler and Rohde-Brandenburger developed a useful average of 0.35 liters gasoline saved per (100 km driven×100 kg weight reduction) [18]. When multiplied by $4.76×10^{12}$ km/year driven by U.S. vehicles [19], 100 kg reduction in weight reduces vehicle fleet gasoline consumption by $1.7×10^{10}$ liters/year, saving 580 PJ/year energy and reducing CO_2 emissions by 39 MMT/year [20] due to reduction in fuel use. This analysis is only "Pump-to-wheels"; "well-to-pump" oil drilling and refining activities add about 25% to both energy and GHG emissions [21].

The best light metal substitution opportunities are in stiffness-limited components, particularly the body-in-white (BIW) and closures. New aluminum-intensive designs can reduce weight by 37-40%, reaching 250 kg for the BIW and 69 kg for closures [22], for total direct weight savings of about 200 kg and 1.3 EJ in system-wide energy savings. And part-by-part substitution can provide substantial weight reduction: one study proposed replacing 286 kg of steel and aluminum parts per vehicle with 156 kg of magnesium equivalents which have already been demonstrated in existing vehicles [23], reducing weight by 130 kg and energy consumption by 800 PJ/year. These are direct weight savings, and enable smaller, lighter engines and chassis as well.

The main obstacles to achieving these goals are the cost and environmental impact of primary aluminum and magnesium metal production. In particular, European auto makers have been reluctant to increase use of magnesium, as very high production process carbon emissions outweigh lower emissions due to light weight for 150,000 km or more vehicle use [3, 24].

129

(Plug-In) Hybrid and Electric Vehicles

Hybrid vehicles (HEV) with rare earths in permanent magnet motors and NiMH batteries use just under 70% of the fuel/km of conventional internal combustion engine (ICE) vehicles [21], potentially saving 8-10 EJ/year of energy and several hundred MMT of GHG emissions, but at considerable additional powertrain and battery cost, *e.g.* $5300 extra for the Ford Fusion.

Battery electric vehicles (BEV) internally use half the energy/km of HEVs, but "well-to-pump" electricity production from fossil fuels and distribution are much less efficient than gasoline production [21], making apples-to-apples comparisons difficult. One comprehensive model [21] indicates that GHG emissions/km are a strong function of location: in Illinois, BEV emissions/km are nearly 90% of non-hybrid ICE vehicles, (*i.e.* over 25% *higher* than HEVs), whereas in New York that figure is about 50%, and in California close to 35% [25]. As the grid reduces emissions by shifting from coal to natural gas and increasing renewables, the BEV fleet will continue to reduce its impact. Plug-in hybrids (PHEV) perform roughly as a weighted average of BEVs and HEVs depending on battery range and driving style. Although most HEVs use NiMH batteries, PHEVs and BEVs generally use lithium batteries, and new magnesium battery chemistries [26, 27] could simplify anode design and improve energy density. And the technology presented here can potentially provide purer low-cost magnesium for this application.

With the exception of lithium batteries, **molten salt electrolysis produces all of the above critical metals** for transportation energy and GHG reduction technology. Pure Oxygen Anode technology can reduce production cost, energy use, and GHG emissions of all of those metals.

Recent Developments in Pure Oxygen Anode™ Technology

As a new process technology, processes using Pure Oxygen Anodes require new supporting and balance-of-plant innovations for commercialization. This is particularly true for magnesium, as producing metal vapor enables in-line distillation, but requires new gas handling arrangements.

Inert Anode/Current Collector

Boston University has developed an oxygen-stable inert anode and current collector with very low resistance [28]. Figure 2 shows its configuration, consisting of a central metal rod *e.g.* Inconel in an outer ceramic tube *e.g.* alumina, with LSM or similar electronically conducting oxide at the tip, and an alumina/LSM seal preventing oxygen from entering the inner region with the metal rod. A droplet of liquid metal such as silver connects the metal rod to the LSM tip. This can connect to a liquid silver anode in the zirconia tube, creating a complete Pure Oxygen Anode assembly immersed into the molten salt.

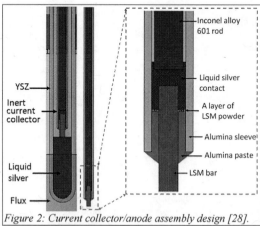

Figure 2: Current collector/anode assembly design [28].

Initial trials indicated excellent performance, with just 0.1Ω resistance which is stable over many hours. This indicates that the materials are stable in this environment, and the alumina/LSM seal prevents oxygen entry and Inconel rod corrosion. Reducing resistance is possible by changing the geometry of the LSM rod, which is the highest-resistance component.

Natural Gas-Fueled Anodes

Fuel injection (Figure 1) reduces costs, energy use and emissions of Pure Oxygen Anodes:
- Material costs are lower because the reducing environment stabilizes low-cost metals like copper and tin instead of silver [14], and a simple nickel tube current collector.
- Energy cost is lower because the fueled anode substitutes chemical energy for 15-25% of the electrical energy. It is like a built-in solid oxide fuel cell (SOFC) with heat recovery, which is the most energy-efficient and lowest-emissions use of fossil fuels.
- The GHG emissions rate of 0.4 kg CO_2e/kWh is much lower than the 0.7 kg CO_2e/kWh U.S. grid average, and lower than average emissions in 42 of 50 U.S. states [25].
- The oxidant is pure oxygen instead of air, so fueled anodes produce no NO_x, making emissions potentially less harmful than those of typical natural gas combustion.

INFINIUM has patented multiple anode-current collector materials systems which work well, and is continuing their development and optimization [29].

Magnesium Condenser

Condensing magnesium vapor to liquid for ease of withdrawal and handling is difficult because its melting point vapor pressure at 400 Pa is among the highest of all metals. Thus even with perfect mass transfer, significant magnesium remains in the vapor phase at its melting point, making it difficult to avoid either leaving significant magnesium in the vapor phase or producing solid metal particles. Furthermore, it reacts zirconia when its partial pressure is above 16 kPa, so we run around 10 kPa. These constraints lead to theoretical best condensation of 96% as a liquid with the remainder condensing as solids, in practice it is hard to exceed 90% liquid condensation.

To overcome this limitation, INFINIUM has invented a continuous two-stage condenser [30]

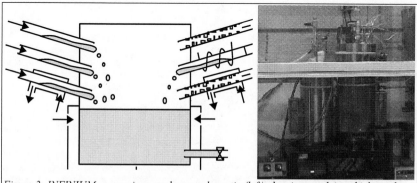

Figure 3: INFINIUM magnesium condenser schematic (left) showing conduits which condense liquid magnesium on the left and solid on the right; and implementation (right).

which condenses most of the magnesium vapor as a liquid in a temperature controlled vessel near the melting point, and traps the remainder as a solid in multiple conduits, then periodically reroutes flow from one or more conduit(s) and drives solids into the liquid for remelting. Figure 3 shows a schematic of this condenser, and its implementation in a small-scale magnesium cell.

Magnesium and Aluminum Energy Balances

For a self-heated electrolysis cell, energy modeling begins with two energy balances: one for free energy, the other for heat. The Nernst equation: $\Delta G = -nFV$ indicates that voltage and free energy are equivalent but use different units: free energy is the energy per mole of reaction extent, and voltage is the energy per coulomb of charge passed. At the process temperature of 1150° C, reduction enthalpy and free energy correspond to 3.81 V and 2.27 V respectively.

The reaction proceeds and current flows if applied voltage is above the 2.27 V dissociation potential, corresponding to reaction free energy. This and other overpotentials in the cell sum to the total voltage, as shown on the left side of Figure 4. The free energy budget consists of the following components:

- Cathode and anode resistances are minimum possible lead energy losses.[1]
- 1.5 V for the zirconia tube is based on 1 A/cm² current density through 3 mm thick YSZ with conductivity 0.2 S/cm.
- Molten salt has much higher conductivity than YSZ, but the anode-cathode distance and effective area are larger, making for half the resistance and voltage drop.
- Oxygen ion mass transfer number is based on concentration (~1/10 of the anions) and boundary layer thickness, and is about 70-80% of molten salt resistance.
- MgO dissociation $V = \Delta G/nF$ (see above).

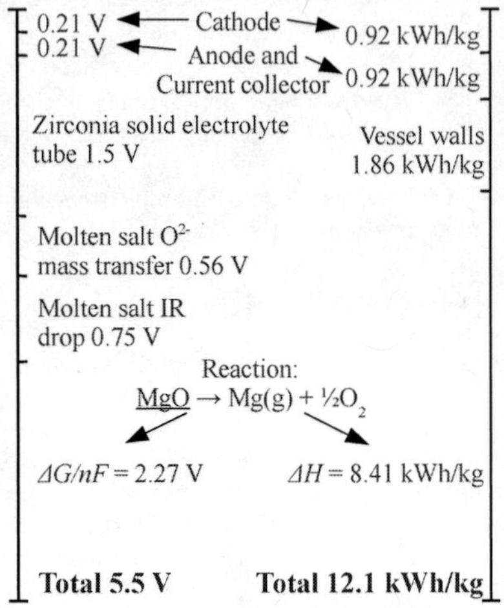

0.21 V ← Cathode → 0.92 kWh/kg
0.21 V ← Anode and → 0.92 kWh/kg
Current collector

Zirconia solid electrolyte tube 1.5 V

Vessel walls 1.86 kWh/kg

Molten salt O^{2-} mass transfer 0.56 V

Molten salt IR drop 0.75 V

Reaction:
$MgO \rightarrow Mg(g) + \frac{1}{2}O_2$

$\Delta G/nF = 2.27$ V $\Delta H = 8.41$ kWh/kg

Total 5.5 V   **Total 12.1 kWh/kg**

Figure 4: Free energy and thermal energy budgets for magnesium oxide electrolysis.

The thermal energy budget on the right side of Figure 4 sums thermal energy use of the cell. The enthalpy ΔH of MgO dissociation to Mg vapor and oxygen consumes most of the heat. If one supplies sufficient free energy but not sufficient enthalpy, then the reaction proceeds but cools the cell and shuts it down. Resistance in the cell or an external heater provides this heat, and excess heat beyond the enthalpy leaves the cell through the cell leads and vessel walls. The thermal energy budget consists of the following:

- Losses through cell leads mentioned above.
- Furnace vessel walls and gas flows, which we estimate will lose about as much heat as the total through electrical leads. Good insulation and heat exchangers could lead to half

---

1 The Wiedmann-Franz law indicates that good electrical conductors are good thermal conductors, so low-resistance leads lose a lot of heat and vice versa. There is thus an unavoidable minimum energy loss, and at process temperature, this minimum is ~0.2 V (0.46 kWh/kg Mg, 0.5 kWh/kg Al) I-R drop in each lead lost to the environment as heat, plus equal heat conducted from the crucible through each lead to the environment.

of this value (0.8 kWh/kg), or various contingencies could make it as high as 3.8 kWh/kg.

- Enthalpy of magnesium oxide dissociation ΔH as mentioned above, note this is above the 1090°C Mg boiling point and includes 1.5 kWh/kg ΔH_v, cf. Table 1).

Based on these energy balances and the range of heat loss values, INFINIUM believes that operating at a cell potential of 5-6.5 V, corresponding to 11-14 kWh/kg, will achieve self-heating and 1 A/cm² zirconia tube current density. Overall vapor generation energy efficiency (ΔH/input energy for the reaction in Figure 4) is then between 63% and 76%. Because the enthalpy of dissociation to liquid is lower, with no condenser heat recovery, energy efficiency is 52-63%, though it may be possible to use condenser heat e.g. for raw material pre-heating and calcining.

The aluminum energy balance is similar, though without a condenser the cell design is much simpler, leading to lower losses. Based on the ability to dramatically reduce or eliminate side-wall and anode gas losses as described above, INFINIUM expects to use about half the energy of today's Hall-Héroult cells with either oxygen-generating or fueled anodes, as shown in Table 1.

Conclusions

The aluminum industry has recognized for decades some of the benefits of zirconia solid electrolytes in molten salt electrolysis. Based on new advances in the science of molten salt-zirconia interactions, and the development of new balance-of-plant technologies, we are close to making Pure Oxygen Anode™ technology using zirconia economically scale for industrial use. In addition to the innovations described above, recent inventions and custom zirconia chemistries have dramatically improved current efficiency and tube lifetime under electrolysis. New on-line diagnostics are helping us to monitor chemical changes in the cell in real time. And INFINIUM has demonstrated the ability to reliably hot-swap zirconia tubes through a steep thermal gradient of a hot furnace. These developments and scale-up activities have brought Pure Oxygen Anodes to the threshold of industrial production of the metals presented above, with tantalum [15], titanium [16], and solar-grade silicon [17] not far behind.

Beyond the direct energy and GHG savings common to all metals, this technology enables major changes in primary production practice for magnesium and rare-earth metals in particular. Direct electrolysis of magnesium oxide completely changes the process flow-sheet, cutting out the very high-cost and often dangerous synthesis of high-purity anhydrous MgCl₂ via carbochlorination or hydrochlorination with chloride dehydration. And eliminating carbon from the process can increase neodymium and lanthanum metal batch size by an order of magnitude or more, and eliminates the anode effect which is very common in today's practice, reducing GHG impact and making it economically feasible to produce these metals in OECD countries.

The reactive metals represented here, which are aluminum, magnesium, neodymium, and lanthanum, are some of the most critical for a new generation of ultra-efficient light-weight vehicles with hybrid and/or electric powertrains. Efficient low-emissions primary production is thus essential for reducing energy consumption and greenhouse emissions of the transportation sector, including meeting new U.S. fuel economy standards, which will save exajoules of energy and millions of metric tonnes of CO_2 emissions in both production and operation of motor vehicles. Lower material costs also reduce barriers to efficient vehicle technologies, particularly in markets with less stringent emissions regulations. INFINIUM keeps these goals in mind as we continue to make advances toward industrial deployment of Pure Oxygen Anodes.

Acknowledgments

This report is based upon work supported by the U.S. Department of Energy under Award No. DE-EE0005547, and by the National Science Foundation under Grants 1026639 and 1322498.

References

1. U.S. Department of Energy (DOE) Energy Information Administration (EIA) Annual Energy Outlook 2013 Report DOE/EIA-0383ER(2013) Tables A2, A18.
2. "U.S. Energy Requirements for Aluminum Production," U.S. Department of Energy, EERE Report, Feb. 2007. Tables A and E.2.
3. International Magnesium Association, *Life Cycle Assessment of Magnesium Components in Vehicle Construction* (Wauconda, Illinois: International Magnesium Association), 2013.
4. Georges Kipouros and Donald Sadoway, "A thermochemical analysis of the production of anhydrous MgCl₂," *J. Light Metals* 1(2):111-117 (2001).
5. Guihua Wang, Xiangsheng Wang and Hongmin Zhu, "Perfluorocarbon Generation During Electrolysis in Molten Fluoride," N.R. Neelameggham *et al.,* eds., *Energy Technology 2011* (Warrendale, PA: TMS).
6. Data source for Al and Mg: NIST Chemistry WebBook, http://webbook.nist.gov/chemistry/ .
7. Data source for Nd and La: HSC Chemistry, Version 6.0.0.
8. Borut Marincek, *Electrolytic cell,* U.S. Patent 3,562,135 February 9, 1971; div. *Electrolytic Process,* U.S. Patent 3,692,645 September 19, 1972.
9. W. Schmidt-Hatting and S. Huwyler, *Electrolytic Cell Apparatus,* U.S. Patent 3,578,580 May 11, 1971.
10. Hans Ginsberg and Siegfried Wilkening, *Verfahren und Vorrichtung zur schmelzflußelektrolytischen Reduktion von Metalloxiden zu ihren Metallen,* German Patent 19 48 462 January 22, 1975.
11. Robert Minck, *Method and apparatus for separating a metal from a salt thereof,* U.S. Pat. 4,108,743 8/22/1978.
12. Robert Rapp, *Method and Apparatus Featuring a Non-Consumable Anode for the Electrowinning of Aluminum,* U.S. Patent 5,942,097 August 24, 1999.
13. Alfred LaCamera and Siba Ray, *Electrolysis in a Cell having a Solid Oxide Ion Conductor,* U.S. Patent 6,187,168 February 13, 2001.
14. Ajay Krishnan, Xiong Gang Lu and Uday B. Pal, "Solid Oxide Membrane Process for Magnesium Production Directly from Magnesium Oxide," *Metall. Mater. Trans.* 36B:463-473, 2005.
15. Ajay Krishnan, Xiong Gang Lu & Uday Pal, "Solid Oxide Membrane (SOM) technology for environmentally sound production of tantalum metal and alloys from their oxide sources," *Scand. J. Metall.* 34(5):293-301, 2005.
16. M. Suput, R. DeLucas, S. Pati, G. Ye, U. Pal and A. Powell, "Solid Oxide Membrane Technology for Environmentally Sound Production of Titanium," *Min. Proc. Extract. Metall.* 117(2):118-122, June 2008.
17. Y. Jiang, J. Xu, B. Lo, U.B. Pal, S.N. Basu, "Production of Silicon from Silica: Solid-Oxide-Membrane Based Electrolysis Process", in M. Free and A. Siegmund eds., EPD Congress 2013, (Hoboken, NJ: Wiley), 2013.
18. Christoph Koffler and Klaus Rohde-Brandenburger, "On the calculation of fuel savings through lightweight design in automotive life cycle assessments," *Int. J Life Cycle Assess.* 15(1):128-135, 2010.
19. U.S. Department of Transportation Bureau of Transportation Statistics Reports: National Transportation Statistics July 2013 Table 1-35.
20. U.S. EPA References Sheet: Gasoline: 0.125 mmBTU/gal = 34.8 MJ/liter, 8.92 kg CO₂/gal = 2.36 kg CO₂/liter.
21. A. Elgowainy *et al.,* "Well-to-Wheels Analysis of Energy Use and Greenhouse Gas Emissions of Plug-In Hybrid Electric Vehicles," Argonne National Laboratories ANL/ESD/10-1, June 2010, Table 3.3 and Fig. 6.6.
22. Tim Skszek and Jeff Conklin, "Multi-Material Lightweight Prototype Vehicles Demonstration," U.S. DOE Vehicle Technologies Program Annual Merit Review 2013.
23. U.S. Automotive Materials Partnership, *Magnesium Vision 2020,* USCAR Technical Report, Nov. 2006.
24. Karl Kainer, "Global Magnesium Research: State-of-the-Art and What's Next?" Presentation at the TMS Annual Meeting, February 28, 2011, and extended abstract in Wim Sillekens, ed., *Magnesium Technology 2011* TMS, Warrendale, Pennsylvania, 2011, p. 5.
25. The US EPA eGRID report shows GHG emissions rates for each state: http://www.epa.gov/egrid/ .
26. D. Aurbach *et al.,* "Prototype systems for rechargeable magnesium batteries," Nature 407:724-727, 2000.
27. Pellion Tecnologies, "Rechargeable Magnesium Batteries," ARPA-E BEEST project 9/1/2010-12/31/2012.
28. X. Guan, U. Pal, S. Gopalan and A.C. Powell, "LSM (La₀.₈Sr₀.₂MnO₃₋δ)-Inconel Inert Anode Current Collector for Solid Oxide Membrane (SOM) Electrolysis," *J. Electrochem Soc.* 160(11):F1179-F186, 2013.
29. A. Powell, S. Pati, U. Pal, S. Derezinski and R.S. Tucker, *Liquid Anodes and Fuels for Production of Metals from their Oxides by Molten Salt Electrolysis with a Solid Electrolyte,* U.S. Patent Application 13/592,211.
30. A. Powell *et al., Apparatus and Method for Condensing Metal Vapor,* U.S. Patent Application 13/543,975.

Energy Technology 2014: Carbon Dioxide Management and Other Technologies
Edited by: Cong Wang, Jan de Bakker, Cynthia K. Belt, Animesh Jha, Neale R. Neelameggham,
Soobhankar Pati, Leon H. Prentice, Gabriella Tranell, and Kyle S. Brinkman
TMS (The Minerals, Metals & Materials Society), 2014

Electrochemistry of Fe(III) in Molten Salt $CaCl_2$-KF and $CaCl_2$-CaF_2-KF

Li Li[1,3], Xuan Liu[2], Shulan Wang[3*]

[1]Department of Materials Science and Engineering, Cornell University, Ithaca, NY, 14850, USA
[2]Department of Materials Science and Engineering, Carnegie Mellon University, Pittsburgh, PA 15213, USA
[3*]Department of Chemistry, School of Science, Northeastern University, Shenyang, 110004, China, Email: slwang@mail.neu.edu.cn

Abstract: Electrochemistry of Fe(III) in $69CaCl_2$-$12CaF_2$-19KF and $81CaCl_2$-19KF (mole percentage, %) was investigated at 827 °C on molybdenum electrode by cyclic voltammetry, chronoamperometry and galvanostatic electrolysis using Fe_2O_3 as the Fe(III) source. The reduction of Fe(III) to Fe was conducted in one step and controlled by diffusion process. Fe(III) shown greater reduction rate and diffusion coefficient in molten $81CaCl_2$-19KF than in $69CaCl_2$-$12CaF_2$-19KF. Galvanostic electrolysis of Fe was performed in both molten salts with a cylindrical cathode rotating at 200 r/min and dendritic iron deposit was obtained.

Keywords: iron; electrolysis; electrochemistry; molten salt

Introduction

The production of iron and steel causes huge emissions of CO_2. Iron electrowinning in molten salts with dissolved Fe_2O_3 using an inert oxygen evolving anode is an alternative process to reduce or eliminate the formation of CO_2. Research on novel process of producing metal has been reported recently [1-2] by using $CaCl_2$ as the main electrolyte and dissolving CaO in $CaCl_2$ due to the high solubility of CaO in the melt. This process has been used in preparing many active metals and alloys [3-14]. In the present work, the electrochemical behavior of iron in $CaCl_2$-KF was studied and transport property of iron ion in the melt was discussed.

Experimental

$CaCl_2$, CaF_2 and KF (in analytical purity) were dried in air for 48 h. Mixed salts used in experiments were placed in a graphite crucible (ϕ 58 mm inner diameter). The graphite crucible together with the salts was heated at 473 K for 12 h in a furnace with flowing Ar (99.99%) before further heating. Temperature inside the furnace was measured by an S type thermocouple. Electrochemical measurements, such as cyclic voltammetry and chronocoulometry, were performed in a three electrode cell by EG&G PAR273A potentiostat/galvanostatand. Galvanostatic electrodepositon of iron was conducted in a two electrode cell. Molybdenum (ϕ 2 mm) and iron (ϕ 5 mm) were used as the working electrode (WE) and the reference electrode, respectively. A graphite crucible and a magnetite rod (ϕ 25 mm) consisted of 100% magnetite and sintered at 950 °C in argon atmosphere for 2 h with electrical conductivity of 366 $\Omega^{-1}m^{-1}$ at room temperature were used as the counter electrodes in the galvanostatic electrolysis experiment. Carbon (φ 10 mm) was used as the counter electrode in electrochemical measurement. The deposits were washed in distilled water bubbled with argon and dried in air before being characterized by X-ray diffraction (PW3040/60 diffractometer) and scanning electron microscope (SEM, SSX-550).

Results & Discussion

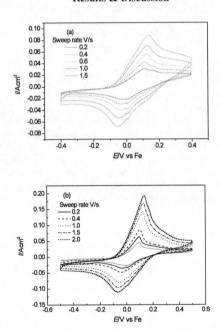

Fig. 1 Cyclic voltammograms at a molybdenum electrode in 0.3 mol% Fe_2O_3 contained molten salts at 827 °C in (a) $69CaCl_2$-$12CaF_2$-19KF, (b) $81CaCl_2$-19KF.

Fig. 1 shows the cyclic voltammograms of a molybdenum electrode at 827 °C with different potential sweep rates in 0.3 mol% Fe_2O_3 contained molten salts of $69CaCl_2$-$12CaF_2$-19KF and $81CaCl_2$-19KF. A cathodic and an anodic current peak at close potentials around -0.05 and 0.1 V in both molten salts were observed. This suggests that the dissolving state of Fe(III) in both molten salts is similar and that the current peaks are ascribed to the redox reaction of Fe(III)/Fe in the molten salts. The cyclic voltammograms with all sweep rates have greater current density in molten $81CaCl_2$-19KF than in molten $CaCl_2$-$12CaF_2$-19KF. This indicates that the former has lower viscosity and higher electrical conductivity than the latter. With increase in sweep rate, the peak current densities increased and the peak potentials shifted.

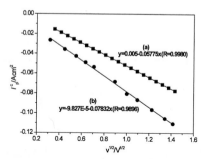

Fig. 2 Voltammetric peak current density versus square root of sweep rate at a molybdenum electrode in 0.3 mol% Fe_2O_3 contained molten salts at 827 °C in (a) $69CaCl_2$-$12CaF_2$-$19KF$, (b) $81CaCl_2$-$19KF$.

Plots of the cathodic peak current density as a function of the square root of the sweep rate in both molten salts show straight lines, indicating that the deposition of Fe(III) was controlled by diffusion of Fe(III) in the molten salts[15]:

$$I^c_p = 0.61(nF)^{3/2}[D/RT]^{1/2}C^0 v^{1/2}$$

The diffusion coefficient of Fe(III) at 827°C in molten $81CaCl_2$-$19KF$ is 9.7×10^{-5} cm/s. This value is greater than that of 5.3×10^{-5} cm/s in $69CaCl_2$-$12CaF_2$-$19KF$ due to the lower viscosity and better electrical conductivity of the former salt. As a result, the slope of the former salt is also greater than that of the latter.

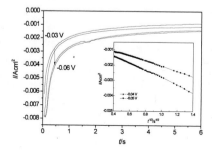

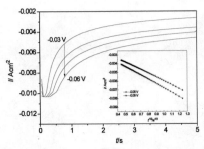

Fig. 3 Potentiostatic current transients recorded at a molybdenum electrode in 0.3 mol% Fe_2O_3 contained molten salts at 827 °C in (a) $69CaCl_2$-$12CaF_2$-19KF, and (b) $81CaCl_2$-19KF. Inset figures are plots of current density vs $t^{-1/2}$ at potentials -0.05 and -0.06 V around the reduction peak potential of Fe^{3+}.

The current-time transients obtained at the molybdenum electrode by the potentiostatic method from a slightly positive potential of Fe deposition to a potential where deposition occurs in the applied potential range of -0.03 to -0.06 V were shown in Fig. 3. No sign of iron nucleation on the electrodes can be seen. The current increases monotonically with time in molten $69CaCl_2$-$12CaF_2$-19KF and dropped initially to a minimum and then rose monotonically with time in molten $81CaCl_2$-19KF. It is assumed that there is a double layer charging process in molten $81CaCl_2$-19KF at the initial stage. Current response in each potential step is more obvious in $81CaCl_2$-19KF than in $69CaCl_2$-$12CaF_2$-19KF. The current-time transients obtained in molten $69CaCl_2$-$12CaF_2$-19KF and $81CaCl_2$-19KF follow Cottrell's equation: the potentiostatic current decays linearly with the inverse of the square root of time due to diffusion control, as shown in the inset figure of Fig. 3.

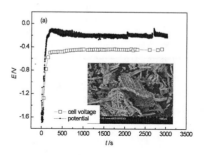

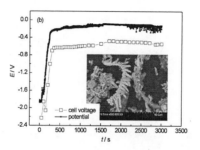

Fig. 4 Galvanostatic electrodeposition of Fe(III) in 0.3 mol%Fe_2O_3 contained molten salt in (a) $69CaCl_2$-$12CaF_2$-$19KF$, (b) $81CaCl_2$-$19KF$.

Several phenomena of galvanostatic electrodeposition of Fe(III) at a stirring rate of 200 rmp in $69CaCl_2$-$12CaF_2$-$19KF$ at a current density of 0.5 Acm^{-2} and at 890°C with the magnetite anode and in $81CaCl_2$-$19KF$ at 827°C at a current density of 0.4 Acm^{-2} with the graphite anode were observed. First, a similar electrode potential of -0.2 V is found in both cases. Secondly, a lower cell voltage in molten $69CaCl_2$-$12CaF_2$-$19KF$ than in $81CaCl_2$-$19KF$ has been observed even though the latter is at a higher electrolysis temperature. This is mainly due to a lower electrical conductivity of magnetite than graphite and a higher Gibbs free energy of production of O_2 in magnetite anode than that of production of CO_2 in graphite. Thirdly, a great difference in the current efficiency of 98% in $69CaCl_2$-$12CaF_2$-$19KF$ and 62% in $81CaCl_2$-$19KF$ has been found. Finally, both products of deposited Fe are dendritic, as shown in the inset of Fig. 4. This suggests that the electrodeposition process of Fe (III) to Fe was controlled by the diffusion step.

Conclusions

Electrochemistry and electrodeposition of Fe(III) in $69CaCl_2$-$12CaF_2$-$19KF$ and $81CaCl_2$-$19KF$ (both in mole percent, %) were investigated by cyclic voltammetry, chronoamperometry and galvanostatic electrolysis using Fe_2O_3 as the Fe(III) source. The electrochemical study shows that the electrodeposition of Fe(III) to Fe was conducted in one step and was controlled by the diffusion process. Fe(III) was found to have a greater diffusion coefficient in molten $81CaCl_2$-$19KF$ than in $69CaCl_2$-$12CaF_2$-$19KF$. This is caused by the lower viscosity and higher electrical conductivity of the former than the latter. The current efficiency of electrodeposition of iron in $81CaCl_2$-$19KF$ at $827^{\circ}C$ is only 62% and increased to 98% in $69CaCl_2$-$12CaF_2$-$19KF$ at $890^{\circ}C$. Both deposits are dendritic Fe. The electrodeposition of Fe(III) was controlled by the diffusion step even at $890^{\circ}C$ under agitation by a rotating cylinder of 200 r/min.

Acknowledgements

This work was supported by National Natural Science Foundation of China No.51274058.

References

[1] G. Z. Chen, D. J. Fray, and T. W. Farthing, 1999, Patent WO9964638
[2] G. Z. Chen, D. J. Fray, and T. W. Farthing. "Direct Electrochemical Reduction of Titanium Dioxide in Molten Calcium Chloride", *Nature*, 407 (2000), 361-364.
[3] X. Y. Yan, and D. J. Fray, "Using Electro-deoxidation to Synthesize Niobium Sponge from Solid Nb_2O_5 in Alkali-Alkaline-Earth Metal Chloride Melts", *Journal of Materials Research*, 18 (2003), 346-356.
[4] G. Z. Chen, E. Gordo, and D. J. Fray, "Direct Electrolytic Preparation of Chromium Powder", *Metallurgy and Materials Transaction B*, 35 (2004), 223-233.
[5] G. H. Qiu *et al.*, "A Direct Electrochemical Route from Oxide Precursors to the Terbium-Nickel Intermetallic Compound $TbNi_5$", *Electrochimica Acta*, 51 (2006), 5785-5793.
[6] S. Q. Jiao *et al.*, "Production of NiTi Shape Memory Alloys via Electro-Deoxidation Utilizing an Inert Anode", *Electrochimica Acta*, 55 (2010), 7016-7020.
[7] S. L. Wang, and Y. J. Li, "Reaction Mechanism of Direct Electrochemical Reduction of Titanium Dioxide in Molten Calcium Chloride", *Journal of Electroanalytical Chemistry*, 571 (2004), 37-42.
[8] C. Schwandt, and D. J. Fray, "Determination of the Kinetic Pathway in the Electrochemical Reduction of Titanium Dioxide in Molten Calcium Chloride", *Electrochimica Acta*, 51 (2005), 66-76.
[9] D. T. L. Alexander, C. Schwandt, and D. J. Fray, "Microstructural Kinetics of Phase Transformations during Electrochemical Reduction of Titanium Dioxide in Molten Calcium Chloride", *Acta Materialia*, 54 (2006), 2933-2944.
[10] D. H. Wang *et al.*, "Electrochemical Metallization of Solid Terbium Oxide", *Angewandte Chemie International Edition*, 45 (2006), 2384-2388.
[11] C. Schwandt, D. T. L. Alexander, and D. J. Fray, "The Electro-Deoxidation of Porous Titanium Dioxide Precursors in Molten Calcium Chloride under Cathodic Potential Control", *Electrochim Acta*, 54 (2009), 3819-3829.
[12] M. F. Liu, Z. C. Guo, and W. C. Lu," Process of Direct Electrochemical Reduction of TiO_2", *The Chinese Journal Nonferrous Metals*, 14 (2004), 1752-1758.
[13] Z. Y. Cai *et al.*, "Electrochemical Behavior of Silicon in the ($NaCl$-KCl-NaF-SiO_2) Molten Salt", *Metallurgy and Materials Transaction B*, 41 (2010), 1033-1037.
[14] S. L. Wang, and X. Y. Chen,"Study on the electro-refining Silicon in Molten Salt $CaCl_2$-$NaCl$ –CaO", *Acta Metallurgica Sinica*, 48 (2012), 183-186.
[15] T. Berzins, and P. Delahay, "Oscillographic Polarographic Waves for the Reversible Deposition of Metals on Solid Electrodes", *Journal of the American Chemical Society*, 36 (1953), 555-559.

Energy Technology 2014: Carbon Dioxide Management and Other Technologies
Edited by: Cong Wang, Jan de Bakker, Cynthia K. Belt, Animesh Jha, Neale R. Neelameggham,
Soobhankar Pati, Leon H. Prentice, Gabriella Tranell, and Kyle S. Brinkman
TMS (The Minerals, Metals & Materials Society), 2014

Novel LiNO$_3$–NaNO$_3$–KNO$_3$–NaNO$_2$ Molten Salts for Solar Thermal Energy Storage Applications

T. Wang[1] and R. G. Reddy[2]*

[1] Graduate Student, [2] ACIPCO Endowed Chair Professor

Department of Metallurgical and Materials Engineering
The University of Alabama, Tuscaloosa, AL 35487-0202, USA.

Abstract

The corrosion of stainless steel SS316L in a low melting point novel LiNO$_3$-NaNO$_3$-KNO$_3$-NaNO$_2$ eutectic molten salt mixture was investigated at 695K. After long-term isothermal dipping corrosion experiments, the SS316L samples were analyzed using scanning electron microscopy (SEM) equipped with energy dispersive spectrometer (EDS) to determine the topography and corrosion products. Dense and protective lithium iron oxide scale formed on the surface and prevented the SS316L sample from further severe corrosion. Therefore, SS316L is considered to be suitable construction material for solar thermal energy storage materials containers.

Keywords: Solar Thermal Energy Storage, corrosion, stainless steel, molten salts.

Introduction

Molten salt mixtures have been considered as potential candidates for thermal energy storage and heat transfer media in solar energy applications. To apply the molten salt in the energy transfer process, the standard pipes made of stainless steel must be utilized to carry the molten salt mixtures in parabolic trough system of solar power production. The pipes encounter continuous flow of hot and cold molten salt every day. Thus corrosion of stainless steel in molten salt mixtures becomes extremely essential for the long-term utilization of the thermal energy storage media. The corrosion of passivity materials inside high temperature molten salt systems have been studied which gives fundamental understanding of basic mechanism about the corrosion process [1-8].

The corrosion behavior of Fe-Ni-Cr alloys in an equi-molar LiNO$_3$–NaNO$_3$–KNO$_3$ mixture was studied [9]. According to the XRD analyses, the major phases in the oxide scales were identified as LiFeO$_2$, Fe$_3$O$_4$ and Cr$_2$O$_3$. A HITEC® salt mixture (NaNO$_3$–NaNO$_2$–KNO$_3$) was also tested as a corrosion medium for stainless steel [10] and the corrosion rate was found to be

---

* Corresponding Author, E-mail: rreddy@eng.ua.edu

affected by the protective oxide layer. In most cases, the alloys having more than 10wt% chromium show high resistance to corrosion in molten salts.

Novel low melting point molten salt mixtures are being developed over the past few decades that can successfully replace the currently used $NaNO_3$-KNO_3 solar salt and HITEC® salt whose melting point are 494K and 415K, respectively. Besides, the $LiNO_3$-$NaNO_3$-KNO_3-$NaNO_2$ [11] was developed using thermodynamic modeling in our research group. In the present work, the corrosion behavior of SS316L stainless steel was investigated at 695 K using the electrochemical method and isothermal dipping method in the $LiNO_3$-$NaNO_3$-KNO_3-$NaNO_2$ quaternary system. This eutectic salt mixture have been studied and found to be very suitable to be used in solar energy receiver plant due to its high energy storage density [11]. After corrosion experiments, the test samples were analyzed using scanning electron microscopy (SEM) equipped with energy dispersive spectrometer (EDS) and x-ray diffraction (XRD) techniques to determine the topography and corrosion products.

Experimental

Materials and salt preparation

Eutectic salt mixture in the $LiNO_3$-$NaNO_3$-KNO_3-$NaNO_2$ system is used as the electrolyte in the present study and composition is given in Table. 1. As-received (99% pure) salt components purchased from Alpha Aesar® Company were used without further purification. The eutectic salt mixture was prepared by mixing weighed amounts of the constituent components in a Pyrex beaker. The mixture was then heated to 473K and kept at that temperature for 1 hour before being heating it to the desired temperature.

Table 1 Composition of $LiNO_3 - NaNO_3 - KNO_3 - NaNO_2$ quaternary eutectic melt

Component	Composition (wt. %)
$LiNO_3$	17.5
$NaNO_3$	14.2
KNO_3	50.5
$NaNO_2$	17.8
Melting point	372K

In the present work, SS316L coupons were used as test samples and the composition ranges are described in Table 2. The test rod was used to determine corrosion behavior with electrochemical method and cut from the stainless steel bar to the dimension of 12mm diameter with 30mm length and spot-welded to a copper wire of approx. 0.5mm diameter. To simplify the analysis, only the bottom surface of the rod was exposed to the molten salt medium, leaving the side and upper surface wrapped by Teflon tape and covered by Pyrex glass tube which has the same inner diameter as that of the test rod. Stainless steel coupons with dimensions of 13mm×10mm×1mm were used to investigate the corrosion behavior and products after certain time of immersion in the $LiNO_3$-$NaNO_3$-KNO_3-$NaNO_2$ quaternary system.

Table.2 Composition ranges for SS 316L

Element	C	Mn	Si	P	S	Cr	Mo	Ni	N
Min(wt%)	–	–	–	–	–	16.00	2.00	10.00	–
Max(wt%)	0.03	2.00	0.75	0.05	0.03	18.00	3.00	14.00	0.10

Isothermal dipping test

The dipping corrosion experiments were carried out on stainless steel (SS316L) coupon in the eutectic $LiNO_3$-$NaNO_3$-KNO_3-$NaNO_2$ quaternary molten salt. The salt mixture was contained in Pyrex crucible and heated using hot plate or box furnace. To achieve uniform temperature distribution inside the molten salt, the entire set-up was hung in the middle of the furnace and covered by insulation materials as illustrated in Fig. 1. The temperature of the molten salt was measured at 695K using K-type thermocouples from 3-4 spots of the bottom of crucible and the deviation was maintained within ±1.2% using the data acquisition software Chartview®. After certain immersion time, the samples were taken out of the crucible and the corrosion products covering the surface of the sample extracted from the molten salt were analyzed by using X-ray diffraction (XRD) and Scanning Electron Microscope (SEM) techniques. Also, the corroded samples were cast into conductive chlorine-fee epoxy and polished with 180-, 320-, 600- and 1200- grit SiC - grinding paper to study the corrosion attack from the scale thickness in the cross sections.

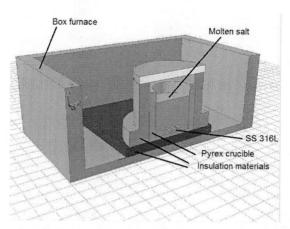

Fig. 1 Experimental set-up for isothermal dipping corrosion test

Results and Discussion

Study of as-received SS316L sample

Fig. 2 shows the SEM image of the sample surface before the experiment. Smooth and uniform colored surface can be observed without any obvious defects. The elemental composition of the alloy before the corrosion experiment was detected using the EDS and listed in Table 3. The composition of the alloy is in excellent agreement with the composition of the SS316L stainless steel coupon shown in Table 1.

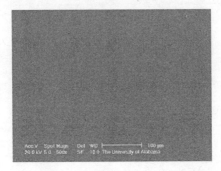

Fig. 2. Scanning electron micrograph of surface of SS316L before corrosion

Table 3 Elemental composition of the as-received SS316L stainless steel coupon surface

Element	wt %	at %
Mo	2.17	1.27
Cr	16.55	17.85
Mn	0.22	0.22
Fe	70.80	71.91
Ni	10.26	9.75
Total	100	100

Scanning Electron Microscope/ Energy Dispersive Spectrometer analysis

The samples extracted from the molten salt at each time interval were investigated using scanning electron microscope (SEM) and energy dispersive spectrometer (EDS). Fig. 4(a) shows the morphology of the corroded sample at 695K after 748hr immersion in molten salt. It is evident that the oxide layer formed directly onto the surface of SS316L and the surface was roughened. High magnification micrograph of corroded sample is shown in Fig. 4(b) which illustrates that the oxide crystals pack together and form oxide film. The elemental composition distribution was determined using the EDX spot analysis on the corroded sample surface. The atomic percentage of each element was listed in Table 4 which reveals that the sample surface was corroded uniformly according to the very similar elemental compositions between any two arbitrarily selected testing spots. By comparing the cation distributions before and after the 748hr immersion with the average element ratio of Fe:Cr:Ni = 8.03:2.02:1.00 and Fe:Cr:Ni = 11.25:2.64:1.00, respectively, it is indicated that more Fe and Cr elements diffuse outward and form oxides covering the surface. Moreover, the higher Fe/Cr ratio after the immersion is possibly attributed to the formation of lithium iron oxides and iron oxides which contain no chromium element.

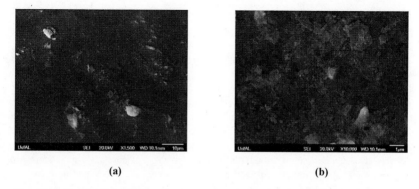

(a) (b)

Fig. 4 Scanning electron micrograph of products adherent on the surface of SS316L after immersion in $LiNO_3 - NaNO_3 - KNO_3 - NaNO_2$ molten salt after 748hrs (a) 1500× magnification (b) 10000× magnification

Table 4 Elemental composition of the SS316L stainless steel coupon surface after 748 hrs immersion in $LiNO_3$-$NaNO_3$-KNO_3-$NaNO_2$ molten salt

Element	wt.%	at.%
O	34.69	62.69
Si	0.51	0.54
Cr	11.1	6.37
Mn	1.19	0.64
Fe	47	27.11
Ni	4.73	2.41
Mo	0.79	0.25
Totals	100	100

With longer immersion time, the morphologies of the corroded sample surfaces are observed to be rougher. As illustrated in Fig. 5(a), the sample after 5088hrs immersion exhibits white color humps which resulted from the coalescence and enrichment of oxide crystals. Fig. 5(b) presents the higher magnification image of the corroded surface. It is viewed that the surface is composed of coarse octahedral oxide crystals which are densely stacked. The elemental composition distribution over the sample surface is given in Table 5. It is found that at 20kV accelerating voltage, O and Fe are the major elements detected at each selected spot. However, only trace amounts of Cr and Ni were measured. The drastic decrease in amount of Cr and Ni compared to that shown in Table 4 reveals the thickening of the lithium iron oxide and iron oxide layers which were formed on the sample surface. When the accelerating voltage was increased to 30kV, the detected amount of Ni and Cr also increased as shown in Table 5, which verifies the thickness effect of lithium iron oxide and iron oxide layers discussed above.

(a)   (b)

Fig. 5 Scanning electron micrograph of products adherent on the surface of SS316L after immersion in $LiNO_3 - NaNO_3 - KNO_3 - NaNO_2$ molten salt after 5088hrs (a) 2000× magnification (b) 10000× magnification

Table 5 Elemental composition of the SS316L stainless steel coupon surface after 5088 hrs immersion in $LiNO_3$-$NaNO_3$-KNO_3-$NaNO_2$ molten salt detected under 20kV and 30kV accelerating voltage

Element	Composition (at %)	
	at 20kV	at 30kV
O K	69.54	68.02
Si K	0.14	0.27
Cr K	1.03	3.43
Fe K	28.56	26.38
Ni K	0.72	1.9
Totals	100	100

Conclusions

The corrosion of stainless steel SS316L in a low melting point novel eutectic $LiNO_3$-$NaNO_3$-KNO_3-$NaNO_2$ salt mixture was investigated at 695K using static isothermal exposure method. Long-term immersion of SS316L coupons in the $LiNO_3$-$NaNO_3$-KNO_3-$NaNO_2$ salt mixture resulted in the formation of oxide scales on the surface. Dense and protective lithium iron oxide prevents the SS316L from severe corrosion. Hence, the SS316L is considered to be compatible to work with the novel quaternary eutectic molten salt mixture. Additional studies will be conducted in the future to determine whether similar results of mixed lithium iron oxide compound film will occur under dynamic molten salt flow conditions such as in the pipes used in solar storage system. The study will define whether the oxide scales peel off under dynamic conditions, or still provide added protection.

Acknowledgements

The authors are pleased to acknowledge the financial support from Department of Energy (DOE) Grant No. DE-FG36-08GO18153 for this research project. We also thank The University of Alabama for providing the experimental facilities.

References

[1] A. Nishikata, H. Numata, T. Tsuru, Electrochemistry of molten salt corrosion, Mater. Sci. Engg. A146 (1991) 15–31.
[2] B. P. Mohanty, D. A. Shores, Role of chlorides in hot corrosion of a cast Fe–Cr–Ni alloy. Part I: Experimental studies, Corr. Sci. 46 (2004) 2893–2907.
[3] L.C. Olson et al., Materials corrosion in molten LiF–NaF–KF salt, J. Fluorine Chem. 130 (2009) 67–73.

[4] T-H. Lim, E. R. Hwang, H. Y. Ha, S. W. Nam, I-H. Oh, S-A. Hong. Effects of temperature and partial pressure of CO2/O2 on corrosion behavior of stainless-steel in molten LiNa carbonate salt, J. Power Sources 89 (2000) 1–6.

[5] C.S. Ni, L.Y. Lu, C.L. Zeng, Y. Niu, Electrochemical impedance studies of the initial-stage corrosion of 310S stainless steel beneath thin film of molten $(0.62Li, 0.38K)_2CO_3$ at 650 °C, Corr. Sci. 53 (2011) 1018–1024.

[6] A. Hendry, D.J. Lees, Corrosion of austenitic steels in molten sulphate deposits, Corr. Sci. 20 (1980) 383–404.

[7] J. Kolchakov, T. Tzvetkoff, M. Bojinov, In situ and ex situ characterization of the passive film on a ferritic stainless steel in molten sodium hydroxide, Appl. Surf. Sci. 249 (2005) 162–175.

[8] S.H. Goods, R.W. Bradshaw, M.R. Prairie, J.M. Chavez, Corrosion of stainless and carbon steels in molten mixtures of industrial nitrates, SAND–94–8211, Sandia National Labs., Livermore, CA, Mar 1994.

[9] D.R. Boehme, R.W. Bradshaw, X-ray diffraction analysis of corrosion products of Fe-Ni-Cr alloys formed in molten nitrate salts, High Temp. Sci. 18 (1984) 39–51.

[10] J. R. Keiser, J. H. Devan, E. J. Lawrence, Compatibility of molten salts with type 316 stainless steel and lithium, J. Nucl. Mater. 85&86 (1979) 295–298.

[11] T. Wang, D. Mantha, R. G. Reddy, Novel low melting point quaternary eutectic system for solar thermal energy storage, Appl. Energ. (2012) in press, http://dx.doi.org/10.1016/j.apenergy.2012.09.001

Energy Technology 2014: Carbon Dioxide Management and Other Technologies
Edited by: Cong Wang, Jan de Bakker, Cynthia K. Belt, Animesh Jha, Neale R. Neelameggham,
Soobhankar Pati, Leon H. Prentice, Gabriella Tranell, and Kyle S. Brinkman
TMS (The Minerals, Metals & Materials Society), 2014

LIFE CYCLE ASSESSMENT OF DIFFERENT GOLD EXTRACTION

PROCESSES

Chao Li[1, 2], Hongxu Li*[1, 2], Meng Wang[1, 2], Xie Yang[1, 2], Xiangxin Hao[1, 2]

1 Beijing Key Lab of Green Recycling and Extraction of Metals, University of Science and
Technology, 30# Xueyuan Road, Beijing, 100083,China
2 School of Metallurgical and Ecological Engineering, University of Science and Technology
30# Xueyuan Road, Beijing, 100083, China
* Corresponding Author

Keywords: Gold extraction process; Resource consumption; LCA; Greenhouse Gas

Abstract

With the development and utilization of resources of gold, high quality gold mine resources
gradually exhausted, and refractory gold ore has become the main raw material for gold
smelting, while the smelting pollutants impacting on the ecological environment is growing,
in recent years there have appeared several methods of green extraction of gold such as
bio-oxidation. In this paper, the life cycle assessment models about three different gold
extraction technologies have been put forward. Material consumption, energy consumption
and emissions during each gold production process are analyzed by using GaBi 6 software,
and the environmental impact of each process is also evaluated by means of CML2001
methodology. Calculation and analysis results show that the energy consumption and material
consumption are the main factors causing the environmental load; and the environmental
impacts of the production process mainly attribute to the category of greenhouse effect,
resources consumption and environmental acidification. Based on the evaluation, some
optimization tips are proposed.

Introduction

Over the past decades, as the industry policy environment continues to improve as well as
gold exploration and mining smelting capacity continues increasing, gold production shows a
prospecting tendency. With the wide use of gold, the demand for gold is growing, while easy
smelting gold mine resources become exhausted and refractory gold ore has become the main
raw material for gold smelting. In addition, pollutants impact on the ecological environment
is also growing year after year.

LCA (Life Cycle Assessment) makes evaluation of product "from cradle to grave" as its
whole process, up to finding the sources of environmental impacts and solutions, to
comprehensively account the use of resources and lower emissions. In the industrial field, life
cycle can compare different technologies from different aspects of resource use, energy
consumption and environmental emissions [1]. In recent years, using LCA to evaluate
metallurgical industry process also has more development. The paper mainly uses LCA and
scenario analysis for the gold metallurgical process by GaBi 6, to investigate the
environmental impact of three different technologies of gold extraction metallurgical process.
The established three evaluation models include the hydrometallurgical, pyro-metallurgical

processes and bio-oxidation pretreatment methods respectively. Several environmental indexes would be selected to evaluate the environmental impact in order to find which process has smallest environmental load and some suggestion of reasonable way gold resource utilization.

Gold extraction process

CIP process

CIP (carbon-in-pulp process) process is one of the most widely used methods in gold production [2]. It is a no-filtering process using activated carbon or 1 shell charcoal directly recovering gold from cyanide pulp [3]. The method mixes the finely ground ore or concentrate with water, lime and cyanide, then stirring and leaching in a series of tank, and finally putting the pulp to a series of absorbers by adding activated carbon to perform adsorption.

Roasting oxidation pretreatment process

Hydrometallurgy has many advantages compared to pyro-metallurgy, such as high metal recovery rate, much more likely to comprehensive use of raw materials by directly dissolving refractory grade ore, with low energy consumption, and the production process is easy to be controlled. It reduces the pollution of waste to the environment compared with the original pyro-metallurgy process [4]. Roasting pretreatment before cyanide leaching can improve cyanide leaching rate but arsenic environment stress still exists.

Bio-oxidation pretreatment process

The main advantages of bio-oxidation pretreatment of concentrate for gold extraction are: little environmental pollution compared with the traditional roasting process, it is more environmentally friendly compared with the pressure oxidation, and it is more moderate and secure. The whole process is simpler and need less investment, runs more stable and reliable, requires lower level operation and low operating cost, and is easy to put into industrial scale. Therefore, the process is more targeted to deal with the complex arsenic, sulfur, micro-wrapped gold ore, and has better applicability of resources utilization [5].

By virtue of some advantages including low investment, low production cost, environmental friendly, easy operation and high efficiency; bio-oxidation technology will make great contribution to the sustainable development of mining industry [6].

LCA Process

Goal and Scope

This paper lists three different gold extraction processes, Carbon-in-Pulp (CIP) process, roasting oxidation pretreatment process and bio-oxidation pretreatment process. The paper compares the three classical processes with each other from different aspects including energy input, energy consumption, environmental impact and resource consumption to quantify the environmental impact during the gold smelting process. The production data is derived from different processes in Hetai CIP process, Jinge roasting oxidation pretreatment process, and bio-oxidation process in Muli County, Sichuan province. Combined with the actual condition of our country, LCA methodology is used for making quantitative evaluation of the environmental load of different gold smelting processes. The purpose is to seek out the

150

most optimal process by comparing the environmental load of three kinds of gold smelting processes.

The scope of evaluation system solution in this paper is from mining, beneficiation, grinding, and production of materials to the waste emissions during the smelting process. Gold per ton was chosen as the production functional in this research.

<u>Inventory Analysis</u>

Inventory analysis is the basis work of LCA, which provide the data foundation for an evaluation of the gold extraction process. The material and energy consumption is the average of actual production from some major gold mines in recent years.

The CIP process includes mining, beneficiation, cyanide leaching, carbon adsorption, desorption, and electro-winning processes. Roasting oxidation pretreatment process includes mining, beneficiation, concentrate roasting, cyanide leaching, carbon adsorption, desorption, and electro-winning processes. Bio-oxidation pretreatment process mainly includes mining, beneficiation, bio-oxidation concentrate, cyanide carbon adsorption, and desorption electro-winning processes.

Table 1 Life cycle inventory of product 1t gold

Process	CIP process	Roasting pretreatment process	Bio-oxidation pretreatment process
Mining	Crude ore 273,224t Power $4.07*10^6$kwh	Crude ore 218,723t Power $3.26*10^6$kwh	Crude ore 208,307t Power $3.1*10^6$kwh
Beneficiation	Concentrate 13,649t Power $7.4*10^6$kwh Water consumption $1.5*10^6$t New water $2.2*10^5$t Sodium butyl xanthate 20.6t Butylamine aerofloat 10.9t No. 2 oil 13.1t Calcium oxide 2,732t	Concentrate 16,388t Sodium butyl xanthate 20.8t Butylamine aerofloat 11t No. 2 oil 13.3t Calcium oxide 2187t Total water $1.52*10^6$t Power $5.9*10^6$kwh	Concentrate 13,658t Total water $1.45*10^6$t Sodium butyl xanthate 19.8t Butylamine aerofloat 10.5t No. 2 oil 12.7t Calcium oxide 2,083t Power $5.64*10^6$kwh
Pretreatment	————	Coal 6,293t Diesel 3.9t	Power $1.25*10^5$kwh Medium 33.4t Flocculant 0.95t Calcium oxide 341t
cyanide leaching	Sodium cyanide 10^9t Calcium oxide 341t Activated carbon 1.64t Power $3.52*10^5$kwh	Sodium cyanide 50.9t Calcium oxide 44.5t Activated carbon 1.5t Power $3.28*10^5$kwh Calcine 12717t	Power $2.46*10^5$kwh Sodium cyanide 38.1t Calcium oxide 33.4t Defoamers 1.43t Activated carbon1.14t Oxide slag 9533t

Table 2 Material and energy consumption of 3 processes

	Crude ore(t)	Power (10^6kwh)	Water (10^6t)	Calcium oxide(t)	Sodium cyanide(t)	Activated carbon(t)
CIP process	273,224	11.8	1.72	3,073	109	1.64
Roasting oxidation pretreatment process	218,723	9.5	1.52	2,231.5	50.9	1.5
Bio-oxidation pretreatment process	208,307	9.11	1.66	2,457.4	38.1	1.14

Gold recovery rate in the order of high to low: bio-oxidation pretreatment process > roasting pretreatment process > CIP process. Material and energy consumption in beneficiation stage increased the burden on the environment because grinding and mixing requires a large amount of water. The lower water recycling rate of gold extraction production causes greater water consumption. It also increases power consumption. Acidic waste water produced in biological oxidation and adsorption and desorption process in CIP gold extraction both require a lot of lime to adjust pH and has indirectly led to increased environment effects.

LCA Model
The established LCA models of three gold extraction processes are shown as Fig. 1, which demonstrates the process of CIP, bio-leaching and roasting pretreatment process respectively. Blue arrows and red arrows in the figure represent the material flow and energy flow respectively.

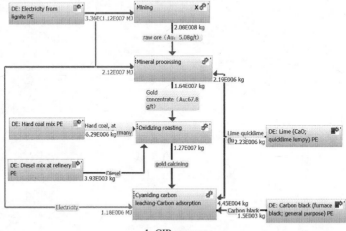

1. CIP process

152

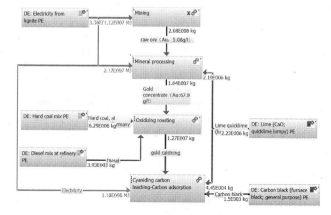

2. Roasting pretreatment process

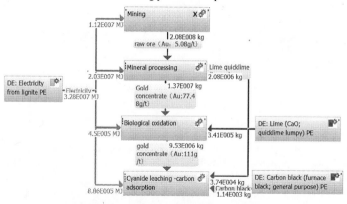

3. Bio-oxidation pretreatment process

Figure1. Evaluation model of the three gold extraction process

Life cycle assessment

The environmental impact of product life cycle was divided into six parts: resource consumption index(ADP), acidification potential(AP), eutrophication potential index(EP), the global warming index(GWP), ozone depletion potential index(ODP), and the formation of ozone potential index(POCP). LCA indexes for the entire life cycle of the three processes are shown in Figure 2. From Figure 2 we can see the power consumption accounted for the vast majority of the environmental load in the process of biological oxidation and CIP process, but the hard coal and diesel fuel that consumed in the production process play a larger role in the process of roasting oxidation pretreatment.

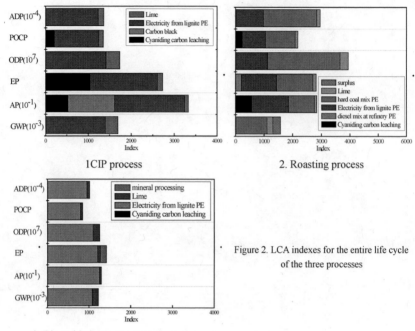

1CIP process

2. Roasting process

3. Bio-oxidation pretreatment process

Figure 2. LCA indexes for the entire life cycle
of the three processes

The relative value of environmental load is shown in Figure 3.   It shows that bio-oxidation
has the smallest impact on the environment, and the roasting process has the largest among
the three. The specific numerical data of AP and GWP are indicated in Figure 4. Evaluation
results are as follows:

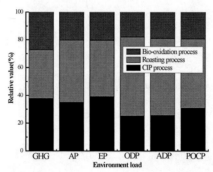

Figure 3. Relative value of environmental load

1) The degree of impact on the global warming index in order is: CIP, roasting process, and
bio-oxidation process. Electrical energy consumption accounts for the majority of its whole
life cycle.

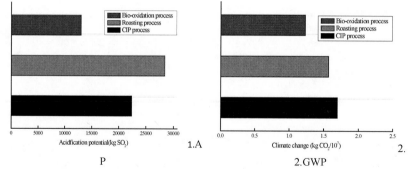

P   2.GWP

Figure 4 Acidification potential index and the global warming index of three processes
(Conversion to SO_2 and CO_2 equivalent)

2) The degree of impact on the acidification potential index in order is: roasting oxidation pretreatment process, CIP, and bio-oxidation pretreatment. In the roasting process, large amounts of SO_2, NO_2 and exhaust residue that contained in flue gas emissions lead to the high level of the acidification potential index (AP). Part of the flue gas can't be recycled which has a very bad impact on the atmospheric environment. It is imperative to introduce new technology on gas treatment or shift to more environmentally friendly process.

Optimization Tips
1) CIP process. Control all kinds of pollutants emissions from the whole production process. Accurately add suitable amounts of lime and do not waste lime while decreasing the cyanide consumptions. Add sodium cyanide to the process of cyanide leaching separately, which is in favor of cyanide leaching for the oxidizing slag.
2) Roasting oxidation pretreatment process. Inlet the air during the process of roasting, and develop a method of rich-oxygen burning while assuring the optimum oxygen concentration. In addition, adopt a new type of circulating fluid-bed roaster to avoid influencing cyanide leaching of gold. Other methods such as using two-stage roasting to fully oxidize sulfide and adopting new roasting pretreatment technologies including alkaline solid sulfur and arsenic roasting are put forward.
3) Bio-oxidation pretreatment process. Strengthen to control effective factors of bio-oxidation process as well as breed, cultivate and domesticate a kind of mineral leaching bacteria with strong adaptability and higher oxidation efficiency. In addition, strengthen the development and introduction of the high and new technology as well as new-style equipment, thus reducing the energy consumption of material and its environmental impact.

Conclusion
This paper has evaluated the potential impact on the environment of raw materials, energy consumption and pollutant emissions in the process of extracting 1 t gold at its entire life cycle from the "cradle" to the product by using the GaBi 6 software. This analysis gives the benefits for selecting the best gold production technology and management system. The main conclusions are as follows:
- The environmental problems are mainly caused by energy and material consumption;
- Water recycling rate is very low and a large number of freshwater resources were

wasted;

- The environmental problems in the process are mainly caused by the greenhouse effect and environmental acidification.

In order to reduce the influence of the environment in the process of gold production, the following measures can be taken:

1) Adjusting the production process;

2) Using economic and efficient technologies for beneficiation.

3) Optimizing energy use, such as, reducing the consumption of power and material; using new methods for saving energy.

4) Developing and introducing advanced technology and equipment to achieve the comprehensive utilization of waste.

REFERENCES

[1]   Song Qingshuang, Fu Yan. Gold and Silver Extraction Metallurgy [M]. Beijing: Metallurgical Industry Press,2012:35-155

[2]   Xue Guang, Ren Wensheng, Xue Yuanxin. Gold and Silver Hydrometallurgy and Analytical Test Methods [M].Beijing: Science press, 2009.

[3]   Gao Bingpeng, Nan Xinyuan. .Research and Application of Bio-oxidation Pretreatment Control System [J].Gold, 2011(1).

[4]   Li Suqin, Cang Daqiang, Li Hong. Industrial Ecology [M].Beijing: Metallurgical Industry Press, 2012:78-93.

[5]   Chen Fangfang, Zhang Yifei, Xue Guang. Disposal to Pollution and Resource Utilization to Waste for Gold Smelting [J].Gold Science and Technology, 2011, (2).

Energy Technology 2014: Carbon Dioxide Management and Other Technologies
Edited by: Cong Wang, Jan de Bakker, Cynthia K. Belt, Animesh Jha, Neale R. Neelameggham,
Soobhankar Pati, Leon H. Prentice, Gabriella Tranell, and Kyle S. Brinkman
TMS (The Minerals, Metals & Materials Society), 2014

PERFORMANCE EVALUATION, TECHNICAL AND ENVIRONMENTAL ASPECTS OF BIOMASS COOKSTOVES: AN EXERGY APPROACH

S. K. Tyagi [1], A. K. Pandey, Kunwar Pal

Sardar Swaran Singh National Institute of Renewable Energy, Kapurthala 144601 (Punjab) India

Keywords: Improved biomass cookstove, emission reduction, energy and exergy analysis

Abstract

This study deals with the performance evaluation, technical and environmental aspects of different cookstove models (NIRE-02, NIRE-03 & NIRE-04). These cookstove models were designed, fabricated and evaluated following Bureau of Indian Standards (BIS) and their performance was compared from technical and environmental points of views. The theoretical efficiencies of all the cookstove models are found to be in good agreement with the experimental one. Also, the overall performance of these models is found to be in the order of (NIRE-04 > NIRE-03 > NIRE-02) which is also observed to be much better than that of the traditional cookstoves being used by the people in most of the developing countries around the world. The CO_2 emission reduction from these models is found to be between 2.0-3.0 tons per household annually which is ultimately in good agreement with the experimental values.

Introduction

More than 2.5 billion people around the world depends on the biomass based resources for cooking and heating applications which includes wood, cattle dung, charcoal and agriculture residue [1,2]. Generally, people of rural area cook their food on unvented open fire, which burn biomass fuel inefficiently besides it creates many thermal and environment pollution hazardous causing health problems in the women and girl children due to more exposure of emissions. Also the consumption of biomass is very high as compared to the fossil fuels due to the fact that the large population of the developing countries lives in the rural area than that of in the urban area. It is estimated that about 3% of the diseases are caused due to incomplete combustion of biomass, which results in the around 1.8 million premature death every year including around 0.9 million death under five year of their age [3,4].

Ramanathan and Carmichael [5] recently found that the black carbon of unburned smoke is playing a major role in global warming. The emission factors for different combinations of fuel and stove tested in China for direct and indirect GHGs as well as other airborne pollutants such

---

[1] Corresponding Author: E-mail(s): sudhirtyagi@yahoo.com

as, CO_2, CO, CH_4, TNMHC, N_2O, SOx, NOx, TSP etc. for a typical set of operating parameters [6]. The improved biomass cookstove technology came into assistance which is technically feasible, socially reliable and economically viable [7, 8]. The improved biomass cookstove not only reduces pressure on the biomass resource, but also reduces the environmental pollution and also minimize the time required for collection of fuel wood [9, 10]. Sharma et al. [11] emphasized on the adoption and large-scale propagation of efficient stoves to improve the health of rural women and to reduce the consumption of firewood and environmental pollution.

Very recently Tyagi et al., [12] presented the experimental study of various improved biomass cookstove models based on energy and exergy analysis. They found that the exergy efficiency is generally, lower than that of the energy efficiency for all type of cookstove model studied by them. Kumar et al. (2013) [13] presented the thorough review on biomass cookstove based on various performance parameters. Improved biomass cookstoves also have the higher potential to get carbon credits [14], not only because of their contribution to climate-change mitigation but also they can yield major co-benefits in terms of energy access to the mass populations around the globe. Carbon trading offsets is growing source of interest comes from improved biomass cookstove program on carbon market for both voluntary and certified emission reduction, under clean development mechanism (CDM) of the UNFCCC following Kyoto Protocol [14,15].

Materials and Methods

In the present article a modified traditional cookstove with different modifications was designed and fabricated using locally available materials. Apart from modifications in the traditional cookstove two improve biomass cookstove models (NIRE-03 and NIRE-04) were also design and fabricated. These improved cookstove work on the principle of down-draft gasifier where pyrolysis, gasification and combustion of biomass are taking place simultaneously. These cookstoves were fabricated using mild steel whereas for insulation clay and wheat straw mixture was used. The photographic view of NIRE-02, NIRE-03 and NIRE-04 are shown in the figure 1. The water boiling test was performed for all the cookstove models and the thermal efficiency of each cookstove was calculated according to BIS standard (IS 13152). The temperature of water, flame and cookstove body was measured with the help of digital temperature sensors, whereas, the temperature of pot, plate and ambient was measured with mercury-glass thermometers. As per the detailed reports available in the literature [16, 17], the thermal efficiency of traditional cookstove is less than 10% and can be verified by following the similar testing protocols as in the present case. Sheesham (Dalbergia sissoo) wood was used as the fuel and prepared as per BIS protocol for cookstove. The fuel wood was cut from the same log into pieces of 3x3 cm square cross-section and length of half the diameter/length of combustion chamber so as to be housed inside the combustion chamber easily. For experimental analysis, the wood samples were crushed into powder form. Calorific value of wood was estimated with Bomb calorimeter. Finally The performance of different types of cookstoves has been evaluated out using energy and exergy analysis following earlier authors [10, 12]. The modified traditional cookstove NIRE-02 is a portable model made up on mud platform and suitable for indoor and outdoor according to the weather condition. This model was tested in three different modes such as, tested with grate (G), tested with grate and top spacing (S) and tested only with top space (S). However, the IBC model NIRE-03 was tested with and without insulation and with.

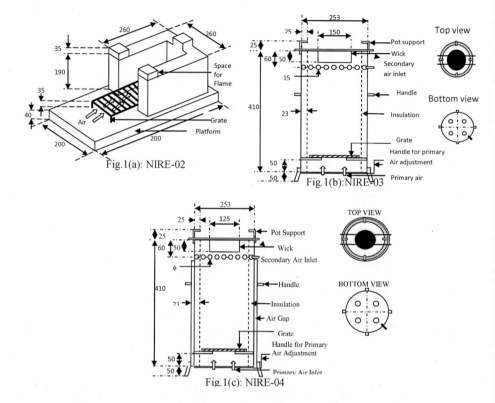

Fig.1: Schematic diagram of the different cookstove models evaluated in the present study

Analysis of Cookstove

Improved biomass cookstoves can reduce indoor air pollution, deforestation, climate change, and therefore, the quality of life can be improved on a universal scale. The better design of these cookstoves can significantly impact their performance and emissions. Although these improved biomass cookstoves have been studied for a long time however, a theoretical understanding of their operating behaviour and the development of engineering tools for an improved cookstove based on natural convection is still missing.

Energy and Exergy Analysis

Energy efficiency of a cookstove may be defined as the ratio of energy utilized to the energy produced by complete combustion of a given quantity of fuel based on the net calorific value of the fuel and this can be written as below [10,12]:

$$\eta = \frac{\text{Energy output}}{\text{Energy input}} = \frac{E_o}{E_{in}} \equiv \left(\frac{m_w C_p (T_{fw} - T_{iw}) + m_{pot} C_{pA1}(T_{fp} - T_{ip})}{m_{wd} c_1 + x \times d \times c_2} \right) \qquad (1)$$

However the exergy efficiency may be defined as the ratio of output exergy to the input exergy and given by [10,12]:

$$\psi = \frac{\text{Exergy output}}{\text{Exergy input}} = \frac{Ex_o}{Ex_{in}} = \left(\frac{m_w C_p (T_{fw} - T_{iw})(1 - \frac{T_a}{T_{fw}}) + m_{pot} C_{pA1}(T_{fp} - T_{ip})(1 - \frac{T_a}{T_{fp}})}{m_{wd} c_1 (1 - \frac{T_a}{T_{fuel}}) \times \eta + x \times d \times c_2} \right) \qquad (2)$$

where m_w is the mass water in the pot, C_p is the specific heat of water, T_{fw} is the final temperature of water, T_{iw} is the initial temperature of water, m_{pot} is the mass of pot, C_{pA1} is the specific heat of Aluminium pot, T_{fp} is the final temperature of pot, T_{ip} is the initial temperature of pot, m_{wd} is the mass of wood, c_1 is the calorific value of wood, T_a is the ambient temperature, T_{fuel} is the flame temperature, η theoretical efficiency, x is the volume of kerosene, d is the density of kerosene and c_2 is the calorific value of kerosene.

Emission reduction Calculation

The emission reduction from each cookstove is based upon fraction of biomass that can be saved during the project year and the calorific value of the biomass used and this can be calculated according to the formula given below [14]:

$$ER_y = B_{y,savings} \times f_{NRB,y} \times NCV_{biomass} \times EF_{projected\_fossilfuel} \qquad (3)$$

where ER_y is the emission reductions during the year y in tCO_2, $f_{NRB,y}$ is the fraction of woody biomass saved by the project activity in year y, $NCV_{biomass}$ is the net calorific value of the non-renewable woody biomass that is substituted (IPCC default for wood fuel, 0.015 Tj/tonne), $EF_{projected\_fossilfuel}$ is the emission factor for the substitution of non-renewable woody biomass by similar consumers and one can use of value of 81.6 tCO_2/TJ. $B_{y,savings}$ is the quantity of woody biomass that is saved in tonnes which can be calculated by the following formula:

$$B_{y,savings} = B_{old} \cdot \left(1 - \frac{\eta_{old}}{\eta_{new}} \right) \qquad (4)$$

where B_{old} is the quantity of woody biomass used in the absence of the project activity in tonnes, η_{old} is the efficiency of the baseline system, η_{new} is the efficiency of the system being deployed as part of the project activity (fraction), as determined using the water boiling test (WBT) protocol.

Results and Discussion

Based on the biomass characteristics, the performance of different cookstoves has been carried out using energy, exergy and carbon emission reduction analysis at a typical location in India, while the discussion of results is given as below:

Modified Traditional Cookstove

The energy and exergy efficiencies were evaluated for NIRE-02 and plotted against the quantity of wood as shown in the Fig. 2. From the figure 2, it is found that as the quantity of wood is increases from 1- 5 kgs, both the energy and exergy efficiencies also increase and found to be highest for 5 kg of fuel wood in the case of with top space. On the other hand, in case of grate only, the energy efficiency is found to be increasing in nature for 1-3 kgs of wood. However, for a given capacity, there should be an optimum time of operation and amount of fuel wood required to get the best possible performance of a cookstove. This due to the fact that the overall heat loss from the cookstove increases as the cooking time and the amount of fuel wood is increased. In other words, the increase in the energy efficiency from 1-3 kgs is found to be significant, while it is almost constant for 4-5 kg of wood consumption. However, the change in the energy efficiency is found to almost constant as the fuel wood supply is increased from 4-5 kgs. Again in the case with top spacing nature of variation in energy efficiency is found to be same as that of the case with grate and space while the average energy efficiency with grate is found to be higher than that of the other two cases mentioned above. The maximum energy and exergy efficiency is found to be 29.28% and 4.79% respectively in the case with top space and with 5 kg of wood. However, the optimum performance of the NIRE-02 model is found for the case with grate only.

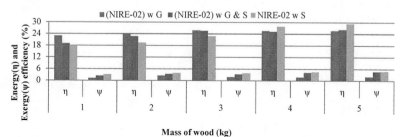

Fig-2: Energy and exergy efficiency of NIRE-02 with different modification parameter and with varying quantity of wood. (G= Grate, S= Spacing between upper edge of cookstove and pot)

The variation of CO_2 emission reduction values of NIRE-02 cookstoves is shown in the figure 3. The value of emission reduction is continuously increases as the fuel wool increases from 1-5 kgs for the case when top space is used only. The emission reduction from NIRE-02 with grate only and with grate and top space is similar in nature. From these modification the value of emission reduction is first increasing as the fuel wood increases from 1-2 kgs and then almost constant with varying quantity of fuel wood from 3-5 kgs. It is clearly understood for the figure 3 that the NIRE-02 model performs good and gives a large amount of emission reduction value i.e. 2.26 tonnes of CO_2 reduced per household per year when it is use with top space only.

161

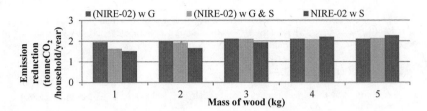

Fig-3: Emission reduction from NIRE-02 with different modification parameter and with varying quantity of wood.

Improved Cookstove Models

The variation in energy and exergy efficiency against the quantity of the wood of NIRE-03 cookstove model with different variations in the design is shown in Fig. 4. From Fig. 4, it is found that as the quantity of wood increases from 1-5 kgs, the energy efficiency of NIRE-03 cookstove model is almost constant without insulation. In the case with insulation and without wick, it was observed that the energy and exergy efficiency first increases with increasing the wood from 1-3 kgs but decreases with further increase in wood from 3-5 kg. Again, it was observed that the energy efficiency first decreases with the increase in the wood quantity from 1-2 kgs and then increases as the quantity of wood increases from 2-5 kgs in the case with wick. Thus it is concluded from this study that the use of wick should be used to regulate and control the fire for specified cooking requirement only.

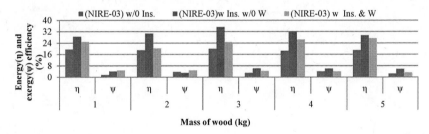

Fig-4: Energy and exergy efficiencies of NIRE-03 with different modification parameter and with varying quantity of wood. (Ins= insulation, w/0 = without, W= wick)

Figure 5 shows the variation of emission reduction potential (tonneCO$_2$/household/year) of NIRE-03 cookstoves. It is clear from the Fig. 5 that there is good potential for annual CO$_2$ emission reduction with improved cookstove. However, the highest potential in emission reduction is found for the case where insulation without wick was applied, and the similar case was also found for the thermal efficiency as shown in the Fig. 4. The maximum value of emission reduction 2.46 tonne CO$_2$/household/year was found with 3 kg of wood. It is clearly understood from the results that the NIRE-03 model performs best and gives a large amount of emission reduction value i.e. 2.46 tonnes of CO$_2$ reduced per household per year. This means if

162

approximately 10,000 of NIRE-03 cookstoves are provided to the consumers it will save approximately 24,600 tonnes of CO_2 in one year.

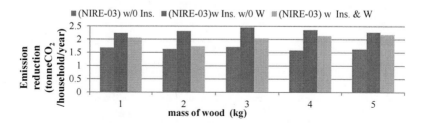

Fig-5: Emission reduction from NIRE-03 with different modification parameter and with varying quantity of wood.

The NIRE-04 cookstove model is the modified version of NIRE-03 in which an air is provided to utilize the heat which is going to waste through the stove body. The combustion chamber as shown in the Fig.1. The performance of NIRE-04 model without insulation has been evaluated and shown in the Fig. 6, further modifications are underway and enhancement in the efficiency is also expected by 10-12% as compared with present case with similar trend for emission reduction potential.

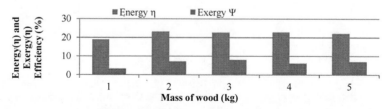

Fig-6: Energy and exergy efficiencies of NIRE-04 without insulation and with varying quantity of wood.

Conclusions

In the present study comparative exergetic, energetic efficiency and CO_2 emission potential of different designs of modified traditional and improved cookstove models have been presented. Based on the experimental study and observations, the following conclusions are drawn:

- Based on the experimental observations the energy and exergy efficiency and as well as the emission reduction potential for NIRE-02 model was found to be best when both grate and top space was provided.
- The performance of NIRE-03 cookstove was found to be best with insulation for 3kg of fuel wood. Similar results were also observed for the emission reduction potential.
- Again, the performance and the CO_2 emission reduction potential of NIRE-03 model is found to be higher than that of NIRE-02 for all set of operating parameters. In general the

efficiencies and CO_2 emission reduction potential of improved cookstoves models to be higher than that of traditional cookstoves i.e., without grate and top space for all set of operating parameters.
- Also energy efficiency was found to be always higher than that of exergy efficiency which is due to the energy gained by the hot water at that particular temperature.

References

1. A. Chaurey et al., "New partnerships and business model for facilitating energy access," *Energy Policy,* 47 (2012), 48-55.
2. I.R. Mercado, O. Mesera, H. Zamora, K.R. Smith, "Adoption and sustained use of improved cookstove," *Energy Policy,* 39 (2011), 7557-7566.
3. E. Rehfuess, "Fuel for Life: household energy and health" (Geneva: WHO, 2006).
4. WHO, "World health report: reducing risks, promoting health life" (Geneva: World Health Organization, 2002).
5. V. Ramanathan, G. Carmichael, "Global and regional climate changes due to black carbon," *Nature Geo-science,* 1 (2008), 221–227.
6. J. Zhang et al., "Greenhouse gases and other airborne pollutants from household stoves in China: a database for emission factors," *Atmospheric Environment,* 34 (2000), 4537–4549.
7. J. Szargut, *Exergy method: technical and ecological applications* (WIT Press, 2005).
8. Y. Kalinci, A. Hepbasli, I. Dincer, "Efficiency assessment of an integrated gasifier/boiler system for hydrogen production with different biomass types," *International Journal of Hydrogen Energy,* 35 (2010), 4991-5000.
9. K. Ojeda, E. Sanchez, V. Kafarov, "Sustainable ethanol production from lignocellulosic biomass-application of exergy analysis," *Energy* 36(4) (2011), 2119-2128.
10. Ibrahim Dincer and Marc A. Rosen, *Exergy, energy, environment and sustainable development* (Elsevier, 2007).
11. R. Saidur, G et al., "A review on exergy analysis of biomass based fuels," *Renewable and Sustainable Energy Reviews,* 16(2) 2012, 1217-1222.
12. S.K. Tyagi, A.K. Pandey, S. Sahu, V. Bajala, J.P.S. Rajput, "Experimental study and performance evaluation of various cookstove models based on energy and exergy analysis," *Journal of Thermal Analysis and Calorimetry,* 111 (3) (2013), 1791-1799.
13. M. Kumar, S. Kumar, S.K. Tyagi, "Design, development and technological advancement in the biomass cookstoves: A review," *Renewable and Sustainable Energy Reviews,* 26 (2013), 265-285.
14. http://cdm.unfccc.int/ accessed from internet on 12-08-13.
15. Global Alliance for Clean Cookstoves; 2011 (http://www.cleancookstoves.org)
16. Small Scale Methodology : Energy efficiency measures in thermal applications of non renewable biomass, United Nations Framework Convention on Climate Change (http://cdm.unfccc.int/filestorage/7/m/24G3EKN6PT0QJ1BHRICMYDX97OW8UF.pdf/EB 70_repan30_AMS-II.G_ver05.0.pdf?t=Q1J8bXZkYnQzfDBezeLfxv3DxRrllGKyy7nq) (accessed from internet on 12-10-13)
17. V.V.N. Kishore, P.V. Ramana, "Improved cookstoves in rural India: how improved are they?: a critique of the perceived benefits from the National Programme on Improved Chulhas (NPIC),". Energy, 27 (2002), 47–63.

Energy Technology 2014
Carbon Dioxide Management and Other Technologies

SYMPOSIUM: ENERGY TECHNOLOGIES AND
CARBON DIOXIDE MANAGEMENT

Energy Efficiency and Furnace Technologies

Session Chairs:
Li Li
Tao Wang

Energy Technology 2014: Carbon Dioxide Management and Other Technologies
Edited by: Cong Wang, Jan de Bakker, Cynthia K. Belt, Animesh Jha, Neale R. Neelameggham,
Soobhankar Pati, Leon H. Prentice, Gabriella Tranell, and Kyle S. Brinkman
TMS (The Minerals, Metals & Materials Society), 2014

AN EXPERIMENTAL INVESTIGATION OF A FLUE GAS RECIRCULATION SYSTEM FOR ALUMINUM MELTING FURNACES

James T. Wiswall[1], Mark Kruzynski[1], Srinivas Garimella[1]

[1]Alcoa
100 Technical Drive, Alcoa Center, PA 15069, USA

Keywords: Flue Gas Recirculation, Melting, Combustion, Furnace, Turbulent Jet, Entrainment

Abstract

A flue gas recirculation (FGR) system for a combustion fired aluminum melting furnace is experimentally investigated. The FGR system flows combustion air in a turbulent jet within the flue gas stream, and the turbulent jet flow is collected in a tube at a distance from the jet exit. Entrainment of the flue gases as well as heat transfer between the flue gases and the jet occurs over the length of the exposed jet. Heat transfer between the flue gases and the jet, and the amount of flue gas entrained into the jet (derived from temperature, flow rate and oxygen fraction measurements) were measured as a function of nozzle exit velocity and the distance between the nozzle exit and collection tube. In addition to flue gas entrainment, a fuel savings of up to 7%, due to heat transfer over the length of the jet, was projected from the measurements obtained.

Introduction

Flue gas recirculation can be applied to aluminum melting furnaces to control combustion temperature. When coupled with a device to improve furnace efficiency by recovering heat from the furnace exhaust, flue gas recirculation provides a control technique that is integral to achieving the goals of reducing fuel consumption, increasing product yield, increasing the longevity of infrastructure, and reducing pollutant emissions. In particular, combustion temperature control can help reduce reactivity of the products of combustion with the molten metal surface and furnace refractory materials. Combustion temperature control can also help improve the life of heat exchangers used to recover energy lost to the environment in the flue gases, and reduce NOx formation in the combustion process. Flue gas recirculation can also dilute oxygen used in oxy-fuel combustion to reduce reactivity.

Flue gas recirculation has been used since the 1970s to control flame temperatures in the automotive industry [1-3] and has successfully reduced NOx emissions. Many different flue gas recirculation approaches for high temperature melting such as melting aluminum and glass have been patented since the 1980s [4-12]. There has also been academic interest to determine the fundamental performance characteristics of flue gas recirculation in high temperature furnaces and in combustion [13-15].

Techniques used to mix flue gas with reactant flow vary. Dinicolantonio [6] and Schindler [10] describe a Venturi method where mixing is driven by the pressure difference between an accelerated flow and more stagnant gases. Joshi [11] and Atreya [12] employ jet entrainment to mix gases. Many others use high temperature ducting and valves to mix flue gases. The location where flue gas is mixed also varies. Most designs mix the flue gas in the burner itself. Syska [7], Moreland [8], and Kobayashi [9] describe staged combustion where a portion of the fuel is oxidized in a fuel rich environment to promote decomposition of the fuel to hydrogen and carbon monoxide prior to combustion in the furnace. Joshi [11] also describes entrainment within the furnace itself. Finally Neville [4] and Atreya [12] describe flue gas mixing prior to the burner.

Although the techniques to circulate flue gas into the reactant flow vary, the intention to reduce NOx production by controlling flame temperature is the major stated goal common to all approaches.

This study experimentally investigates a turbulent air jet within an exhaust flow with a design basis derived from Atreya 2008. Products of combustion (POC) are entrained with the air allowing for both heat transfer and dilution. Heat transfer between the jet and the POC stream can recover some waste heat while dilution due to mass transfer can maintain low NOx production. The hypothesis for this work is that a turbulent jet can control flame temperature by entraining flue gases into the reactant air flow and can also increase furnace efficiency due to heat transfer over the length of the turbulent jet.

Results

The experimental setup, developed at Alcoa [16], passes combustion air flow through a turbulent jet exposed to the exhaust stream. A schematic of the setup is shown in Figure 1. Air for combustion is injected using a high pressure nozzle (location 1 in the figure) located within the products of combustion (POC) exit stream of the furnace (location 3 in the figure). The turbulent air flow exiting the nozzle entrains POC gases and is collected at location 2. The gas mixture at location 2 is higher in temperature than the injected air and also has a lower O_2 fraction. Finally, POC gases which are not entrained in the turbulent jet are exhausted to the atmosphere through location 4.

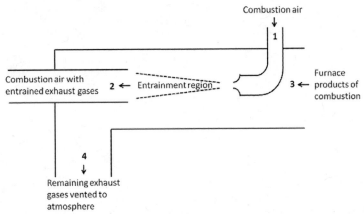

Figure 1. Schematic of the flue gas entrainment system. Experimental measurements were obtained at each of the numbered locations

Table 1 lists the measurements obtained from the experimental setup. Temperature was measured using k-type thermocouples inserted into the flow stream. Flow rate was measured using a Pitot tube, oxygen mole fraction was measured using a non-dispersive infrared sensor

(NDIR) and data is presented on a wet basis. Note that only oxygen fraction was measured at location 4.

Table 1. The measurements at each location shown in Figure 1.

location	Description	measurements
1	combustion air	temperature, flow rate, oxygen fraction
2	combustion air with entrained flue gas	temperature, flow rate, oxygen fraction
3	furnace products of combustion	temperature, flow rate, oxygen fraction
4	remaining exhaust gases that are vented to the atmosphere	oxygen fraction

Figures 2 through 4 plot the data listed in Table 1 for locations 2, 3, and 4 as a function of air flow rate (location 1). Air to the nozzle was supplied by a compressor and was cooled to room temperature (20 °C +- 5 °C) before being injected. Due to fluid mixing and heat transfer within the turbulent jet, the temperature at location 2 is between room temperature and the POC temperature at location 3.

The POC flow rate and location 2 flow rate as a function of the nozzle flow rate is shown in Figure 3. As the nozzle flow rate increases, the POC flow rate at location 3 increases. Such behavior is most likely due to the decrease in pressure in the entrainment zone for increasing nozzle flow rate. The decrease in pressure in the turbulent jet region essentially draws more combustion gases from the furnace, and therefore increases the flow rate at location 3.

Figure 4 shows the oxygen fraction, measured on a wet basis, for locations 2, 3, and 4. The oxygen fraction of the nozzle flow was assumed to be that of ambient air. The oxygen fraction at location 3 and 4 are within the experimental uncertainty of the measurement. This result indicates that the nozzle air flows in its entirety into location 2. In other words there is no air that bypasses the collection tube and is vented to the atmosphere through location 4.

Control of the entrainment length of the turbulent jet was attempted by altering the length between the nozzle exit and entrance to the collection tube at location 2. The results shown in Figure 5 indicate that the jet entrained flue gases into the tube even when the distance between the nozzle exit and collection tube increased. A different method such as setting the turbulent jet in counter flow with the flue gases or varying collection tube diameter must be used to control the entrainment length of the turbulent jet.

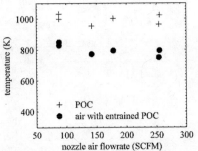

Figure 2. Temperature measurements at location 3, furnace products of combustion (POC), and location 2 combustion air with entrained POC. The ambient temperature was 20 C +- 5 C.

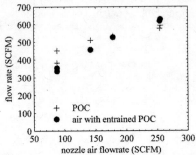

Figure 3. Flow rate at location 3, furnace products of combustion (POC), and location 2, combustion air with entrained POC.

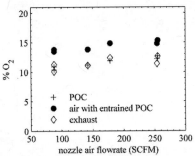

Figure 4. Oxygen fraction (wet basis) for locations 2, 3 and 4.

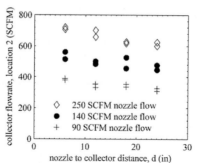

Figure 5. Increasing nozzle to collector distance slightly decreases the flow rate in the collector (location 2).

Discussion

The measurements, presented in Figures 2 through 4, are used to estimate the amount of fluid mixing and energy transfer within the turbulent jet entrainment. Conservation of mass and conservation of oxygen (equations 1 and 2) are used to determine the fraction of nozzle flow (ε_1) and furnace products of combustion (ε_3) entering location 2. The symbols $\dot{n}$ and χ_{O_2} represent the molar flow rate and oxygen mole fraction.

$$\varepsilon_1 \dot{n}_1 + \varepsilon_3 \dot{n}_3 = \dot{n}_2 \tag{1}$$
$$\left(\varepsilon \chi_{O_2} \dot{n}\right)_1 + \left(\varepsilon \chi_{O_2} \dot{n}\right)_3 = \left(\chi_{O_2} \dot{n}\right)_2 \tag{2}$$

Conservation of energy is used to determine the energy flow due to mass transfer and heat transfer (equation 3 and 4). The fuel savings is calculated as the ratio of heat transfer energy flow to the total energy flow of the gases (equation 5). The symbols h, P, and HHV represent the specific enthalpy, energy flow rate, and higher heating value. The specific enthalpy was calculated using thermodynamic tables for the gas mixtures at each location, and the molar flow rate of methane was calculated from oxygen fraction measurements at location 2 and assuming a stoichiometric combustion mixture.

$$(\varepsilon \dot{n} h)_1 + (\varepsilon \dot{n} h)_3 = P_{\text{mass transfer}} \tag{3}$$
$$(\varepsilon \dot{n} h)_2 - P_{\text{mass}} = P_{\text{heat transfer}} \tag{4}$$

$$f_{\text{fuel savings}} = \frac{P_{\text{heat transfer}}}{P_{\text{heat transfer}} + (HHV \times \dot{n})_{\text{CH}_4}} \tag{5}$$

Figure 6 shows the fraction of the flow entering location 2 from the nozzle and POC. Similar to Figure 4, Figure 6 indicates that nearly 100% of the nozzle flow enters location 2 implying that the collection tube is larger than the expansion of the turbulent jet, and indicates the system entrains POC gases as intended.

Figure 7 shows the enthalpy flow of the POC, collection tube, and the difference between these values. This difference is the energy flow due to heat transfer. In other words, deviation from the enthalpy due to the POC mass flow entrained within the turbulent jet is due to heat transfer. As expected, there is additional heating of the flow entering location 2 due to heat transfer.

Finally, Figure 8 shows the fuel savings expected due to heat transfer between the POC flow and the turbulent jet. The largest calculated fuel savings was 7% and is a function of nozzle air flow velocity. It is the energy transferred due to heat transfer that ultimately manifests itself in reduced fuel consumption when coupling the turbulent jet entrainment system to a furnace.

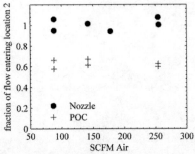

Figure 6. The fraction of the nozzle flow and POC flow that flow into location 2.

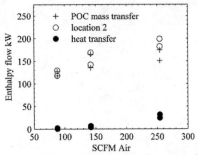

Figure 7. Enthalpy flow due to POC mass entrained in the nozzle flow, total enthalpy flow at location 2, and enthalpy flow at location 2 due to heat transfer.

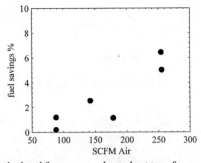

Figure 8. Fuel savings calculated from energy due to heat transfer.

Implementation challenges

The experiments presented in this work are intended to be a proof of concept for the flue gas recirculation device performance. Reduced combustion stability due to diluted reactant mixtures, reactivity of heated oxidizer with flow control and burner components, and inadvertent mixing of fuel and oxidizer in the event of combustion extinction are among the most important technical issues to be solved before implementation.

The technical issues, although non-trivial, have been solved for other combustion systems. Extinction limits in the published literature can help determine when combustion stability is an issue. Gas composition and temperature sensors can determine if extinction occurred flue gas recirculation system must be shut down. Operating conditions of the equipment can be designed such that the gas temperature remains with acceptable limits, governed by the materials used.

Summary and Conclusions

The turbulent air jet within the flue gas stream entrained flue gas. The control technique of varying distance between the nozzle exit and collection tube entrance did not have the intended influence on entrainment length, and additional techniques must be tested to properly control the amount of flue gas entrained with the air in the turbulent jet. Heat transfer between the hot flue gases and cooler jet is projected to increase furnace efficiency, and a fuel savings of 7% was projected for the current configuration. Heat transfer increased for increase air velocity. Further studies can determine how much the heat transfer between the flue gas and jet can be enhanced.

References

1. D. Hill. "Exhaust Gas Recirculation", *U.S. Patent 3,641,989*, filed November 16, 1970, and issued February 15, 1972

2. R. Bolton. "Exhaust Gas Recirculation", *U.S. Patent 3,799,131*, filed April 19, 1972, and issued March 26, 1974

3. S. Mick. "Exhaust Gas Recirculation", *U.S. Patent 3,868,934*, filed March 21, 1973, and issued March 4, 1975.

4. W. Neville. "Furnace System with Reheated Flue Gas Recirculation", *U.S. Patent 4,358,268*, filed December 15, 1980, and issued November 9, 1982.

5. T. Rampley and P. Hoffarth. "Flue Gas Recirculation System", *U.S. Patent 4,995,807*, filed March 20, 1989, and issued February 26, 1991.

6. A. Dinicolantonio. "Flue Gas Recirculation For NOx Reduction in Premix Burners", *U.S. Patent 5,092,761*, filed November 19, 1990, and issued March 3, 1992

7. A. Syska, J. Beer, M. Togan, D. Moreland, and C. Benson. "Staged Air, Recirculating Flue Gas Low NOx Burner", *U.S. Patent 5,269,679*, filed October 16, 1992, and issued December 14, 1993.

8. D. Moreland, "Staged Air, Low NOx Burner with Internal Recuperative Flue Gas Recirculation", *U.S. Patent 5,413,477*, filed December 13, 1993, and issued May 9, 1995.

9. W. Kobayashi, A. Francis, and H. Kobayashi. "Staged Combustion for Reducing Nitrogen Oxides", *U.S. Patent 5,601,425*, filed June 13, 1994 and issued February 11, 1997.

10. E. Schindler, L. Tsirulnikov, J. Guarco, J.Moore, and M. Gamburg. " Low NOx and Low CO Burner and Method For Operating Same", *U.S. Patent 6,347,935*, filed June 17, 1999 and issued February 19 2002.

11. M. Joshi, K. Heier, and G. Slavejkov "Ultra Low NOx Burner For Process Heating", *U.S. Patent 6,773,256*, filed February 5, 2002, and issued August 10, 2004.

12. A. Atreya. "Method of Waste Heat Recovery From High Temperature Furnace Exhaust Gas", *U.S. Patent Application Publication US 2008/0014537*, filed July 12, 2007 and issued January 17, 2008.

13. J.Baltasar, M.G. Carvalho, P. Coelho, and M. Costa, "Flue gas recirculation in a gas-fired laboratory furnace: Measurements and modeling", *Fuel*, Volume 76, Issue 10, August 1997, Pages 919-929.

14. J.J. Feese, and S.R. Turns, "Nitric Oxide Emissions from Laminar Diffusion Flames: Effects of Air-Side versus Fuel-Side Diluent Addition", *Combustion and Flame*, Volume 113, Issues 1-2, April 1998, Pages 66-78.

15. H.K. Kim, Y. Kim, S.M. Lee, and K.Y. Ahn, "NO reduction in 0.03–0.2 MW oxy-fuel combustor using flue gas recirculation technology", *Proceedings of the Combustion Institute*, Volume 31, Issue 2, January 2007, Pages 3377-3384.

16. Private communication between S. Garimella, M. Kruzynski, and A. Atreya, January – July 2009

Energy Technology 2014: Carbon Dioxide Management and Other Technologies
Edited by: Cong Wang, Jan de Bakker, Cynthia K. Belt, Animesh Jha, Neale R. Neelameggham,
Soobhankar Pati, Leon H. Prentice, Gabriella Tranell, and Kyle S. Brinkman
TMS (The Minerals, Metals & Materials Society), 2014

RESEARCH ON COMMON BIOMASS PYROLYSIS PRODUCTION OF BIOMASS CARBON, PYROLYSIS GAS, AND BIOMASS TAR

Li Yang [1,2], Yonggang Wei [1,2*], Hua Wang [1,2], Jianhang Hu [1,2], Kongzhai Li [1,2]

[1]Faculty of Metallurgy and Energy Engineering, Kunming University of Science and Technology;
Kunming, 650093, China
[2]Engineering Research Center of Metallurgical Energy Conservation and Emission Reduction
Ministry of Education; Kunming University of Science and Technology, Kunming, 650093,
China

Keywords: Pyrolysis, Biomass Carbon, Pyrolysis Gas, Biomass Tar.

Abstract

In this paper, the pyrolysis of biomasses including bamboo, sawdust and walnut shell were performed in a tube furnace under temperatures between 400℃ and 600℃ to obtain biomass carbon, gas, and tar. Textural parameters analysis revealed the caloric value of biomass carbons between 32 MJ/kg and 34 MJ/kg. It also indicated that the surface of biomass carbon consists of mesopore and macropore structures. By using GC analysis, it was determined the main ingredients of pyrolysis gas are CO, H_2, CO_2, CH_4, C_2H_4 and C_2H_6. Calculation states that the density of pyrolysis gas is roughly between 1.0 kg/m^3 and 1.7 kg/m^3, while the calorific value of pyrolysis gas is between 12000 kJ/m^3 and 16500 kJ/m^3, illustrating that pyrolysis gas is a median gas. The calorific value of biomass tar was found to be between 32 MJ/kg and 34 MJ/kg, with more than 80 compounds in the biomass tar identified by GC-MS method.

Introduction

Biomass as a renewable source is the world's largest and most sustainable energy resource [1].The conversion of biomass has the potential to be a renewable and carbon neutral method of producing fuels and chemicals [2-4]. Pyrolysis is a thermo-chemical process defined as applying heat to degrade biomass in an inert atmosphere. It transforms unstable biomass and other organic materials into biomass carbon, gas, and tar [5]. Biomass carbon can be stored or transported much easier as compared to parent biomass or be converted to value added products such as activated carbons[6]; gas is a relatively low-energy-density fuel; tar can be considered as a source of useful chemicals[7].The characterization of biomass pyrolysis will lead to its increased understanding and utilization as a source of solid, liquid and gaseous products which can be used as fuels and petrochemical feedstock[8, 9].In recent years, many researchers studied the effect of temperature on biomass pyrolysis product distribution characteristics in a variety of pyrolysis reactor[10]. Ramin Azargohar's paper [11] presents the results of characterization for biomass and investigates the effects of pyrolysis temperature on the pyrolysis products obtained by fast pyrolysis of four different types of biomass.

This paper takes bamboo, sawdust and walnut shell as raw material to obtained biomass carbon, gas, and tar by pyrolysis. Through analysis of products, comprehend the properties of biomass

---

* Corresponding author. E-mail address: weiygcp@yahoo.com.cn

pyrolysis. These results expand the comprehensive utilization range of biomass, as well as provide basic data with use of biomass waste.

Experiment

Experimental instrument

The pyrolysis of biomass wastes was performed in pipe resistance furnace (SGM68) at variant temperature (400°C~600°C), which power is 4kW, and the highest temperature can reach 1200°C. Textural parameters were got by automatic calorimeter (5E-AC). The main ingredients of pyrolysis were determined by gas chromatograph (Agilent 7890A). The biomass tars were identified by GC-MS (TRACE DSQ).

Raw material and method

The experimental raw materials are comparatively common to be seen as crude biomass. Table 1 is Proximate and ultimate analysis of raw material.

Table 1 Proximate, calorific value and ultimate analysis of biomasses

Material	Proximate analysis [%]				HHV /MJ·kg^{-1}	Ultimate analysis [%]				
	M	V	FC	A		C	H	O	N	S
Bamboo	7.10	77.78	19.92	2.16	19.728	51.54	5.25	42.65	0.62	0.19
Sawdust	9.60	79.54	17.22	3.24	18.521	51.41	5.85	42.46	0.18	0.10
Walnut shell	10.20	80.88	18.44	0.68	17.599	51.11	5.80	42.84	0.20	0.05

The pyrolysis of biomass wastes were performed in pipe resistance furnace. Taking raw materials in drying oven at 105°C for 2 hours before experiment, then weight adequate material sample to the intermediate position of resistance furnace, and setting carbonization temperature, heating rate and carbonization time, pass into nitrogen as carrier gas all the time, determine the main ingredients of pyrolytic gas and tars by gas chromatograph and GC-MS, take out biomass carbon to determinate textural parameters in the end.

Carbon product analysis

Physical properties of biomass carbon

The results of the proximate analysis of biomass carbons as shown in table 2.

Table 2 Proximate analysis of biomass carbons

Biomass	M/ %	V/ %	FC/ %	A/ %	HHV /MJ·kg^{-1}
Bamboo	3.5	18.64	76.58	3.8	31.174
Sawdust	4.0	19.91	72.58	4.2	27.196
Walnut shell	2.7	16.77	80.58	2.2	31.154

Compared with biomass, carbon content of biomass carbons is improved greatly with volatile sharply reduces, That is, biomass carbons are more flammable. Calorific value of biomass carbon has increased considerably (about 30 MJ/kg) account for the combustion properties better than raw.

Pore structure analysis of biomass carbon

The nitrogen adsorption-desorption isotherm of biomass carbons under the best pyrolysis conditions are shown in figure 1.

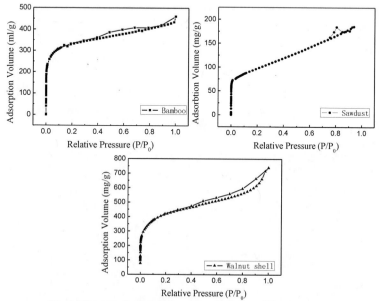

Fig. 1 Nitrogen adsorption-desorption isotherm of biomass carbon

In the Inception phase of adsorption, Adsorption capacity is sharp rise with relative pressure (P/P0) and then gradually slow down present a certain slope called "tail" appearance which caused by the existence of mesopore and macropore[12]. When the relative pressure (P/P0) is close to 1, adsorption has a faster rise rate due to capillary condensation.

Textural parameters of biomass carbons are obtained by analyzed the biomass carbon structural, the specific surface area, pore volume, the average pore size as shown in table 3.

Table 3 Textural parameters of biomass carbons

Sample	specific surface area (m²/g)	pore volume (ml/g)	average pore size (nm)
Bamboo	270	0.37	50.65
Sawdust	174	0.11	37.18
Walnut shell	331	0.33	40.55

Gas-phase product analysis

GC analysis

The chemical composition of the pyrolysis gas was determined by an Agilent 7890A GC gas chromatographer, the flow rate of the carrier gas (helium) was 40 mL/min, injector temperature was 100 ℃. The effect of pyrolysis temperature on gas constituents is shown in Fig. 2.

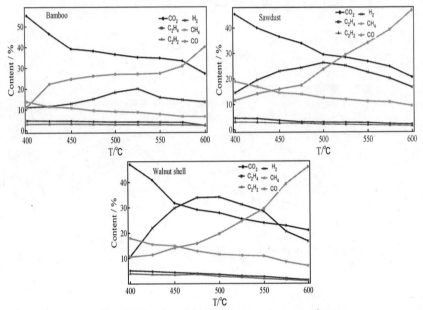

Fig. 2Effect of pyrolysis temperature on gas constituents

Pyrolysis gas is one of the main products of the biomass pyrolysis. The composition of the gas would be helpful to investigate the reactions in the reactor, the uncondensed pyrolysis gas was mainly composed of CO_2, C_2H_4, C_2H_2, H_2, CH_4, and CO. temperature plays a decisive role that high pyrolysis temperature is advantageous to the gas generation. The content of CO_2 decreases with temperature, then flatten out, C and water vapor reaction generate H_2 and CO_2 in low temperature while H_2 and CO in high temperature, under high temperature ,reduction reaction of CO_2 belong to endothermic reaction will occur, and reaction increased with the temperature lead to CO_2 content decreased dramatically after 400℃ and then slow down due to the shrinking of reactants and the increase of products.H_2 come from endothermic reaction on water vapor and C,CH_4 so H_2 content increases before 525℃, and then decreases by the water vapor content. CO mainly comes from the reaction of C with H_2O and CO_2, CO content increase with temperature due to elevated temperature is propitious to promote the reaction. The content of CH_4, C_2H_4 and C_2H_2 decreased with temperature and CH_4 will be decomposition under high temperature.

Calculate the density (γ_m) and calorific value (H_m) of pyrolysis gas under the temperature between 400℃ and 600℃, according to the data get diagrams of effect of temperature on density and calorific value of pyrolysis gas, as shown in Fig. 3 and Fig. 4.

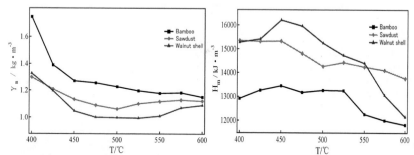

Fig. 3 Effect of pyrolysis temperature
on density of gas

Fig. 4 Effect of pyrolysis temperature
on heat value of gas

The density of biomass pyrolysis is between 1.0kg/m³ and 1.7kg/m³ and decrease (25% ~ 35%) with the temperature, it may associated with the change of gas composition that H_2 (low density) content increase while other compositions(high density) content decrease with temperature. Calorific value of pyrolysis gas is between 12000 kJ/m³ and 16500 kJ/m³ (median fuel gas) and rise with temperature and then fall, it may be caused by the content change of H_2, CO, CH_4, C_2H_4 and C_2H_2 that H_2 and CO (high calorific value gas) content increase with temperature before 450℃ while CH_4, C_2H_4 and C_2H_2 (high calorific value gas) content decreases sharply after 550℃,the calorific value of bamboo, sawdust and walnut shell reached the maximum at 450℃, respectively is 13443.56 kJ/m³,15324.661 kJ/m³ and 16208.24 kJ/m³.

Liquid product analysis

Physical properties of biomass tar

Biomass tar is a black flammable liquid has high sticky, high viscosity and special excitant smell, its specific gravity bigger than water and become gas after 300 ℃. Yield and physical properties of different kind of tars are shown in Tab.4.

Table 4 Yield and physical properties of different kind of tars

Sample	Yield %	HHV /J·kg⁻¹	pH	Density /kg·m⁻³
Bamboo	30.46	33.678	3.56	1057.9
Sawdust	28.40	32.443	2.55	1085.9
Walnut shell	27.32	34.090	3.49	1029.5

The calorific value of biomass tar is between 32MJ/kg and 34MJ/kg. Nitrogen content in the main components of biomass tar is few and do not contain sulfur, so biomass tar as fuel have not negative impact on the environment can be used as a kind of organic green fuel directly. The calorific value of tar have differences for different biomass raw materials, the tar made from walnut shell has highest calorific value, it is related to the biomass ash that ash is unfavorable for tar production. Morell in his study also point out that ash is beneficial to the formation of small molecular gas product [13]. In addition, the density of biomass tar is about 1000 kg/m^3, much higher than the accumulation density of biomass, it means biomass tar is more convenient for storage and transportation.

GC/MS analysis

The chemical compositions of the tars were determined by TRACE DSQ GC-MS with an ECTM-5 capillary column (30m×0.25mm×0.25um). The GC was first maintained at 60 ℃ for 5 min and then increased at 10 ℃/min to 200 ℃. The injector temperature was 200 ℃ and the injection size was 1 uL. The flow rate of the carrier gas (helium) was 1 mL/min, the ion source is Electron Ionization (EI). The compounds were identified by comparing the spectral data with the NIST Mass Spectral library[14]. TIC chromatogram from GC/MS of biomass tars are shown in Fig. 5.

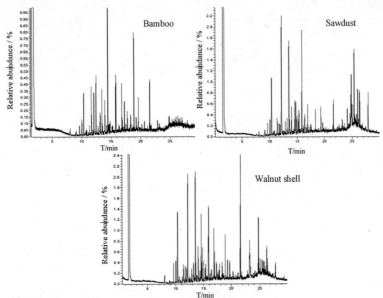

Fig. 5 TIC chromatogram from GC/MS of biomass tars

From Fig.5, Tar is a complex mixture of anhydrosugar, esters, furans, acids, ketones, alcohols, phenols, etc. In order to determine the chemical composition of the tars from pyrolysis of bamboo, sawdust and walnut shell, GC-MS was used to characterize the tar chemical compounds which were categorized into eight functional groups (Tab.5).

180

Table 5 contents of organics in tar of three biomasses

Composition	relative peak area /%		
	Bamboo	Sawdust	Walnut shell
acids	9.98	14.84	10.7
aromatic hydrocarbons	48.75	56.53	48.2
aldehyde and ketone	14.36	13.91	15.08
furans	4.22	2.00	2.46
anhydrosugar	1.25	1.20	1.46
alcohols	6.45	1.82	8.59
esters	12.24	4.92	9.83
pyran	0.47	1.99	3.46

Tars from the pyrolysis of bamboo, sawdust and walnut shell were a mixture compound. Aromatic hydrocarbons were the main composition of the tars, these aromatic hydrocarbons were mainly composed of benzene, toluene, xylene, phenols and their derivatives. Phenols became the highest amount of compounds. Pyrolysis favored the production of biomass as aromatic hydrocarbons, which were cracked by lignin to produce aromatic hydrocarbons. Lignin was depolymerized and dehydrated to produce guaiacols and guaiacols were mainly composed of 2-methoxy-phenol, 2-methoxy-4-methyl-phenol and 4-ethyl- 2-methoxy-phenol.Organic acid would affect the pH of biomass tar, high content of organic acid was relatively low pH value of pyrolysis tar. The organic acid content of sawdust tar is highest, its pH value is lowest, conform to the inversely proportional relationship of organic acid content and pH value. In addition, the esterification reaction of organic acids and alcohols lead to poor stability. Aldehyde and ketone produced by the pyrolysis of cellulose. Influence of the carbonyl group，chemical character of aldehyde and ketone are very active can produce a series of reaction illustrates that the stability of biomass tar is poor. And condensation reaction not only affect the stability of biomass tar but also increase the viscosity of biomass tar. Cellulose was decomposed and dehydrated to form furans and anhydrosugar, such as levoglucosan[15]. Levoglucosan with low thermal stability easily occurs secondary thermal cracking under the high temperature and furan may react with acid and results in the poor stability of the tar.

Conclusion

In this paper，bamboo, sawdust and walnut shell were pyrolysis to obtain biomass carbon, gas, and tar. Carbon content of biomass carbons is improved greatly with volatile sharply reduce by proximate analysis. Calorific value of biomass carbon is about 30 MJ/kg, determine the surface of biomass carbon is mesopore and macropore by analyzed the structural parameters of biomass carbon. Determined the pyrolysis gas was mainly composed of CO_2, C_2H_4, C_2H_2, H_2, CH_4, and CO and known temperature plays a decisive role that high pyrolysis temperature is advantageous to the gas generation by GC analysis. Through calculate the density and calorific value of pyrolysis gas, found the density of biomass pyrolysis is about 1.5kg/m^3 and decrease with the temperature; Calorific value of pyrolysis gas is about 15000 kJ/m^3.The calorific value of biomass tar is between 32MJ/kg and 34MJ/kg. Nitrogen content in the main components of biomass tar is few and do not contain sulfur. The result of GC-MS analysis show that biomass tar is a complex mixture of acids, aromatic hydrocarbons, aldehyde and ketone, furans, anhydrosugar, alcohols, esters and pyran.

References

1. Siwei Luo et al., "Conversion of Woody Biomass Materials by Chemical Looping Process Kinetics, Light Tar Cracking and Moving Bed Reactor Behavior," *Ind. Eng. Chem. Res.*, 10(2013), 1-35.

2. Arthur J. Ragauskas et al., "The Path Forward for Biofuels and Biomaterials," *Science*, 27(2006), 484-489.

3. Y.-C. Lin and G.W. Huber, "The critical role of heterogeneous catalysis in lignocellulosic biomass conversion," *Energy Environ. Sci*, 2 (2009), 68-80.

4. G.W. Huber, S. Iborra and A. Corma, "Synthesis of Transportation Fuels from Biomass: Chemistry, Catalysts, and Engineering," Chem. Rev., 106 (2006), 4044-4098.

5. D.A. Laird et al., "Review of the pyrolysis platform for coproducing bio-oil and biochar," Biofuels, Bioproducts, and Biorefining, 3(2009), 547-562.

6. R. Azargohar and A.K. Dalai, "Steam and KOH activation of biochar: experimental and modeling studies," Microporous and Mesoporous Materials, 110(2008), 413-421.

7. P.K. Rout et al., "Supercritical CO2 fractionation of bio-oil produced from mixed biomass of wheat and wood sawdust," Energy and Fuels, 23(2009), 6181-6188.

8. D. Mohan, C.U. Pittman, P.H. Steele, "Pyrolysis of wood/biomass for bio-oil: a critical review," Energy Fuels, 20(2006), 848-889.

9. A.V. Bridgwater, D. Meier and D. Radlein, "An overview of fast pyrolysis of biomass," Organic Geochemistry, 30(2011), 1479-1493.

10. Yan R et al., "Influence of temperature on the distribution of gaseous products from pyrolyzing palm oil wastes," Combustion and Flame, 142(2005), 25.

11. Ramin Azargohar et al., "Evaluation of properties of fast pyrolysis products obtained, from Canadian waste biomass," Journal of Analytical and Applied Pyrolysis, 6(2013), 1-11.

12. Kruk et al., " Adsorption study of porous structure development in carbon blacks," J. Colloid Interface Sci, 182(1996), 282-288.

13. Morell J.I, Amundson N R and Park S.K, "Dynamics of a single particle during char gasification," Chemical Engineering Science, 45(1990), 387-401.

14. Ren S et al., "The effects of torrefaction on compositions of bio-oil and syngas from biomass pyrolysis by microwave heating," Bioresour Technol, 135(2013), 659-645.

15. Carlson, T. R. et al., "Catalytic Fast Pyrolysis of Glucose over ZSM-5," Catalysis, 270(2010), 110-124.

Energy Technology 2014: Carbon Dioxide Management and Other Technologies
Edited by: Cong Wang, Jan de Bakker, Cynthia K. Belt, Animesh Jha, Neale R. Neelameggham,
Soobhankar Pati, Leon H. Prentice, Gabriella Tranell, and Kyle S. Brinkman
TMS (The Minerals, Metals & Materials Society), 2014

OPTIMIZATION THE PREPARATION OF ACTIVATED CARBON FROM

WALNUT SHELL WITH MICROWAVE HEATING USING RESPONSE

SURFACE METHODOLOGY

Zheng Zhao-qiang[1,2,3], Xia Hong-ying[1,2,3*], Peng Jin-hui[1,2,3], Zhang Li-bo[1,2,3]

[1] Yunnan Provincial Key Laboratory of Intensification Metallurgy,
Kunming, Yunnan 650093, China;
[2] Key Laboratory of Unconventional Metallurgy, Ministry of Education, Kunming University of
Science and Technology, Kunming, Yunnan 650093, China;
[3] Faculty of Metallurgical and Energy Engineering, Kunming University of Science and
Technology, Kunming, Yunnan 650093, China;

Keywords: Walnut shell; Activated carbon; Microwave heating; Response Surface Methodology;

Abstract

Preparation of activated carbon from the walnut shell by microwave assisted steam activation was studied. Influences of the three parameters, activation temperature, activation duration and steam flow rate on the adsorption capacity and yield of walnut shell activated carbon(WSAC) were investigated, optimization process was obtained by the Response Surface Methodology (RSM). The optimum preparation conditions were as follows: activation temperature of 980°C, activation duration of 41 min and steam flow rate of 1.66 ml/min. The optimum conditions resulted in an activation carbon with an iodine number of 1025 mg/g and a yield of 52.17%, while the BET surface area, total pore volume and average pore diameter of WSAC were 1328m^2/g, 0.72ml/g and 2.2nm respectively. The study supported the potentiality of comprehensive utilization of Walnut shell and the benefits of the microwave heating.

Introduction

Activated carbons are one of the most important adsorbent materials in modern society because of its abundantly developed pore structure, strong adsorption ability, high surface area and thermo stability. They have extensive application in many industries fields which include gas purification [1], catalyst supports [2] and other pollutants [3]. However, in recent years, the

---

Corresponding author. E-mail address: hyxia81@hotmail.com

industrial production of activated carbon is facing the problem of raw material scarcity and high cost, which limit to be widely used. Therefore, the selection of an appropriate precursor plays an important role and it has become hot sport to seek cheap raw material to lower preparation cost. As a kind of reproducible resource, the waste from agriculture and forestry is a high-grade raw material for preparing activated carbon due to is high carbon content, such as palm shell [4], rice hull [5], and coconut shell [6]. At present Walnut shell is widely accepted to be a renewable source of biomass. According to statistics, there is more than 1,000,000 tonnes of walnut shells are produced in China annually [7]. As a good precursor for activated carbon production, Walnut shell is a major agricultural waste, to make better use of the cheap and abundant agricultural waste, it is proposed for activated carbon production.

Compare with the conventional heating, microwave heating is increasingly utilized in various technological, due to its advantage of faster and uniform heating rate. The energy transfer is not based on the effect of conduction or convention [8-9].Response surface methodology (RSM) is a statistical technique for modeling and analysis of problems to study the interactions of two or more factors [10-11]. The aim of the present work is to explore the possibility of converting the Walnut shell into value-added activated carbons and optimize the WSAC preparation variable: activation temperature, activation duration and steam flow rate utilizing the microwave heating.

Materials and methods

Materials

Walnut shell was obtained from Kunming, Yunnan Province of China. The raw materials were oven-dried at 105°C and stored in moisture free environment for utilization in the experiments. The carbonization of Walnut shell carried out at a carbonization temperature of 600°C for 60 min, under N_2 gas flow (100cc/min). The proximate analysis of chars represented in weight percent, shows a volatile matter of 10.24%, fixed carbon of 80.75% and an ash content of 9.19%.

Experimental Device

The carbonized product is activated using a self-made microwave tube furnace, which utilizes a single-mode continuous controllable power for the experiments and is shown in Fig. 1.

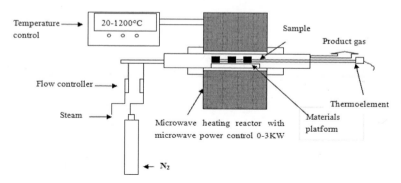

Fig.1 The modified diagram for the preparation of microwave heating equipment

Result and Discussion

Response analysis and interpretation

Table 1 shows the experimental conditions of preparation of WSAC generated by the Design Expert Software covering the parameters such as activation temperature(X_1), activation duration (X_2) and steam flow rate (X_3). The WSAC are characterized for iodine number, yield and results are listed as well in Table 1. This iodine number was tested for the regenerated samples according to the National Standard Testing Methods of PR China(GB/T12496.8-1999) [12].

Table 1 Experimental design matrix and result

Run	X_1(°C)	X_2(min)	X_3(ml/min)	Y_1(mg/g)	Y_2(%)
1	920	30	0.75	852	79.49
2	980	30	0.75	920	74.58
3	920	90	0.75	938	66.52
4	980	90	0.75	989	57.63
5	920	30	1.75	900	70.49
6	980	30	1.75	947	64.48
7	920	90	1.75	956	56.84
8	980	90	1.75	1142	37.55
9	900	60	1.25	915	62.41
10	1000	60	1.25	1102	46.84
11	950	9.55	1.25	801	74.24

12	950	110	1.25	985	45.42
13	950	60	0.41	883	73.87
14	950	60	2.09	1089	43.58
15	950	60	1.25	1012	50.38
16	950	60	1.25	1008	50.47
17	950	60	1.25	1015	51.34
18	950	60	1.25	1001	49.85
19	950	60	1.25	1024	50.28
20	950	60	1.25	1019	51.27

<u>Activated carbon iodine number</u>

All three parameters chosen in the present study are found to have significant effect on iodine adsorption capacity. Fig.3 shows the effect of activation temperature and activation duration on Iodine number while the third parameter steam flow rate held constant at 1.25 ml/min. Fig. 4 shows the effect of activation temperature and steam flow on Iodine number while the third parameter activation duration held constant at 60min. The Iodine number gradually increased with the increasing of the three parameters within the range studied. An increase in the activation temperature leads to increases the rate of C-H$_2$O reaction. The higher rate of reaction results in a better porous carbon, which in turn would increase the porosity of the activated carbon.

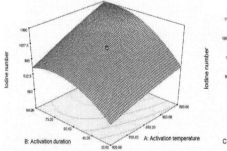 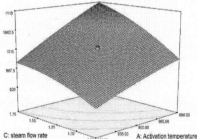

Fig. 3 Three-dimensional response surface
plot of iodine number: effect of X_1 and X_2
(X_3: 1.25ml/min)

Fig. 4 Three-dimensional response surface
plot of iodine number: effect of X_1 and X_3
(X_2: 60min)

Activated carbon yield

The yield of activation carbon is also influenced by the activation process and the activation conditions. Fig.6 shows the combined effect of activation temperature and activation duration with the steam flow rate of 1.25 ml/min while Fig.7 shows the combined effect of activation temperature and steam flow rate with the activation duration of 60 min. The yield was found to decrease with increasing activation temperature, activation duration and steam flow. The increase in the three parameters would resulting in decreased yield and increased carbon burn-off which was found to depend on the extent of $C-H_2O$ reaction.

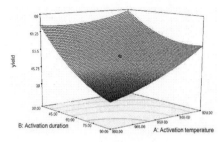 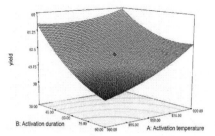

Fig. 3 Three-dimensional response surface plot of yield: effect of X_1 and X_2 (X_3: 1.25ml/min)   Fig. 4 Three-dimensional response surface plot of yield: effect of X_1 and X_3 (X_2: 60min)

Process optimization

For the economical practicability, the WSAC should have both a high carbon yield and a high adsorption capacity. Hence it is desirable to prepare activated carbon with high yield and a high adsorption capacity. However, the Iodine number (Y_1) and yield (Y_2) were competitive, when Y_1 increases, Y_2 will decrease, and vice versa. In order to compromise these two values, the Design Expert software version 7.15 (STAT-EASE Inc., Minneapolis, USA) was applied. The WSAC was preparation under the experimental condition shown as follows: activation temperature of 980°C, activation duration of 41 min and steam flow rate of 1.66 ml/min. The predicted Iodine number and yield were 1032 mg/g and 54.18%, three repeat experiments were carried out to justify the accuracy of the predicted result which showed an average Iodine number of 1025 mg/g and average yield of 53.25%, respectively. It was observed that the experimental value obtained was in good agreement with the value calculated from the model.

Characterizations of pore structure

Fig. 8 shows the nitrogen adsorption isotherm estimated using the Autosorb instrument exhibited by the char and WSAC under the optimum condition. The shapes of these isotherms are the intermediates between types I and II of the referred classification [13-14]. It can be seen that, the N_2 adsorbed volume of WSAC is higher than that of char. The cumulative pore volume plot shown in Fig. 9 substantiates the amount of pore in the mesoporous range. Table 5 provides the pore structural parameters of WSAC and char. The SEM of the char and WSAC were shown in the Fig. 10. For Fig. 10(a), we can find the pores of char is planar, covering by impurities and without any obvious porous. After the activation progress the char transform into activated carbon with a significant increase in the micro and mesopores, as shown in Fig. 10(b).

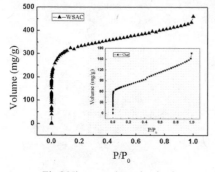 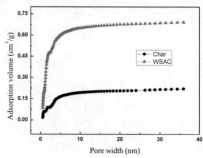

Fig.8 Nitrogen adsorption isotherm of the WSAC and Char

Fig. 9 Cumulative pore volume distribution chart for WSAC and char

Table 5 Pore structural parameters of WSAC vs. chars

Properties		WSAC	Char
Pore volume	(ml/g)	0.72	0.17
Average pore diameter	(nm)	2.2	2.7
BET Surface area	(m^2/g)	1328	175

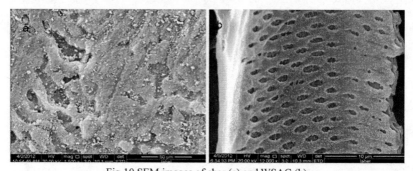

Fig.10 SEM images of char (a) and WSAC (b)

Conclusion

This research was carried out to determine optimum conditions for the activation of activated carbon from walnut shell based on output characteristics. The process parameters were optimized utilizing the Design Expert software and were identified to be an activation temperature of 980°C, activation duration of 41 min and steam flow rate of 1.66 ml/min. The optimum conditions resulted in an activated carbon product with the Iodine number of 1025mg/g and a yield of 52.17%. The results indicate that walnut shell has a potential to be a good precursor material for the production of activated carbon with superior adsorption capacity.

Acknowledgements

The authors would like to express their gratitude to the Specialized Research Fund for the Doctoral Program of Higher Education of China (NO.20115314120015) and the Kunming University of science and technology personnel training fund (NO. KKSY201252077) for financial support.

References

1. Nieto-Delgado C, Terrones M, and Rangel-Mendez J R, "Development of highly microporous activated carbon from the alcoholic beverage industry organic by-products" *Biomass and Bioenergy*, 35(2011), 103-112.

2. Mudoga H L, Yucel H, and Kincal N S, "Decolorization of sugar syrups using commercial and sugar beet pulp based activated carbons" *Bioresource technology*, 99(2008), 3528-3533.

3. Hejazifar M et al., "Microwave assisted preparation of efficient activated carbon from grapevine rhytidome for the removal of methyl violet from aqueous solution" *Journal of Analytical and Applied Pyrolysis*, 92(2011), 258-266.

4. Issabayeva G, Aroua M K, and Sulaiman N M N, "Removal of lead from aqueous solutions on palm shell activated carbon" *Bioresource technology*, 97(2006), 2350-2355.

5. Rahman I A et al., "Adsorption characteristics of malachite green on activated carbon derived from rice husks produced by chemical–thermal process" *Bioresource technology*, 94(2005) , 1578-1583.

6. Deng H et al., "Preparation of activated carbons from cotton stalk by microwave assisted KOH and K_2CO_3 activation" *Chemical Engineering Journal*, 163(2010), 373-381.

7. Yang J and Qiu K, "Preparation of activated carbons from walnut shells via vacuum chemical activation and their application for methylene blue removal" *Chemical Engineering Journal*, 165(2010), 209-217.

8. Venkatesh M S and Raghavan G S V, "An overview of microwave processing and dielectric properties of agri-food materials" *Biosystems Engineering*, 88(2004), 1-18.

9. Yuen F K and Hameed B H, "Recent developments in the preparation and regeneration of activated carbons by microwaves" *Advances in colloid and interface science*, 149(2009), 19-27.

10. Tan I A W, Ahmad A L, and Hameed B H, "Optimization of preparation conditions for activated carbons from coconut husk using response surface methodology" *Chemical Engineering Journal*, 137(2008), 462-470.

11. Azargohar R and Dalai A K, "Production of activated carbon from Luscar char: experimental and modeling studies" *Microporous and mesoporous materials*, 85(2005), 219-225.

12. Liu X et al., "Temperature measurement of GAC and decomposition of PCP loaded on GAC and GAC-supported copper catalyst in microwave irradiation" *Applied Catalysis A: General*, 264(2004), 53-58.

13. Rouquerol J et al., "Recommendations for the characterization of porous solids (Technical Report)" *Pure and Applied Chemistry*, 66(1994), 1739-1758.

14. Smâiések M and éCernây S, *Active carbon: manufacture, properties and applications* (Amsterdam and New York, Elsevier Pub. Co. 1970).

Energy Technology 2014: Carbon Dioxide Management and Other Technologies
Edited by: Cong Wang, Jan de Bakker, Cynthia K. Belt, Animesh Jha, Neale R. Neelameggham,
Soobhankar Pati, Leon H. Prentice, Gabriella Tranell, and Kyle S. Brinkman
TMS (The Minerals, Metals & Materials Society), 2014

RESEARCH ON USING CARBIDE SLAG TO MINERALIZE THE CARBON DIOXIDE IN ELECTROLYTIC ALUMINUM WASTE GAS

Yan Liu, Yu Fang, Zimu Zhang*, Guanting Liu, Ting'an Zhang,Zhihe Dou, Xiaoli Jiang Dianhua Zhang*

Key Laboratory of Ecological Utilization of Multi-metal Intergrown Ores of Education Ministry, * State Key Laboratory of Rolling and Automation, Northeastern University, Shenyang, 110004, China

Keywords: venturi; carbide slag; electrolytic aluminum; local gas holdup; fiber optic probe

Abstract

With the idea of "waste control by waste itself", we come up with the crafts of using waste calcium carbide to mineralize CO_2 in electrolytic aluminum waste gas, design and make out "Venturi gas-liquid-solid three stage reactor", do experiments under the system of CO_2-Ca(OH)$_2$, investigate the influence on gas absorption by pH meter; investigate the influence on diameter distribution , rise velocity of bubble and local gas hold-up by dual conductivity probes; investigated the absorption of CO_2 in every stage reactor under different initial proportion of CO_2 by intelligent gas measurement system. The results showed that the reactor can fully absorb the CO2 in waste gas, and it has largely shorten the time of absorption and improve the CO_2 utilization compared with the single stage reactor.The residual proportion of CO_2 in the gas is below 0.08% through the absorption of three stage reactor,which show that absorption is extremely improved.

Introduction

Carbide slag is a kind of polluting hydrolysis reaction byproducts in producing carbide in wet reactor, and it is mainly comprised by Ca(OH)$_2$, It's estimated that there are about million tons carbide slag are piled up per year. Carbide slag is a kind of industrial wastes that hard to deal with ,which are treated by land filled in the past. It would not only occupy a lot of land, but also lead to secondary pollution [1].

CO_2 is the main greenhouse gas by now. In 2010, global CO_2 emission has reached 33.5 billion tons, while China has now become the world's second largest emitter of CO_2, emissions accounting for about one-seventh of the world [2].As the experts estimates, the greenhouse gas emission of whole nonferrous metal industry may over 0.6 billion tons per year, while CO_2 emission of electrolytic aluminum is 0.334 billion tons per year [3]. So it is necessary and pressing to discover the reusing CO_2 of electrolytic aluminum. With the idea of "waste control by waste itself", we came up with the crafts of using carbide slag to mineralize the CO_2 in electrolytic aluminum waste gas, which was experimented on the counterflow-three stage-venturi reactor,and studied the local gas holdup, residual concentration of CO_2 and the volumetric transfer coefficient in the reactor.

Experimental Principle and Equipment

The experimental is going under the system of Ca(OH)$_2$-CO_2-Ar. Fig. 1 is the design of the experimental equipment. The main reactor are three vertical Plexiglas column, 300mm in inner diameter and 0.7m in height. Carbide slag solution and different ratios of CO_2 and Ar mixed gas are used as the liquid and gas phase, respectively.

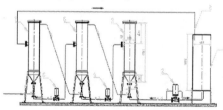

Fig. 1 Design of the experimental equipment

Local gas holdup is an important parameter of gas-liquid-solid phase dispersion in the reactor, which can reflect the local dispersion and mass transfer characteristics, and has great influence on internal flow regime, gas-liquid contact area and other hydrodynamic parameters. Therefore, the gas holdup is of great significance for in-depth study.

Gas holdup is defined as the volume fraction of gas phase. Assuming the local flow are uniform, and the probability of probe piercing any point of bubble's positive projection plane, then in in the statistical sense, a spatial point's time fraction of gas phase is equal to its gas holdup. Therefore local gas holdup can be calculated as follows:

$$\varepsilon = \frac{\sum_{i=1}^{i=n_b} \tau_{1i}}{t} \qquad [1]$$

The value of τ_{1i} is residence time of probe in the ith bubble, n_b is the number of bubbles in sample, and t is the total sampling time. Fig 2 is construction of the dual electric conductivity probe. Due to the turbulence in bubble passing through the probe, there are differences in measurements between two probes. In this experiment, measurements of sensor2 are smaller than sensor1. Taking into account that the top probe's measurements are closer to actual conditions, we chose the data of sensor1 as the measurements [4,5].

Rise velocity of bubble is an important parameter in this experiment, which also can be determined by dual electric conductivity probe.

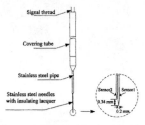

Fig. 2 Construction of the dual electric conductivity probe

The residual concentration of CO_2 is the main reflection of absorption effect. Fig. 3 is the photo of Leibo intelligent gas measurement system. The residual CO_2 in the flue is pumped into sampling tube and pass through the sample-gas-room after dust removal and dehydration. The concentration of CO_2 can be measured by determine the characteristic absorption spectrum under the irradiation of infrared light.

Fig.3 Leibo intelligent gas measurement system

The volumetric transfer coefficient In this experiment pH meter is needed, which is used to measure the pH changes with time during the absorption, and then the volumetric transfer coefficient can be calculated by the curves of pH-time, which is also an important parameter in the process of absorption.

Results and Discussion

Compare of Single-Stage Reactor And Three-Stage Reactor On Absorption

Effect of Single-Stage Reactor And Three-Stage Reactor On Time of Absorption

Fig. 4~6 are the curves of pH –time under different superficial gas velocity. The graphs indicate that the time of three-stage reactor is far less than single-stage reactors', because CO_2 can be totally used because of the counterflow designing.

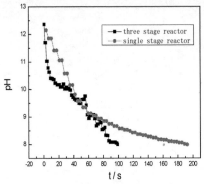

Fig. 4 curve of pH-time when velocity of gas is 10.616 m/s

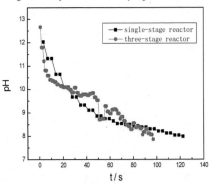

Fig. 5 curve of pH-time when velocity of gas is 14.154 m/s

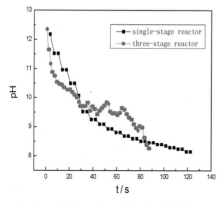

Fig. 6 curve of pH-time when velocity of gas is 17.693 m/s

Efficiency Of CO₂ In Different Reactor

Fig. 7 is the efficiency of CO_2 under different superficial gas velocity. From fig. 7 we can know thar with the increasing of superficial gas velocity ,the efficiency of CO_2 is decreasing. Due to the increasing superficial gas liquid and gas agitation, the gas residence time decrease, and then the efficiency of CO_2 decrease. And efficiency of CO_2 in three-stage reactor is far more than single stages', so the three-stage reactor has great advantages on dealing CO_2.

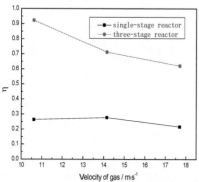

Fig. 7 The efficiency of CO_2 under different superficial gas velocity

Residual Proportion of CO₂ In Different Reactor

Pump CO_2-Ar mixture into reactor under the certain velocity of gas ,velocity of liquid and gas pressure, then measure the residual proportion of CO_2 in different reaction after the absorption, the results is in fig. 8 and fig. 9.

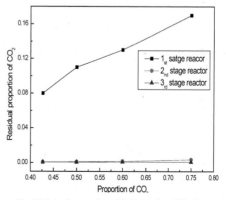

Fig. 8 Relation between the residual proportion of CO_2 in each reactor and initial proportion of CO_2

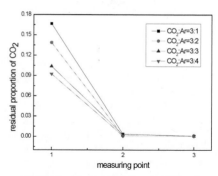

Fig. 9 Relation between residual proportion of CO_2 and measuring point under different initial proportion of CO_2

From fig. 8 and fig. 9 we can know that with the proportion of CO_2 in mixed gas increasing in the first stage reactor, the residual proportion of CO_2 is increasing, while the residual proportions of CO_2 in the second and third stage reactor are nearly unchanged. After the absorption of three reactors, the concentration of CO_2 in residual gas has already decreased below 0.08%, the absorption results are satisfactory.

Effect of Superficial Gas Velocity and Superficial Liquid Velocity on Local Gas-holdup

Fig. 10 indicates that with the increasing of superficial gas velocity and superficial liquid velocity , their influences on local gas-holdup are opposite.

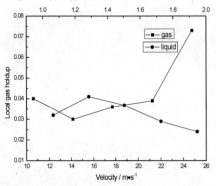

Fig. 10 Effect of superficial gas velocity and superficial liquid velocity on local gas-holdup

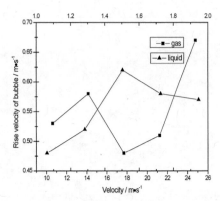

Fig. 11 Effect of superficial gas velocity and superficial liquid velocity on rise velocity of bubble

With superficial gas velocity increasing in 7.077 ~21.231 $m \cdot s^{-1}$ range, local gas holdup was basically stable in 3%~4%. When superficial gas velocity is below 21.231 $m \cdot s^{-1}$, CO_2 is absorbed fairly completely, increasing gas velocity has little effect on local gas holdup. While superficial gas velocity is above 21.231 $m \cdot s^{-1}$, CO_2 is relative overdose, and the number of bubble increase, which lead to local gas holdup rising.

local gas holdup first increased and then decreased with superficial liquid velocity increasing. The maximum value of local gas holdup is achieved when superficial velocity is 1.283 $m \cdot s^{-1}$. When superficial liquid velocity is in low range, gas and liquid are not so completely mixed that capacity of Ca(OH)$_2$ to absorb CO_2 are diminished, and then local gas holdup increase with superficial liquid velocity. While superficial liquid velocity is in high range, due to the good mixing effect and improved absorbing capacity of Ca(OH)$_2$, local gas holdup decrease with superficial liquid velocity increasing.

Effect of Superficial Gas Velocity and Superficial Liquid Velocity on Rise Velocity of Bubble

Fig. 11 shows that with the increasing of superficial gas velocity and superficial liquid velocity , their influences on rise velocity of bubble are also opposite.

With superficial gas velocity increasing in 7.077 ~21.231 $m \cdot s^{-1}$ range, bubble rise velocity is basically stable in 0.5~0.55$m \cdot s^{-1}$, because gas holdup and bubble rising resistance are nearly unchanged.When superficial gas velocity is above 21.231 $m \cdot s^{-1}$, rise velocity of bubble increased with superficial gas velocity increasing. High superficial gas velocity leads to incomplete absorption of CO_2, which causes higher gas holdup and smaller density of solution and bubble rising resistance.

Rise velocity of bubble first increased and then decreased with superficial liquid velocity increasing. The maximum value of rise velocity of bubble is achieved when superficial velocity is 1.504 $m \cdot s^{-1}$. When in the low velocity range, rising velocity of liquid will decrease bubble rising resistance, and then increase the rise velocity of bubble. While in the high velocity range,there is more fierce turbulent flow in the reactor, which leads to higher bubble rising resistance and lower rising velocity.

Grading Analysis of Product

$CaCO_3$ powder product can be obtained by filtering solution after absorption and 24-hour drying. And the $CaCO_3$ powder product from industrial venturi reactor(in Xinyu Iron&Steel Group in Jiangxi province) is to be the contrast test, results are shown in fig. 12.

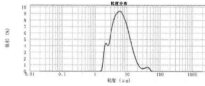

Experimental product

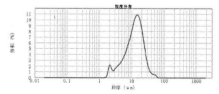

Industrial product

Fig. 12 Grading analysis of $CaCO_3$ product from different reactor

Fig. 12 indicates that grain size of $CaCO_3$ from three stage reactor is 5.9um, which is much smaller than industrial product's 12.6um.

Conclusion

New three-stage reactor has great performance on dealing CO_2 gas. With the analysis and the summary of the results, we can get the conclusions below:

(1)Three-stage reactor has better capacity on dealing CO_2 gas, which would take far less time than single-stage reactor,CO_2 can be totally used because of the counterflow designing.

(2)Through three stages reactors' absorption, the concentration of CO_2 in the residual gas is below 0.08%, the CO_2 is nearly totally absorption, which achieved the best efficiency.

(3)Local gas holdup first decrease and then increase subsequently with the increasing of superficial gas velocity. And first increase and then decrease with the increasing of superficial liquid velocity.

(4)Rise velocity of bubble keep stable basically at first and then decrease with the increasing of superficial gas velocity. And first increase and then decrease with the increasing of superficial liquid velocity.

(5)Grain size of $CaCO_3$ powder product from three stage reactor can reach 5.9um, which is far more better than industrial venturi reactor's $CaCO_3$ powder product.

Reference

1. Wang H Q. Resourcification utilization routes for carbide slag. Chemical production and technology，2007.14(1)：1-4

2. Zheng X Y. Global carbon dioxide emission reduction situation and strategies. International observation，2010.(6)：55-57

3. Chen X P, Li W X, Qiu S L. Study on carbon dioxide emission from primary aluminium industry，Light Metal, 2012 (007): 33-36

4. Gao N. Research of multilayer slurry gas - liquid mixing tank bubble size and local gas holdup. Beijing: Beijing University of Chemical Technology，2010

5. Lv S S, Chen X L, Yu Z H. Application conductivity probe parameters measured bubble bubble column. Beixue Reaction Engineering and Technology，2003.19(4):344-351

Acknowledgement

This research was supported by the National Natural Science Foundation of China (Nos. U1202274, 51004033, 50934005, 50974035, 51204040, 51374064 and 51074047), National 863 Plan (2010AA03A405 and 2012AA062303) and a grant from the National High Technology Research and Development Program of China (No. 2009AA063701)；the doctoral fund of EDU gov (20050145029)

Energy Technology 2014: Carbon Dioxide Management and Other Technologies
Edited by: Cong Wang, Jan de Bakker, Cynthia K. Belt, Animesh Jha, Neale R. Neelameggham,
Soobhankar Pati, Leon H. Prentice, Gabriella Tranell, and Kyle S. Brinkman
TMS (The Minerals, Metals & Materials Society), 2014

STUDY ON THE COMBUSTION CHARACTERISTICS AND KINETICS

OF BLENDING COAL

Xiangdong Xing[1], Jianliang Zhang[1], Shan Ren[1], Xingle Liu[1], Zhenyang Wang[1], Hongen Xie[2],

1 School of Metallurgical and Ecological Engineering, University of Science and Technology Beijing; Beijing 100083, China

2 Pangang Group Research Institute Co.,Ltd.; Panzhihua 617000, China

Key words: Thermogravimetric analysis (TG), Combustion, Blending coal, combustion

Abstract

Non-isothermal combustion experiments of different additive amount of bituminous (0, 20, 40, 60, 80, 100 wt%) were conducted by synthesized thermogravimetry analyzer(STA409PC) from room temperature to 900°C in air. The changes of combustion characteristic parameters o f pulverized coals in different atmospheres are analyzed. The results show that DTG curves of coal combustion move to low temperature zones when the amount of bituminous increases. It indicates that both ignition and burnout temperature are lower, burnout time decreases, combustion characteristic index obviously increases, and combustion performance of blending coal are improved. The iso-conversional method involving Kissinger-Akah-Sunose （KAS） method was used for the kinetic analysis of the main combustion process. The results indicated that, when the additive amount of bituminous varied from 0 to 100 wt%, the value of activation energy which would sharply reduce if the additive amount of bituminous was under 60% increased from 133.94 kJ•mol-1 to 78.03 kJ•mol-1by using KAS method.

Introduction

Pulverized coal injection (PCI) to blast furnace (BF) is an important way for saving energy and reducing resource consumption during the ironmaking process. In order to acquire higher coal injection quantity, the relevant research topics in the field of pulverized coal combustion technology such as enhancing the efficiency of pulverized coal combustion, improving the combustion characteristics and reducing the harmful gas emissions and so on received extensive attention in recent years [1,2]. The combustion of blending coal when the appropriate proportion is selected for coal and also mixed well- distributed could give full play to the advantages of various types to make up the defects of single coal which is widely used throughout the world. So the blending coal injection to BF could produce a good impact on safety and economy for productive process. Therefore, a large amount of experimental researches on the combustion characteristic of blending coal for anthracite and bituminous were carried out by experts and scholars around the world [2-7].

Nevertheless, the systematic research on the combustion behavior of anthracite with different content of bitumite in a thermogravimetric analyzer (TGA) is less. So in this paper, using the method of multiple scanning rate of conversion method, bituminous to promote the pulverized coal combustion reaction was studied to analyze its promotion mechanism of pulverized coal combustion reaction process, and the kinetic parameters of activation energy was also calculated, which to provide theoretical basis for the rationalization of resource utilization.

Experiment

Material

The powder of anthracite and bituminous which were dried at 105℃ for 2 h to remove the free water were provided by one steel enterprise in Hebei province (PR China). The analyses of ultimate, proximate analyses and calorific value of single testing coal sample were carried out according to ASTM standards, and the results were presented in **Table I**.

Table I. Ultimate, proximate analyses and calorific value of single testing coal sample，%

Sample	Proximate analysis, ad				Ultimate analysis, daf					Bomb calorific value /J•g⁻¹
	FC_{ad}	M_{ad}	A_{ad}	V_{ad}	C_{ad}	H_{ad}	O_{ad}	N_{ad}	S_{ad}	
Anthracite	76.09	1.34	13.22	9.32	79.17	3.45	3.51	1.01	0.98	29172.62
Bituminous	45.40	3.13	8.33	42.59	66.58	3.82	19.10	1.06	1.05	25867.58

Coal analysis data of blending coal could be calculated due to the linear weighted of moisture, ash, fixed carbon and volatile in coal [3], as shown in **Table II**.

Table II. The amount of adding bituminous and calculated results of blending coal

Additive amount(%)	FC_{ad}	A_{ad}	V_{ad}	M_{ad}
0%	76.09	13.22	9.32	1.34
20%	69.95	12.24	15.97	1.70
40%	63.81	11.26	22.63	2.06
60%	57.68	10.29	29.28	2.41
80%	51.54	9.31	35.94	2.77
100%	45.40	8.33	42.59	3.13

Thermogravimetric analysis

The experiments were carried out in a NETZSCH instruments STA409PC thermogravimetric analyzer. In each experiment, (10±0.2)mg of anthracite with different additive amount of bituminous (0, 20, 40, 60, 80, 100 wt%) was spread uniformly on the bottom of the alumina crucible of the thermal analyzer. The combustion experiments were performed at heating rates of 5, 10, and 20℃/min in air atmosphere, respectively.

The temperature of the furnace was programmed to rise from room temperature to 900℃. The experiments were replicated at least thrice to determine the irreproducibility, which was found to be very good. The combustion behavior and the kinetic parameters of the process could be obtained by analyzing TG and DTG curves.

Results and discussion

The determination of combustion characteristic value

2.1.1 Ignition and burnout temperature

The initial combustion temperature of blending coal with different additive amount of bituminous (0, 20, 40, 60, 80, 100 wt%) was determined by thermogravimetric analysis which could be described as an vertical line was made from the maximum points in DTG curves to TG curves at point A, and then made a tangent line from A to the parallel lines of initial weightlessness to point C on TG curves. The corresponding temperature C was the ignition temperature T_i, and the burnout temperature T_F was defined as the corresponding value when the weight loss ratio reached 98%. The burning-down period could be determined as the time from ignition temperature rising to burnout temperature.

2.1.2 Comprehensive combustion characteristic index

The comprehensive combustion characteristic index of S reflects the ignition and combustion characteristics of pulverized coal, and the combustion performance would be better with the larger S value. The S is defined as follows,

$$S = \frac{R}{E}\frac{d}{dT}\left|\frac{dw}{dt}\right|_{T-T_i} \cdot \frac{(dw/dt)_{max}}{(dw/dt)_{T=T_i}} \cdot \frac{(dw/dt)_{mean}}{T_F} = \frac{(dw/dt)_{max}(dw/dt)_{mean}}{T_i^2 T_F} \tag{1}$$

where $(dw/dt)_{max}$ is maximum combustion rate, %/min, $(dw/dt)_{mean}$ is average combustion rate, %/min, T_i is the ignition temperature, ℃, and T_f is the burnout temperature, ℃

2.1.3 TG/DTG curves analysis

The curves of the weight loss (TG) and the rate of weight loss (DTG) are obtained during the combustion of blending coal with different additive amount of bituminous under air atmosphere at a heating rate of 20 K/min, as shown in Figure 1 and Figure 2, respectively. It could be seen that three individual stages can be distinguished during the heating process of the different samples. Furthermore, the more content of bituminous, the clearer stages will appear. Furthermore, the TG curve of just anthracite or bituminous has only one obvious mass loss stag. Correspondingly, the DTG curve has also only one obvious the rate of weight loss peak. The first stage (I) goes from room temperature to 315.1~421.2℃, and a slight weight loss in the TG curve as well as a small hump in the DTG curve is observed, corresponded to the loss of water and very light volatile compounds. The second stage (II) is from end temperature of stage I to 373.2-536.9℃. Stage II is characterized by a major weight loss, which corresponds to the main release and combustion of organic components. Most of the combustible materials are combustion in this stage. The mass loss of this stage is more than 90% of total mass. The third stage (III) goes from end temperature of stage II to the final temperature (900℃). In the third stage, a slight continued loss of weight is shown in the TG curve.

As the paper published [1,3-5], the combustion of pulverized coal is a complex chemical process. In the three stages, the reaction rate of the stage II which is the main reaction zone to study on combustion dynamics of anthracite with different additive amount of bituminous. It can be seen from Figure 1 and Figure 2 that TG curves tend to move to low temperature zones with the additive amount of bituminous increase. Three peaks would be found apparently when the additive amount exceeds 60%. This is mainly due to the high surface activity of bituminous which could promote the reaction of pulverized coal burning. Meanwhile, the former peak value in DTG curves becomes larger with the additive amount of bituminous increase, and the latter tends to lessen which reflects the combustion process of bituminous and anthracite, respectively.

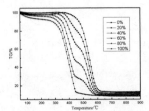 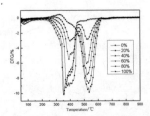

Figure 1. TG curves of different additive amount of bituminous at a heating rate of 20 K•min⁻¹

Figure 2. DTG curves of different additive amount of bituminous at a heating rate of 20 K•min⁻¹

Combustion characteristic parameters of samples at a heating rate of 20 K/min are shown in **Table III**. Figure 3 shows the relationships between the ignition temperature (T_i) and final combustion temperature (T_F) with different additive amount of bituminous for blending coal. It can be found from Table 3 and Figure 3 that both of the T_i and T_F tend to decrease gradually with the additive amount of bituminous increase which mainly because of high violate and porosity of bituminous, all of which are good for combustion. In addition, the average combustion reaction rates also display the same change tendency. The effect of adding bituminous on the final temperature is much more remarkable than that of ignition temperature. Therefore, adding the high reactivity bituminous can promote the combustion of anthracite and cause the change of maximum reaction rate.

Table III. Characteristic parameters of coal combustion under different additive amount of bituminous

Additive amount (%)	$T_i(°C)$	$(dw/dt)_1$ (%•min⁻¹)	$(dw/dt)_2$ (%•min⁻¹)	$T_{max}(°C)$	$(dw/dt)_{max}$ (%•min⁻¹)	$(dw/dt)_{mean}$ (%•min⁻¹)	$T_F(°C)$
0	421.2	-	8.97	536.9	8.97	1.07	608.5
20	391.2	3.08	8.68	532.9	8.68	1.11	600.4
40	355.0	5.48	7.54	524.2	7.54	1.12	593.2
60	361.8	7.62	6.03	391.4	7.62	1,13	583.3
80	338.4	8.88	2.90	380.4	8.88	1.16	577.8
100	315.1	9.89	-	373.2	9.89	1.23	524.5

Notes:

T_i is the ignition temperature.

$(dw/dt)_1$ and $(dw/dt)_2$ are the reaction rates of the first and second, respectively.

$(dw/dt)_{max}$ and $(dw/dt)_{mean}$ are the maximum reaction rate and average reaction rate, respectively.

T_{max} is the temperature for maximum reaction rate.

T_F is the final combustion temperature detected as weight stabilization.

The changes of comprehensive combustion characteristic indexes with the additive amount of bituminous increase are shown in Figure 4.

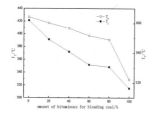 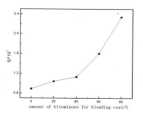

Figure 3. Relation between T_I and T_F with different additive amount of bituminous for blending coal

Figure 4. Relation between combustion indexes with different additive amount of bituminous for blending coal

It can be concluded from Fig.4 that adding bituminous can improve the combustion performance of the blending coal, and the influence on the comprehensive combustion characteristic index is not a linear relationship which the effect would be significantly increase after the additive amount more than 60%.

Kinetic analysis

2.2.1 Kinetic methods

The sample mass is measured as a function of temperature in the non-isothermal experiments carried out with a thermobalance, and the rate of combustion or conversion, $d\alpha/dt$, is a linear function of a temperature-dependent rate constant, $k(T)$, and a temperature-independent function of conversion, $f(\alpha)$:

$$\frac{d\alpha}{dt} = k(T)f(\alpha) \tag{2}$$

The reaction rate constant, $k(T)$, has been described by the Arrhenius expression,

$$k = A\exp\left(-\frac{E}{RT}\right) \tag{3}$$

where A is the pre-exponential factor, E is the activation energy, R is the universal gas constant, 8.314J/(mol · K), and T is the absolute temperature.

$f(\alpha)$ can be described as[5] ,

$$f(\alpha) = (1-\alpha)^n \tag{4}$$

where n is reaction order.

The combustion rate is defined as,

$$\alpha = \frac{m_i - m_t}{m_i - m_\infty} \tag{5}$$

where m_i is the initial mass of the sample, m_t is the mass of the sample at time t, m_∞ the final mass of the sample in the reaction.

By combined Eqs.(2), Eqs.(3) and Eqs.(4), the kinetic equation of Eqs.(6) can be obtained in the following form,

$$\frac{d\alpha}{dt} = A\exp\left(-\frac{E}{RT}\right)(1-\alpha)^n \tag{6}$$

The heating rate is

$$\beta = \frac{dT}{dt} \tag{7}$$

If the temperature of the sample is changed by a controlled and constant heating rate, the rearrangement of Eqs.(6) gives

$$\frac{d\alpha}{d(1-\alpha)^n} = \frac{A}{\beta}\exp\left(-\frac{E}{RT}\right)dT \tag{8}$$

The integrated form of Eqs.(8) is generally expressed as,

$$g(\alpha) = \int_0^\alpha \frac{d\alpha}{(1-\alpha)^n} = \frac{A}{\beta}\int_{T_0}^T \exp\left(-\frac{E}{RT}\right)dT \tag{9}$$

$$T = T_0 + \beta t \tag{10}$$

where $g(\alpha)$ is the integrated form of the conversion dependence function $f(\alpha)$. Based on these equations, different kinetic methods were applied in this study.

It is well known that the iso-conversional method easily gives an estimate of activation energy regardless of the reaction mechanism. A series of experiments of different heating rate were designed in this article, and the iso-conversional method of Kissinger-Akahira-Sunose (KAS) was adopted to calculate the combustion kinetic parameters of activation energy, respectively.

Kissinger-Akahira-Sunose(KAS) equation[5,6],

$$\ln\left(\frac{\beta}{T_\alpha^2}\right) = \ln\left[\frac{AR}{E_\alpha g(\alpha)}\right] - \frac{E_\alpha}{RT_\alpha} \tag{11}$$

The combustion activation energy $E\alpha$ with a given value of conversion rate could also be calculated according to the linear relationship between $\ln(\beta/T_\alpha^2)$ and $1/T_\alpha$.

2.2.2 Analysis of kinetic parameters

According to Eqs.(11), a plot of $\ln(\beta/T_\alpha^2)$ against $1/T_\alpha$ should be a straight line corresponding to the various conversion degrees from the combustion with different additive amount of bituminous for blending coal [7]. From the slope of line, apparent activation energy (E_α) of the dynamic combustion process at various conversions (α) can be estimated. In the present study, three different heating rates (5 K•min^{-1}, 10 K•min^{-1} and 20 K•min^{-1}) are used to evaluate the relationship between E_α, and α by the iso-conversional method. Fig.5 show the linear relationships between $\ln(\beta/T_\alpha^2)$ and $1/T_\alpha$ based on the KAS methods in $0.2 \leqslant \alpha \leqslant 0.8$, respectively.

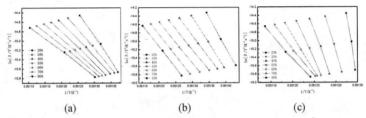

(a)   (b)   (c)

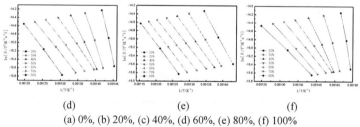

(d)   (e)   (f)

(a) 0%, (b) 20%, (c) 40%, (d) 60%, (e) 80%, (f) 100%

Figure 5. The linear relationship between $\ln(\beta/T_\alpha^2)$ and $1/T$ of KAS

The activation energies obtained by TG data at different heating rates by KAS methods are shown in **Table IV** for each degree of conversion and the corresponding related coefficient (R^2) are also listed.

Table IV. The activation energies obtained by TG data at different rates by KAS methods

Additive amount(%)	α	E (kJ•mol^{-1})	related coefficient R^2	Additive amount(%)	α	E (kJ•mol^{-1})	related coefficient R^2
0%	0.2	87.89	0.9927	60%	0.2	277.14	0.9918
	0.3	82.69	0.9951		0.3	181.18	0.9960
	0.4	61.22	0.9960		0.4	133.39	0.9980
	0.5	56.78	0.9964		0.5	100.24	0.9980
	0.6	52.00	0.9969		0.6	91.81	0.9991
	0.7	47.65	0.9973		0.7	86.57	0.9992
	0.8	44.45	0.9973		0.8	93.46	0.9990
	Average	60.70			Average	137.68	
20%	0.2	120.70	0.9996	80%	0.2	314.27	0.9879
	0.3	106.16	0.9996		0.3	223.02	0.9957
	0.4	89.90	0.9999		0.4	139.77	0.9982
	0.5	83.54	0.9995		0.5	98.46	0.9984
	0.6	83.24	0.9999		0.6	90.10	0.9991
	0.7	85.95	0.9996		0.7	81.96	0.9993
	0.8	86.23	0.9995		0.8	74.31	0.9996
	Average	93.67			Average	145.98	
40%	0.2	216.63	0.9689	100%	0.2	362.72	0.9678
	0.3	196.05	0.9945		0.3	278.09	0.9872
	0.4	141.65	0.9866		0.4	155.55	0.9953
	0.5	92.83	0.9938		0.5	103.01	0.9912
	0.6	83.23	0.9996		0.6	87.14	0.9987
	0.7	60.30	0.9940		0.7	80.95	0.9932
	0.8	56.24	0.9944		0.8	70.30	0.9992
	Average	120.99			Average	162.54	

As shown in Table 4, the apparent activation energy has the high linear correlation coefficients R^2 in the range of 0.9689-0.9999 which guarantee the calculated results reliability, and the E_α are 133.94 kJ•mol^{-1}、122.22 kJ•mol^{-1}、97.52 kJ•mol^{-1}、85.11 kJ•mol^{-1}、85.04 kJ•mol^{-1}、78.03 kJ•mol^{-1} for the combustion process of blending coal with different additive amount of bituminous (0, 20, 40, 60, 80, 100 wt%), respectively. Combined with the cost of bituminous and the combustion reaction rates of pulverized coal with different additive amount of bituminous discussed in Section 2.1.3, the appropriate bituminous content can be maintained about 60%. It should be noted that the average activation energy increased with the content increase of bituminous. The activation energy is affected by decrease of activated molecule concentration, diffusion limitation and organic impurities during the process of combustion of samples [8,9]. As bituminous content increases, the ignition temperature decreases and heat release from anthracite oxidization increases and thus surface temperature of pulverized coal increases. Therefore, the activation energy increases with increased bituminous content.

Conclusions

(1) Comprehensive combustion index of blending coal increases with additive amount of bituminous increase, and the combustion behavior of blending coal is improved which means bituminous has a stimulative effect on blending coal combustion and is beneficial for combustion and burnout of blending coal.

(2) During combustion processes of blending coal with different additive amount of bituminous with a heating rate of 20℃/min, the TG curves change from first stage to third stages gradually along with the increase of bituminous content. Meanwhile, the maximum peak values of DTG curves of blending coal increase from first to second gradually, and the T_F of maximum mass loss approaches to a lower zone.

(3) The average values of activation energy at the bituminous content of 0, 20 %, 40 % , 60%, 80%and 100% are 62.37 kJ•mol^{-1}, 95.79 kJ•mol^{-1}, 119.91 kJ•mol^{-1}, 136.90 kJ•mol^{-1}, 146.57 kJ•mol^{-1}, 164.24 kJ•mol^{-1}, respectively. It is obvious that the activation energy is increased with the increase of bituminous content.

(4) The appropriate bituminous content is about 60% when combined with the cost of bituminous and the combustion reaction rates of blending coal with different additive amount of bituminous.

Acknowledgements

The present work was supported by National Key Technology R&D Program (No. 2011BAC01B02) and the authors would like to sincerely appreciate the fund support.

References

[1] Smoot L D et al., "Coal combustion and gasification," New York: *Plenum press*, 1985.

[2] Nie Q H et al., "Thermogravimetric Analysis on the Combustion Characteristics of Brown Coal Blends," *JOURNAL OF COMBUSTION SCIENCE AND TECHNOLOGY*, 2001, 7(1):72-76.

[3] Qi C L et al., "Characteristics of Qingxu coal applied in the 4350m3 blast furnace of Taigang," *Journal of University of Science and Technology Beijing*, 2011,33(1):80-86.

[4] Zhang J L et al., "Blended Coal Combustion Studied by Thermogravimetry," *Journal of*

Iron and Steel Research, 2009, 21(2):6-10.

[5] Zou S P et al., "Pyrolysis characteristics and kinetics of the marine microalgae Dunaliella tertiolecta using thermogravimetric analyzer," Bioresour Technol, 2010, 101(1):359-365.

[6] Boonchom B et al., "Thermodynamics and kinetics of the dehydration reaction of $FePO_4 \cdot 2H_2O$," *Phys B*, 2010, 405(9):2350-2355.

[7] SEO DONG KYUN et al., "Study of the Pyrolysis of Biomass Using Thermogravimetric Analysis (TGA) and Concentration Measurements of the Evolved Species," *Journal of Analytical and Applied Pyrolysis*, 2010, 89(1):66-73.

[8] Zhang H. Study on the Influence of Mineral Matters on the Combustion Characteristics and Kinetics of Pulverized Coals," *Journal of CUMT Mining Science and Technology*, 2009, 38(3):455-456.

[9] Chunxiang Chen et al., "Thermogravimetric Analysis of microalgae combustion under different oxygen supply concentrations," *Applied Energy*. 2011, 88:3189-3196.

Energy Technology 2014: Carbon Dioxide Management and Other Technologies
Edited by: Cong Wang, Jan de Bakker, Cynthia K. Belt, Animesh Jha, Neale R. Neelameggham,
Soobhankar Pati, Leon H. Prentice, Gabriella Tranell, and Kyle S. Brinkman
TMS (The Minerals, Metals & Materials Society), 2014

EFFECT OF BATCH INITIAL VELOCITY ON THE GLASS FURNACE EFFICIENCY

Nasim Soleimanian[1], Mark Jolly[2], Karl Dearn[3], Oliver Brinkman[4], and William Brinkman[5]

[1,2] Cranfield University, School of Applied Science, Cranfield, Bedfordshire, MK43 0AL, United Kingdom
[3] University of Birmingham, School of Mechanical Engineering, Edgbaston, Birmingham, B15 2TT, United Kingdom
[4,5] Glassworks Hounsell limited, Park Lane, Halesowen, West Midlands B63 2QS, United Kingdom.

Key Words: CFD, energy efficiency, glass furnace, oscillating batch chargers, batch initial velocity.

Abstract

Glass manufacturing is a heat intensive process. There is a direct correlation between the batch distribution techniques and the furnace energy consumption, productivity, and quality of the glass manufactured. All four major segments (float, container, fibre, and specialty glasses) would benefit from using an optimised batch distribution technique where possible. Oscillating batch chargers (OBC) have been in use since the early 70s, despite their superior batch shape, coverage, and in turn positive effects on the energy consumption and productivity of the furnace they are almost exclusively used in container glass manufacturing. The OBC's main difference compared with other charging methods is its ability to directly influence the batch initial velocity. This paper reports on results achieved in CFD models (GFM) used to study effect of the machine on the overall energy consumption in the doghouse and the melt space.

Introduction

Glass manufacturing can be divided to four major segments float glass, container glass, fiberglass, and specialty glasses. The financial cost of manufacture, its impact on the environment, lack of standardisation and mounting pressure imposed by the authorities for reduction in emissions means that the industry and its suppliers are utilising CFD modelling in order to tackle these challenges.

The need for high temperature to create and maintain the viscous flow, homogenisation of the glass melt, and maintenance of the process to allow continuous output of glass of the right quality are the contributors to the energy intensive activities that are involved in continuous glass manufacturing. This in turn has resulted in design of large container glass furnaces which have high energy consumption, low specific performance, and high CO_2 emissions where the most efficient furnaces in the container glass sector have a specific primary energy consumption of 3.8 GJ /tonne of glass at a level of 50% cullet in the batch making even very small improvements in efficiency by operation optimisation a value adding activity [1-14].

The theoretical calculated energy required to melt glass is 2 to 3 times less than the energy actually used to melt glass as a result of glass melt passing a zone 5 to 8 times before leaving the tank.[8]. This causes contamination as the result of mixing between the completely molten glass free from bubbles returning to the batch area where there is freshly molten non-homogeneous glass, still containing very large numbers of bubbles and un-molten batch. At high temperature some components are dissolved rather than melted. Given that an un-molten batch is a good thermal insulator, it can be transported a long way into the furnace, increasing the need for homogenization which also increases the glass residence time of glass melt. This is why large unbroken strings of batch should be avoided, as should an overall blanket without gaps.

Batch distribution technique is one area where each segment tends to have its preference due to technical and historical reasons [6, 15]. This paper focuses on oscillating batch chargers (OBC) which have been in use since the early 70s. Despite their proven superior batch shape, coverage, and in turn positive effects on the energy consumption and productivity of the furnace, they are still almost exclusively used in container glass manufacturing [15]. The introduction of an emerging new generation of OBC (with better seal in the doghouse and an improved control system which enables even more control over the batch velocity in the melt space.) has created an opportunity to develop a CFD simulation to improve and identify possible operating optimisation windows.

OBC's Impact on the Glass Melting Process

Glass Melting process
Glass melting is a multi-phase process. The furnace uses the intense heat to melt the batch into primary glass, where there is a simultaneous flow of materials in different states, i.e. solid, liquid, and gas. Glass melt being the primary phase, batch particles and gases are the secondary phases dispersed within the continuous phase. The batch (raw material) make up consists of collection of particles of various sizes and cullet pieces.

There are several chemical reactions involved in glass making, where reactive dissolution of sand grains in the primary phase is directly related to the heat flux within the batch pile and sand grain size distribution. The melting kinematics of a batch pile is determined by formation of eutectic melts. When the batch pile is exposed to high temperature (at the edges) its viscosity increases, increasing the viscous flow in the primary phase and begins the endothermic processes in the batch to bring it to the reaction temperature [6-8]. In the melt zone of a glass furnace the heat transferred mainly from radiation (achieved by combustion of fuel and air/oxygen) from the combustion space to the glass-melt tank is used to melt the glass batch and heat the liquid glass already in the furnace [6].

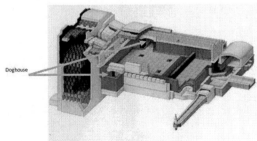

Figure 1. is a end-fired furnace with tow side doghouses.

The shape and pattern of the batch, its initial velocity, and temperature (depending on the presence of a doghouse (see figure 1) when it enters the melt space is influenced by the type of batch chargers in the process [15].

OBCs and Their Effect on the Energy Consumption in the Melt Space

OBC's fall under charging systems that can influence the pattern, shape, and initial velocity of the batch. They are positioned on top of the doghouse and they create batch piles which are then pushed in different directions as the equipment swivels on the surface of the glass melt in the doghouse (see figure 2). In such cases the charging flexibility offered by the pusher design currently cannot be bettered, and today this design of charger is the type most commonly used on such furnaces [16]. The calculated energy required to melt glass is 2 to 3 times less than the energy actually used to melt glass [8]. In most cases the glass melt passes a zone 5 to 8 times before leaving the tank. Most of the completely molten glass free from bubbles from the hot spot area returns to the batch area and is mixed with freshly molten non-homogeneous glass, still containing very large numbers of bubbles and un-molten batch. At high temperature some components are dissolved rather than melted. The un-molten batch is a good thermal insulator making heat transfer within a batch pile difficult. As a result, un-molten batch can be transported a long way into the furnace, increasing the need for homogenization which also increases the glass residence time. This is why large unbroken strings of batch should be avoided, as should an overall blanket without gaps.

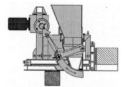

Figure 2. Shows a side profile an OBC [16].

Soleimanian and Jolly in [15] showed that high space utilisation is essential for reducing energy consumption and increasing the melting performance of a furnace, by optimising and manipulating the ratios between the transversal and longitudinal temperature gradients in the melt space using batch piles. They compare batch patterns produced by different charging methods and show use of OBC maximises the melt space utilisation value for dissolution by improving the spiral critical trajectories close to the glass surface.

Polák, and Němec in [17] state for every given process intensity and melting space there is one maximal value of the space utilisation. Soleimanian and Jolly in [15] show that modeling of the batch pattern is also essential in order to identify maximum space utilisation. Their optimal circulation with spiral flow at the lowest theoretical ratio for the average residence time of glass in the melting space to the fraction of utilised space (α) in their specific 3D melt space was achieved with feeding rate of 0.5 kg/s, with use of OBC chargers.

$$\alpha = \frac{\bar{\tau}}{(1-m)} = \frac{[H_M^0 - (H_M^T + C^G(T - T^e))]}{\frac{A\,H_A^L}{\rho V}} \qquad (1)$$

Where H_M^0 is the specific energy consumption, H_M^T is the theoretical heat needed for glass phase transition, chemical reaction and heating to T^e (the exit temperature), C^G is the average heat capacity of glass, T is the melting temperature, H_A^L is the specific average heat flux above the glass which has a total surface area of A, $\bar{\tau}$ is the average residence time of glass in the melting space, ρ is the glass density, V is the volume of the melt space, and m is the fraction of dead space.

The Batch Initial Velocity Sensitivity Study

Furnace Model and Batch Composition
Four different batch velocity combinations were simulated (see table 1). To resemble the OBC batch pattern and its signature random sized batch piles, six inlets with different distances between them, shifted to one side was added to the melt space (see figure 3).

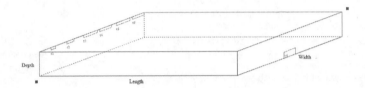

Figure 3. Shows the melt space and the location of 6 batch inlets.

Each inlet dispatches at a different rate (the width of the inlets are corresponding with the rate of dispatch) in order to simulate an oscillating batch charger feeding from the side wall of the furnace.

Table 1. shows the different arrangement batch initial speed for each setup.

	V_{c1} (ms⁻¹)	V_{c2} (ms⁻¹)	V_{c3} (ms⁻¹)	V_{c4} (ms⁻¹)	V_{c5} (ms⁻¹)	V_{c6} (ms⁻¹)
Base line	0.003	0.003	0.003	0.003	0.003	0.003
Case 1	0.005	0.004	0.003	0.003	0.003	0.003
Case 2	0.003	0.003	0.003	0.003	0.002	0.001
Case 3	0.005	0.004	0.003	0.003	0.002	0.001

The furnace that was modelled was an end-fired, regenerative, container glass furnace. The glass tank is 8m long, 4.5m wide and 1m deep (no bubblers or boosting). The combustion space has an arc shape crown with the minimum and maximum heights of 1.2m and 1.45m respectively. There are 6 burners in total, and 2 air ports, 1 air port for each set of 3 burners. The direction of flame is changed every 20 min. Soda-lime silicate glass of composition SiO_2 (74%) -Na_2O (16%) -CaO (10%) (mol) with 50% cullet was selected as the batch. The batch and glass thermo physical data were from [10, 18]. All the variables were kept constant excluding batch input rate for each batch pattern. The basic features of the mathematical model and the numerical method employed to compute the flow and temperature in the glass melt and the combustion space are detailed in [7]. The relevant criteria were studied: 1) The H_M^0, 2) Glass exist temperature, and furnace output rate.

Results
Changing the batch introduction rate as well as adjusting the initial speed has an optimising effect on the temperature gradient across the melt space. By using batch to enhance the stirring in a melt tank it is possible to achieve high pull and relatively low

specific energy consumption, where high temperatures may not be the optimum solution (See figures 4, 5, and 6) [13, 15, 17].

Table 2. The total percentage reduction in H_M^0 for cases, 1, 2, and 3.

Batch rate in (kg/s)	Base line (MJ/s)	Case 1 (%)	Case 2 (%)	Case 3 (%)
0.50	1.1030	0.1238	-0.2515	-0.1798
0.75	1.6774	0.0828	-0.0598	0.0120
1.00	2.2159	0.0627	-0.0299	0.0136
1.50	3.0223	0.0182	-0.0138	0.0044
1.75	3.7819	0.0211	-0.0090	-0.0011
2.00	4.2634	0.0187	-0.0062	-0.0003
2.50	5.2561	0.0000	-0.0033	-0.0008
2.75	5.8517	0.0147	0.0114	-0.0014
3.00	6.3356	0.0124	-0.0025	-0.0017

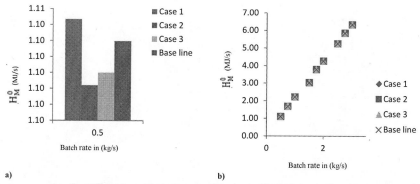

a)   b)

Figure 4. Shows H_M^0 (MJ/s) at total of 0.5 kg/s batch input and b) for different batch inlet (0.5 – 3 kg/s).

The results predicted by show (case 1) has the highest H_M^0 and T^e in the group (see figure 4 and 5: a and b). This is as a result of different velocities along the melt trajectories. The batch pattern produces a broader spectrum of melt residence times in the melt tank, compared to the other cases and temperature differences along the trajectories cause differences of the melting rate. Melting performance is restrict as the result of critical pathways through the melting space increasing the energy demands (see table 2).

As the batch input increases T^e is kept well between the 1000°C-1200°C temperature required for optimal viscosity for cutting the glass to gobs (solid cylinders of glass). By reducing the glass temperature in preparation for the forming process energy loss through water-cooling the feeding channels is also reduced(See figure 5) [6, 15]. These effects are improved by better managing the gaps, and batch velocity in case 2.

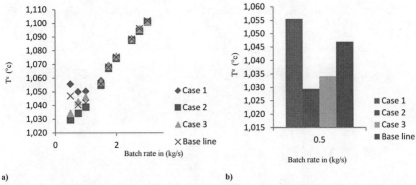

a)   b)

Figure 5. Shows Mean T^e (°c) a) for different batch inlet (0.5 – 3 kg/s) and b) at total of 0.5 kg/s batch input.

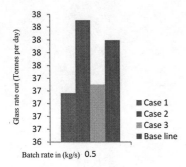

Figure 6. Shows glass rate out (Tonnes per day) for different batch inlet arrangements for 0.5 kg/s input.

The result is showing by controlling the patch coverage on the glass melt surface, the introduction of new control system certainly has a positive impact on the overall energy consumption and performance of the furnace.

Further work is needed to establish the total energy savings incurred by use of new generation of OBCs vs in energy losses due to presence of doghouse, in large, high capacity end-fired furnaces, in particular those with high specific melting rates, are particularly dependent on the batch charging technology studied above.

Acknowledgment

The completion of the above project could not have been accomplished without the support of GLASSWORKS HOUNSELL, and the Funding provided by TSB as part of 8637 KTP project.

References

1.	Abbassi, A. and K. Khoshmanesh, *Numerical simulation and experimental analysis of an industrial glass melting furnace.* Applied Thermal Engineering, 2008. **28**(5-6): p. 450-459.
2.	Beerkens, R., *Analysis of elementary process steps in industrial glass melting tanks-some ideas on innovations in industrial Glass melting* Ceramics – Silikáty 2008. **52**(4): p. 206-217
3.	Beerkens, R.G.C. and J. van der Schaaf, *Gas Release and Foam Formation During Melting and Fining of Glass.* Journal of the American Ceramic Society, 2006. **89**(1): p. 24-35.
4.	Beerkens, R.G.C. and J. van Limpt, *Energy Efficiency Benchmarking of Glass Furnaces,* in *62nd Conference on Glass Problems: Ceramic Engineering and Science Proceedings*2008, John Wiley & Sons, Inc. p. 93-105.
5.	Boyd, G.A. and J.X. Pang, *Estimating the linkage between energy efficiency and productivity.* Energy Policy, 2000. **28**(5): p. 289-296.
6.	C. Philip Ross , G.L.T., *Glass Melting Technology: A Technical and Economic Assessment* M. Rasmussen, Editor 2004, U.S. Department of Energy. p. 292.
7.	Chang, S.L., C.Q. Zhou, and B. Golchert, *Eulerian approach for multiphase flow simulation in a glass melter.* Applied Thermal Engineering, 2005. **25**(17-18): p. 3083-3103.
8.	Charles H. Drummond, I., *72nd Conference on Glass Problems: Ceramic Engineering and Science Proceedings*2012: Wiley.
9.	Choudhary, M.K., *Recent Advances in Mathematical Modeling of Flow and Heat Transfer Phenomena in Glass Furnaces.* Journal of the American Ceramic Society, 2002. **85**(5): p. 1030-1036.
10.	Fluegel, A., *Glass viscosity calculation based on a global statistical modelling approach.* Glass Technology - European Journal of Glass Science and Technology Part A, 2007. **48**(1): p. 13-30.
11.	Jiang, S., *Advances in Glass and Optical Materials: Proceedings of the 107th Annual Meeting of The American Ceramic Society, Baltimore, Maryland, USA 2005, Ceramic Transactions*2012: Wiley.
12.	Levitin, L., et al., *Raw materials with prescribed properties — additional resources for increasing glass-furnace efficiency and float-glass quality.* Glass and Ceramics, 2012. **69**(1): p. 3-8.
13.	mec, L., Jebav, and Marcela, *Analysis of the performance of glass melting processes as a basis for advanced glass production.* Glass Technology - European Journal of Glass Science and Technology Part A, 2006. **47**(3): p. 68-77.
14.	Paramonova, O., et al., *Study of glass batch components segregation.* Glass and Ceramics, 2012. **68**(9): p. 319-322.
15.	Soleimanian, N. and M. Jolly, *Effect of Batch Charging Equipment on Glass Furnace Efficiency,* in *Energy Technology 2013*2013, John Wiley & Sons, Inc. p. 77-84.
16.	Sims, R. *Batch charging technologies - a review.* [Online] 2007 [cited 2007.
17.	Polák, M. and L. Němec, *Glass melting and its innovation potentials: The combination of transversal and longitudinal circulations and its influence on space utilisation.* Journal of Non-Crystalline Solids, 2011. **357**(16–17): p. 3108-3116.
18.	Seward, T.P. and T. Vascott, *High temperature glass melt property database for process modeling*2005: American Ceramic Society.

213

Energy Technology 2014: Carbon Dioxide Management and Other Technologies
Edited by: Cong Wang, Jan de Bakker, Cynthia K. Belt, Animesh Jha, Neale R. Neelameggham,
Soobhankar Pati, Leon H. Prentice, Gabriella Tranell, and Kyle S. Brinkman
TMS (The Minerals, Metals & Materials Society), 2014

KINETIC MODELING STUDY OF OXY-METHANE COMBUSTION AT ORDINARY PRESSURE

Xianzhong Hu, Qingbo Yu, Qin Qin

School of Materials and Metallurgy, Northeastern University, Shenyang, Liaoning, 110819, P.R. China

Keywords: Oxy-fuel combustion, Laminar flame speed, Reaction path, Kinetic simulation

Abstract

Oxy-fuel combustion is one promising option among different CO_2 capture technologies. The object of this work is to show the effect of CO_2 dilution on the laminar flame speed, flame structure and reaction pathway for methane oxidation at O_2/CO_2 atmosphere. In this paper, kinetic simulations were made using CHEMKIN with GRI-Mech 3.0 mechanism, which was validated by experimental data of laminar flame speeds. The reaction pathway of $CH_4/O_2/CO_2$ mixture was derived from reaction flow analysis. Results showed that CO_2 decreased the flame speeds, lowed concentration of major intermediates and increased flame thickness. The major chemical pathways of CH_4 to CO_2 have not changed at O_2/CO_2 atmosphere comparing with O_2/N_2 atmosphere, but contributions to major intermediates formation from each related elementary reaction have changed a lot. In addition, the secondary chemical pathways of CH_3 oxidation became different at O_2/CO_2 atmosphere.

Introduction

Nowadays, as the global economy is developing fast and the pollutant levels (especially global warming gases) are increasing, a lot of researches are focusing on new technologies which can cut greenhouse gas emissions [1]. New technologies have been developed, such as high Mild combustion, flameless combustion and Integrated Gasification Combined Cycle (IGCC) technology. Oxy-fuel combustion using carbon dioxide as dilution of oxygen is considered as a promising approach for reducing CO_2 emission. The properties of CO_2 in flue gas are different from N_2 in air. On one side, the transport, thermal and radiative properties of CO_2 are different from N_2 in air. On the other side, CO_2 is not inert but directly participates in chemical reactions and recent researches [2-5] also present the chemical effects of CO_2 on oxy-combustion. For these differences above, the characteristics of oxy-combustion are not alike with traditional combustion. It is necessary to develop the investigation of oxy-combustion.

In all combustion processes, laminar flame speed is an important parameter, which links to combustion behavior. Analysis of this parameter can provide a better understanding of the effect of CO_2 dilution in the process of oxy-combustion. Several investigations of laminar flame speeds of oxy-fuel flame have been improved, including experimental measurements and numerical computations. A.A.Konnov et al. measured the flame speeds of oxy-methane and oxy-ethane mixture using the heat flux method with a flat burner [6-8]. V. Ratna Kishore et al. used heat flux method to measure the adiabatic laminar flame speeds of hydrogen [9] and H_2/CO mixture [10]. Adiabatic laminar flame speeds of hydrogen in O_2/CO_2 atmosphere are much smaller than in O_2/N_2 or in O_2/Ar. The burning velocity of H_2/CO mixture decreased with increase in CO_2 content in mixture. Jeongseog Oh et al. [11] used Bunsen flame to measure the laminar flame speeds of premixed oxy-methane mixture with angle methods in atmospheric condition (300k, 1atm). The results were S_L=29.1cm/s and S_L=29.5cm/s respectively derived from oxy-methane from CH* images and Schlieren photos.

In those papers, kinetic simulations were also performed to analyze impact factors of laminar flame speeds of oxy-fuel mixture. However, most of researches focused on the observation, measurements of flame speeds of oxy-fuel flame and impact factors analysis, seldom to investigate the impacting mechanism of CO_2 dilution deeply. The object of this study is therefore to investigate the CO_2 dilution effect on flame speeds, flame structure and reaction path.

Simulation Methods

Numerical simulations were performed using CHEMKIN packages and flame models were adopted as the freely-propagating, premixed, one-dimensional unstretched flame based on PREMIX code [12] to obtain the laminar burning velocity and some parameters to analyze the flame structure. The reaction mechanism GRI Mech 3.0 is used, which is very popular and used widely for methane flame in literatures[13].This mechanism was developed and optimized as 53 species and 325 reactions.

For simulation, the initial temperature of gas mixture was from 300K to 800K, and the pressure was fixed at 1 atm. The computations used multispecies and thermal diffusion along with GRAD 0.05-0.08 and CURV 0.01-0.02 leading to grids consisting of about 200 points on average, which showed fine enough to make the simulation grid-independent. In this calculation, the solution domain length is 2cm. It is long enough because most of combustion reactions have finished among the distance of 1cm.

The present calculation results of oxy-methane flame speeds were validated here. Fig.1 shows the comparison between the computed laminar flame speeds (S_L) by using mechanism GRI Mech 3.0 and experimental results from A.A.Konnov's measurement [7]. In literature [7], the flame speeds of oxy-methane mixture were measured using the heat flux method with a flat burner at 300K, 1atm. In Fig1, good agreement is found over a wide range of equivalence ratios (φ), in spite of a slight underestimation in numerical results. The results show that mechanism GRI Mech 3.0 can well produce the laminar burning velocity of $CH_4/CO_2/O_2$ flames.

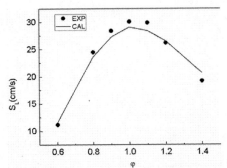

Fig. 1 Comparison of the predicted variation of laminar speed speeds S_L with equivalence ratio φ with experimental measurements

Results and discussion

CO₂ dilution effect on flame speed

In present work, the carbon dioxide concentration in the combustion-supporting gas (the mixtures of CO_2 and O_2)

$$X_{CO_2} = \frac{CO_2}{CO_2 + O_2} \tag{1}$$

is varied from 65% to 75%. The corresponding oxygen concentration

$$X_{O_2} = \frac{O_2}{CO_2 + O_2} = 1 - X_{CO_2} \tag{2}$$

is varied from 35% to 25%. The pressure of mixture is fixed at 1 atm and the initial temperature varied from 300K to 1200K.

Fig. 2(a) shows that the laminar speed speeds S_L varied with equivalence ratio φ in different CO_2 concentrations at 300K. Obviously, it is found that the flame speed decreased as the CO_2 concentrations increased through comparison of the three parallel curves. The same change law can also be found at 800K from Fig. 2(b). As carbon dioxide concentration increased, the oxygen concentration decreased, which meant flue concentration decreased at fixed equation rate. This reduced the concentration of reactant, lowered the flame temperature, which decreased the flame speed. It should be noted that there was an approximately quadratic function relationship between the flame speed S_L and oxygen concentration (expression 1- X_{CO_2}) through present calculation. Fig. 2 also showed that it was certainly not a linear relationship between flame speed S_L and X_{CO_2} expression.

In addition, the flame speeds of mixture $CH_4/CO_2/O_2$ increased with the initial temperature raised and the change law of which is similar to the mixture $CH_4/N_2/O_2$. Initial temperature increase led to the increase of combustion temperature, which would raise the laminar flame speeds. Through analysis of our calculation, the laminar flame speeds of mixture $CH_4/CO_2/O_2$ increased with temperature evidently as quadratic function, which was similar to the empirical correlation of Andrews and Bradley [14] for stoichiometric methane-air flames,

$$S_L = 10 + 3.71 \cdot 10^{-4} [T_u(K)]^2 \tag{3}$$

Fig. 3 shows the variation of laminar flame speeds of mixture $CH_4/CO_2/O_2$ and $CH_4/N_2/O_2$ with initial temperature. Through comparison, the laminar flame speeds of mixture $CH_4/N_2/O_2$ were always higher than mixture $CH_4/CO_2/O_2$ at the same initial temperature. Obviously, the difference between the two become wider as the temperature increased. In other words, the laminar flame speeds of mixture $CH_4/N_2/O_2$ increased faster than mixture $CH_4/CO_2/O_2$. This was due to the greater inhibitory effect of carbon dioxide on combustion than nitrogen.

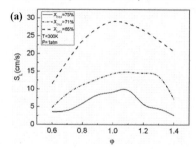 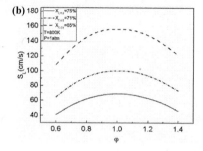

Fig. 2 Variation of laminar speeds S_L with equivalence ratio φ with different carbon dioxide concentration (65%, 71%, 75%) at different initial temperature 300K(a),800K(b).

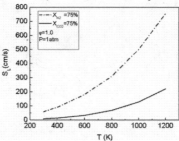

Fig. 3 Variation of laminar speed speeds S_L with different initial temperature T.

CO_2 dilution effect on flame structure

The detailed structure of the oxy-methane flame was also presented to analysis the impacting mechanism of CO_2 dilution. Fig.4(a)-(b) shows the mole-fraction profiles of reactants (CH_4 and O_2) and products (CO and H_2O) in $CH_4/O_2/CO_2$ mixture and $CH_4/O_2/N_2$ mixture respectively at X_{O_2}=31%, φ=1. In $CH_4/O_2/CO_2$ mixture, CH_4 concentration goes to zero at 0.11cm, which locates behind 0.08cm in $CH_4/O_2/N_2$ mixture. The location of approach to equilibrium of O_2, H_2O and CO in $CH_4/O_2/CO_2$ mixture is also behind the $CH_4/O_2/N_2$ mixture. In other words, the reaction zone increased and the species concentration gradients dropped in $CH_4/O_2/CO_2$ mixture compared to $CH_4/O_2/N_2$ mixture. It's worth pointing out that the species equilibrium concentration of products in different mixtures were very close. Thus, CO_2 chiefly slow the rate of chemical reactions, slightly or not change the chemical equilibrium. Calculations of flame thickness were also carried out with equation:

$$\delta = \frac{(T_{max} - T_0)}{(dT/dx)_{max}} \tag{4}$$

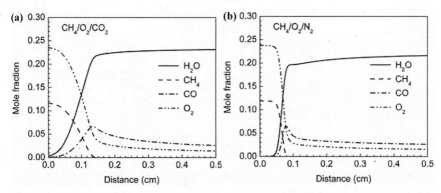

Fig. 4 Profile of H_2O, CH_4, CO, O_2 mole fractions in one-dimensional (a) $CH_4/O_2/CO_2$ and (b) $CH_4/O_2/N_2$ flames, φ=1.0, X_{CO_2} =79%

Where, T_{max} is the highest temperature of flame, T_0 is the temperature of unburned mixture and $(dT/dx)_{max}$ is maximum temperature gradient. It is estimated that flame thickness increase to 0.062cm in $CH_4/O_2/CO_2$ mixture (X_{O_2}=31%, X_{CO_2}=69%,) compared to 0.024cm in $CH_4/O_2/N_2$ mixture (X_{O_2}=31%, X_{N_2}=69%). Fig.5(a)-(b) show the mole-fraction profiles of H, O and OH radicals in $CH_4/O_2/CO_2$ mixture and $CH_4/O_2/N_2$ mixture respectively at X_{O_2}=31%, φ=1. The maximum of O and OH in mixture $CH_4/O_2/CO_2$ is about one third of that in $CH_4/O_2/N_2$ mixture and the maximum of H is near one fifth of that in mixture $CH_4/O_2/N_2$ by comparing with different dilutions. This also indicated that the radicals were restrained seriously by CO_2, hence the flame speed of oxy-methane decreased considerably.

The inhibitory effect of CO_2 on chemical reaction of combustion is due to the high special heat of CO_2, which decrease the adiabatic temperature of flame. Besides, the other reason is the low rate of diffusion of CO_2, which reduces the diffusion velocity of high-energy radicals and heat transferring from reaction zone to unburned zone.

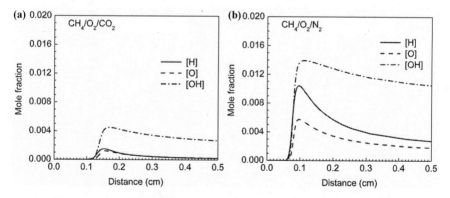

Fig.5 Profile of H, O, OH mole fractions in one-dimensional (a) $CH_4/O_2/CO_2$ and (b) $CH_4/O_2/N_2$ flames, φ=1.0, X_{CO_2}=79%

CO$_2$ dilution effect on reaction pathway

The reaction pathway of mixture $CH_4/O_2/CO_2$ and $CH_4/O_2/N_2$ were analyzed in order to investigate the CO_2 dilution effect on chemical reaction process. In present work, rate of production analysis and sensitivity analysis were used to perform the reaction pathway of the gas mixture based the mechanism of GRI 3.0. Fig. 5 showed the reaction pathway comparison of mixture $CH_4/O_2/CO_2$ and $CH_4/O_2/N_2$ at normal temperature (T=300K).Each arrow represents an elementary reaction, with the primary reactants species at the tail and primary products at the head. Additional reactants species are shown along the length of the arrow, and the corresponding reaction numbers from the mechanism of GRI 3.0 were indicated. The width of the arrow gave a visual indication of the relative importance of a particular reaction path, while the parenthetical numerical values quantify the contribution ratio of the reaction.

As we known, the combustion process of methane could be regard as an oxidation reaction process. Comparing Fig. 6(a) and (b), the major oxidation pathway of mixture $CH_4/O_2/CO_2$ has not changed and the major oxidation reaction pathway still was divide five steps: $CH_4 \rightarrow CH_3 \rightarrow CH_2O \rightarrow HCO \rightarrow CO \rightarrow CO_2$. But contributions to major intermediates formation from each related elementary reaction have changed a lot, which was the reason of CO_2 dilution decreased the flame speeds. There were two changes on the secondary chemical pathways. First, the secondary chemical pathways of CH_3 oxidation became different at O_2/CO_2 atmosphere. The reactions for producing CO from CH_3 brought more effect on CO production, which could not be neglected. Secondly, the major pathway of destruction of $CH_2(s)$ has changed into: $CH_2(S) \rightarrow CO$. Through production analysis, the two changes above decreased the overall reaction rate of CH_4 oxidation, which were due to the effect of high concentration of CO_2 on chemical reaction. In other words, all these changes of reaction pathway decreased the laminar flame speed.

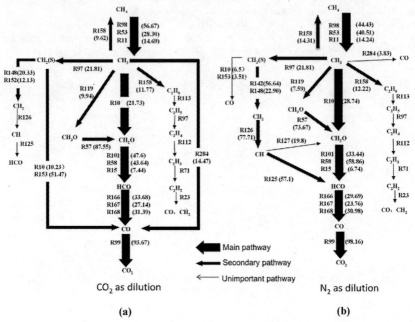

Fig. 6 Reaction pathway diagram for combustion of methane with CO_2 as dilution (a) and N_2 as dilution (b) at T=300K, p=1atm

Conclusions

In this paper, the effect of CO_2 dilution effects on laminar flame speed, flame structure and chemical reaction pathway were studied based on kinetic simulation. The main results are summarized as follows:

(1) High concentration CO_2 in $CH_4/O_2/CO_2$ decreased the flame speeds comparing with $CH_4/O_2/N_2$ mixture.

(2) Compared to N_2 dilution, high concentration CO_2 lowed concentration of major intermediates and increased flame thickness for oxy-methane flame.

(3)The major reaction pathways of methane oxidation in O_2/CO_2 atmosphere at O_2/CO_2 atmosphere have not changed, but the elementary reactions have changed a lot. The secondary reaction pathway including CH3 and $CH_2(s)$ oxidation altered new way.

References

1. Wall, Terry F. "Combustion processes for carbon capture," Proceeding of the combustion institute, 31(2007), 31-47.
2. C. Zhang, A. Atreya, K. Lee. "Sooting structure of methane counterflow diffusion flames with preheated reactants and dilution by products of combustion," (Paper present at the Twenty-Fourth Symposium on Combustion, The Combustion Institute, Pittsburgh, 1992), 1049-1051.
3. A.R. Masri, R.W. Dibble, R.S. Barlow. "Chemical kinetic effects in nonpremixed flames of H_2/CO_2 fuel," Combustion and Flame, 91 (1992), 285-309.
4. F.S Liu, et al. "The chemical effects of carbon dioxide as an additive in an ethylene diffusion flame: implications for soot and NOx formation," Combustion and Flame 125 (2001), 778–87.
5. F.S Liu, et al. "The chemical effect of CO_2 replacement of N_2 in air on the burning velocity of CH_4 and H_2 premixed flames," Combustion and Flame, 133 (2003), 495-497.
6. A.A.Konnov, I.V.Dyakov. "Meaurement of propagation speeds in adiabatic flat and cellular premixed flames of $C_2H_6+O_2+CO_2$," Combustion and Flame, 136 (2004), 371-376.
7. A.A.Konnov, Igor V.Dyakov. Measurement of propagation speeds in adiabatic cellular premixed flames of $CH_4+O_2+CO_2$. Experimental Thermal and Fluid Science, 29 (2005), 901-907.
8. A.A.Konnov, I.V.Dyakov. "Measurement of adiabatic burning velocity and sampling in methane-oxygen-carbon dioxide mixtures," Archivum Combustionis 22 (2002), 13-24.
9. V.Ratna Kishore, Ringkhang Muchahary. "Adiabatic burning velocity of H_2-O_2 mixtures diluted with $CO_2/N_2/Ar$," International Journal of Hydrogen Energy, 34 (2009), 8378-8388.
10. V. Ratna Kishore, M.R. Ravi, Anjan Ray. "Adiabatic burning velocity and cellular flame characteristics of H_2–CO–CO_2–air mixtures," Combustion and Flame, 158(2011), 2149-2164.
11. Jeongseog Oh, Dongsoon Noh. "Laminar burning velocity of oxy-methane flames in atmospheric condition," Energy, 45 (2012), 669-675.
12. Kee RJ, Grcar JF, Smooke MD, Miller JA. A Fortran program for modeling steady laminar one-dimensional premixed flames.(Report SAND 85-8240. Livermore, CA: Sandia National Laboratories; 1985)
13. Smith GP, Golden DM, Frenklach M, Moriarty NW, Eiteneer B, Goldenberg M, et al. GRI Mech 3.0, Http://www.me.berkeley.edu/gri_mech/; 2006.
14. Stephen R. Turns, *An introduction to combustion: concepts and applications* (New York, NY: USA McGraw-Hill, 2002), 275.

Energy Technology 2014
Carbon Dioxide Management and Other Technologies

SYMPOSIUM: ENERGY TECHNOLOGIES AND
CARBON DIOXIDE MANAGEMENT

Poster Session

Energy Technology 2014: Carbon Dioxide Management and Other Technologies
Edited by: Cong Wang, Jan de Bakker, Cynthia K. Belt, Animesh Jha, Neale R. Neelameggham,
Soobhankar Pati, Leon H. Prentice, Gabriella Tranell, and Kyle S. Brinkman
TMS (The Minerals, Metals & Materials Society), 2014

CORROSION BEHAVIOR OF DIFFERENTLY HEAT TREATED STEELS IN CCS ENVIRONMENT WITH SUPERCRITICAL CO_2

Anja Pfennig[1], Phillip Zastrow[1], Axel Kranzmann[2]

[1] HTW University of Applied Sciences Berlin,
Wilhelminenhofstraße 75 A, Gebäude C, 12459 Berlin
anja.pfennig@htw-berlin.de

[2] BAM Federal Institute of Materials Research and Testing,
Unter den Eichen 87, 12205 Berlin

Keywords: steel, heat treatment, corrosion, CCS, supercritical CO_2, CO_2-storage

Abstract

Properties of pipe steels for CCS technology require resistance against the geothermal corrosive environment (heat, pressure, salinity of the aquifer, CO2-partial pressure). To predict the reliability of two different injection pipe steels with 13% chromium and 0.46% carbon (1.4034, X46Cr13) as well as 0.20% carbon (1.4021, X20Cr13) steel coupons were heat treated by comparable routines prior to long term corrosion experiment (60°C, 100 bar, artificial brine close to a CCS-site in the Northern German Basin, Germany). In general higher carbon content exhibits better corrosion resistance with respect to corrosion rate, amount of pits and maximum intrusion depth in the case of local corrosion. At 100 bar hardened steels with martensitic microstructure have the lowest corrosion rates with an average of 0.003 mm/a after 2000 h exposure time. At ambient pressure normalized steels show the lowest corrosion rates, indicating that the corrosion is driven by grain boundaries.

Introduction

Engineering a geological on-shore saline aquifer CCS-site (CCS Carbon Capture and Storage [1-3]) corrosion of injection pipe steels may become an issue when emission gasses, e.g. from combustion processes of power plants, are compressed into deep geological layers [4-8]. Typically 42CrMo4 (1.7225, AISI 4140) is used for casing, and for injection pipe the steel X46Cr13 (1.4034, AISI 420 C) or X20Cr13 (1.4021, AISI 420 J) is used. Saline formations are the most favorable storage sites in Germany [9], because of their large potential storage volume and their common occurrence. Leakage of CO_2 back into the atmosphere may be a problem in saline aquifer storage, especially around the water level within the borehole. Carbon dioxide corrosion may cause failure of pipelines and wells as known from the oil and gas industry [4,6,10,11, 12].

From thermal energy production it is known, that the CO_2-corrosion is sensitively dependent on alloy composition, contamination of alloy and media, environmental conditions like temperature, CO_2 partial pressure, flow conditions and protective corrosion scales [6-8,13-20]. Considering different environments, aquifer waters and pressures, the analyzed temperature regime between 40 °C to 60 °C is a critical temperature region well known for corrosion processes as shown by Pfennig et al. [21-23] and various other authors [10,24-28] (figure 1). Here a maximum around 4.7 mm/year was found for the pit intrusion depth of 13% Cr steel X46Cr13. Still this may be predicted by the rather conservative Norsok-Model used in the oil and gas industry to calculate surface corrosion rates of carbon steels [29].

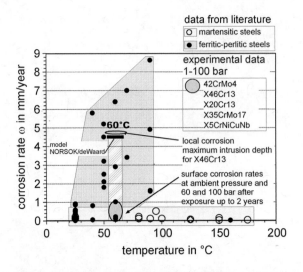

Figure 1. Comparison of measured uniform corrosions rates of 42CrMo-4 and X46Cr13 (also intrusion depth rates for shallow pit and pitting corrosion) with data obtained from literature [21].

Generally steels, that are exposed to CO_2-environment, precipitate slow growing surface layers mainly comprised of $FeCO_3$ (siderite) and ferrous hydroxide [4,8,30]. After the CO_2 is dissolved to build a corrosive environment the solubility of $FeCO_3$ in water is low (p_{Ksp} = 10.54 at 25 °C [28,31]. As a result of the anodic iron dissolution a siderite corrosion layer grows on the alloy surface which is also found in pits of locally corroded samples [7,21,23] and a precipitation model has been introduced by Han et. al [30]. The initiation of pit formation is located at the grain boundaries which act as corrosion catalyzed due to their rather high carbide concentration compared to the base metal.

The influence of heat treatment, that is: temperature and time of austenitisizing, cooling rate as well as temperature and time of annealing, has been shown by various authors. The presence and amount of retained austenite as a microstructural component resulting from the heat treatments applied has a beneficial effect on the pitting corrosion resistance of 13%-chromium steels (13CrNiMo) [32]. A higher Ni and Cr content in the heat treated steels improve the corrosion resistance [32, 33]. Hou et al. introduce a method of empirically calculating the influence of alloying elements in heat treated steels [34]. Cvijović and G. Radenković showed that the corrosion resistance of duplex steels with chromium contents even as high as 22-27% varied with solidification mode and annealing condition [35]. In general raising the annealing temperature lowers the pitting potential of lean duplex stainless steels [35-37]. The lowest potentials, corresponding to the transition from metastable to stable pitting, are observed for annealing at 900 °C while a maximum improvement of corrosion stability can be achieved by annealing at 1200 °C [35]. The better corrosion resistance of martensitic stainless steels with 13% Cr at higher austenitizing temperature (980-1050 °C) is related to the dissolution of carbides [37-39]. The precipitation of Cr-rich $M_{23}C_6$ and M_7C_3 carbides reduced the resistance of passive film and pitting corrosion [37] and has high impact on mechanical properties due to secondary hardening [38]. The influence of heat treatment on the microstructure and mechanical properties is well known [37,40,41]. However for C-Mn (carbon) steels in a H2S-containing NaCl solution the martensitic microstructure has the highest corrosion rate up to two orders of magnitude higher than ferritic or ferritic-bainitic microstructures due to the fact that martensitic grain boundaries are more reactive [41].

This work was carried out to assess the influence of heat treatment and microstructure of steels on the local corrosion behaviour. This knowledge may be used to estimate the corrosion phenomena during CO_2-injection into aquifer water reservoirs and predict the reliability of steels used in on-shore CCS sites.

Materials and Methods

Exposure tests were carried out using samples made of thermal treated specimen of AISI 4140 (1%Cr) and AISI 420 (13%Cr) with 8 mm thickness and 20 mm width and 50 mm length. A hole of 3.9 mm diameter was used for sample positioning. The surfaces were activated by grinding with SiC-Paper down to 120 μm under water. Samples of each base metal were positioned within the vapour phase, the intermediate phase with a liquid/vapour boundary and within the liquid phase. The brine (as known to be similar to the Stuttgart Aquifer [42]: Ca_2^+: 1760 mg/L, K_2^+: 430 mg/L, Mg_2^+: 1270 mg/L, Na_2^+: 90,100 mg/L, Cl^-: 143,300 mg/L, SO_4^{2-}: 3600 mg/L, HCO_3^-: 40 mg/L) was synthesized in a strictly orderly way to avoid precipitation of salts and carbonates.). Austenitizing prior to exposure was done at 950 °C 1000 °C and 1050 °C for 30 min, 60 min and 90 min [43]. Heat treatment of the coupons was done according to protocols used in industrial use (Table 1). The exposure of the samples between 700 h to 8000 h was disposed in a chamber kiln at 60 °C at 100 bar in an autoclave system and for reference at ambient pressure as well. Flow control (3 NL/h) at ambient pressure was done by a capillary meter GDX600_man by QCAL Messtechnik GmbH, München. X-ray diffraction was carried out in a URD-6 (Seifert-FPM) with CoKα-radiation with an automatic slit adjustment, step 0.03 and count 5 sec. Phase analysis was performed by matching peak positions automatically with PDF-2 (2005) powder patterns. Mainly structures that were likely to precipitate from the steels were chosen of the ICSD and refined to fit the raw-data-files using POWDERCELL 2.4 [44] and AUTOQUAN ® by Seifert FPM.

Table 1: heat treatment of samples used in exposure experiments.

material	abbr.	heat treatment	temperature [°C]	hold time [min]	cooling medium
	HT1	normalizing	785	30	air
X20Cr13	HT2	hardening	1000	30	oil
1.4021	HT3	hardening + tempering 1	1000 / 600	30	oil
	HT4	hardening + tempering 2	1000 / 670	30	oil
	HT5	hardening + tempering 3	1000 / 755	30	oil
	HT1	normalizing	785	30	air
X46Cr13	HT2	hardening	1000	30	oil
1.4034	HT3	hardening + tempering 1	1000 / 600	30	oil
	HT4	hardening + tempering 2	1000 / 670	30	oil
	HT5	hardening + tempering 3	1000 / 755	30	oil

Then the samples were embedded in a cold resin (Epoxicure, Buehler), cut and polished first with SiC-Paper from 180 μm to 1200 μm under water and then finished with diamond paste 6 μm and 1 μm. Different light optical and electron microscopy techniques were performed on specimens to investigate the layer structures and morphology of samples 60°C/700 h, 60°C/2000 h, 60°C/4000 h and 60°C/8000 h.

Results and Discussion

Differently heat treated steels with 13% Cr and 0.46% C (X46Cr13 / 1.4034) or 0.2% C (X20Cr13 / 1.4021) that may be used as injection pipe in CCS technology were exposed to CO_2 and saline aquifer water at ambient pressure and 60 °C for up to 8000 h (approximately 1 year) [45]. Similar experiments were performed in supercritical CO_2 at 100 bar also at 60 °C. According to DIN 6601, X20Cr13 and X46Cr13 as well as X5CrNiCuNb16-4 would be unsuitable for use in pressure vessel applications, if the material is surrounded by the CO_2-saturated brine. All heat treatments fail the requirements, because even though the corrosion rates are all tolerable (below 0.1 mm/year), the maximum pit penetration depth exceeds 0.2 mm by far (figure 2).

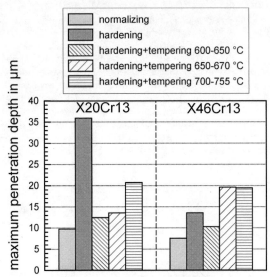

Figure 2. Maximum penetration depth of X20Cr13 and X46Cr13 after 700 h of exposure to artificial geothermal brine condidions.

During the normal storage procedure, the CO_2 is supposedly injected in its supercritical phase. Therefore, experiments at ambient pressure with excess oxygen in the open system can overestimate the pit corrosion because results give higher corrosion rates and greater pit penetration depths at ambient pressure than at 100 bar. However, when there are intermissions of the injection, the water level may rise into the injection pipe. This will then lead to the precipitation of corrosion products and formation of pits as stated. Therefore, long-term exposure experiments at high pressure, where the CO_2 is in its supercritical state, and detailed microstructure analysis were necessary [46].

Taking into account the surface corrosion rate under supercritical CO_2 conditions, a martensitic microstructure of hardened and tempered steel at low temperatures (600-670 °C) offers the best corrosion resistance (figure 3). After 4000 h of exposure, the corrosion rates for uniform corrosion seem to stabilize. This indicates that a certain thickness of the carbonate layer is reached, and mutual diffusion of ionic species into the base materials (CO_3^{2-} and O_2-species) as well as diffusion of Fe-ions towards the outer surface is slow. The possibly low ductility of the steel, which results from tempering at low temperatures and in addition the brittleness caused by internal strain are unfavourable [45]. Tempering at higher temperatures does not offer a solution due to the formation of iron carbides. Especially in 13% Cr steel Cr carbides susceptible to local corrosion precipitate through secondary hardening, which do not dissolve under the CCS conditions given. The increased chromium content in the carbides results then in an insufficient Cr-content to form a passivating layer in the matrix surrounding the carbide. Therefore both, the carbides and the missing passivating layer initiate and enhance local corrosion [45].

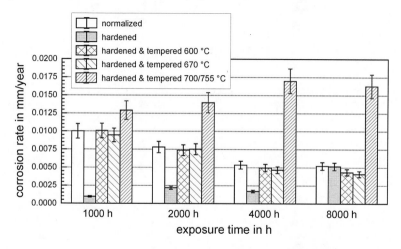

Figure 3. Corrosion rate of X20Cr13 and X46Cr13 combined after 8000 h of exposure to artificial geothermal brine conditions.

Pit growth cannot be calculated as easily as surface corrosion rates because of its little predictability. Therefore, it is not possible to give reliable corrosion rates and lifetime predictions regarding pit corrosion in CCS technology. Independent of the exposure time, the fewest pits are found for steels with martensitic microstructure [45]. Regarding steels with similar Cr content, the higher C content in 1.4034 results in fewer pits and a lower maximum intrusion depth compared to 1.4021 (figure 4). Specimens exposed to the vapour phase, where corrosion rates are rather low, do not show enhanced pit corrosion. Non-uniform corrosion forms carbonate corrosion products on the surface, such as siderite and goethite. Even in the supercritical CO_2 phase a rather thick corrosion layer could form.

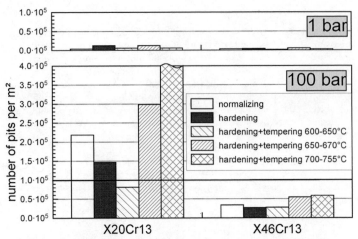

Figure 4. Corrosion rate of X20Cr13 and X46Cr13 combined after 8000 h of exposure to artificial geothermal brine conditions.

Summarizing the kinetic results at ambient pressure, the heat treatment preferred to obtain least corrosive attack is normalizing. If higher tensile strength of the steels is needed and hardening and tempering will be applied, austenitising at low temperature and tempering for a short time gives the best corrosion resistance. At 100 bar, where the CO_2 is in its supercritical state, hardening and tempering at low temperatures (600 to 670 °C) for X20Cr13 and X46Cr13 give best results. Although surface corrosion rates are low, pit growth rates only allow the steel to be suitable for injection pipes in CCS environments if monitored closely. Future investigations will focus on hardening and tempering to determine a combination of austenitising and tempering which optimizes, strength, toughness and corrosion resistance. Of high interest is also the evaluation of carbide distribution within the base material in relation to pit distribution and size to gain knowledge on pit initiation processes.

Conclusion

The following facts observed could be summarized [45]:

- Following from long-term exposure tests in CCS environment, hardening and tempering at 600 °C to 670 °C offers the best corrosion resistance against uniform and pit corrosion at injection conditions of 60 °C and 100 bar.
- Shorter austenitising (best around 60 minutes) at lower austenisation temperature results in better corrosion resistance, especially against pitting corrosion. Note, that the austenisation time has the greater impact.
- The corrosion rate is higher by a factor of 2 to 5 when the steel is kept in the CO2-saturated liquid phase than in the water-saturated supercritical CO2 phase due to the formation of carbonic acid in the liquid phase.
- The higher the carbon content of the steels (X20Cr13<C46Cr13), the lower the amount of pits initiated but the larger the pit diameter. X46Cr13 shows slightly better corrosion resistance than X20Cr13.

Acknowledgement

This work was supported by the FNK (Fachkonferenz für wissenschaftlichliche Nachwuchskräfte) of the Applied University of Berlin, HTW and by IMPACT (EU-Project EFRE 20072013 2/21).

References

[1] D.C. Thomas, Carbon Dioxide Capture for Storage in Deep Geologic Formations - Results from CO2 Capture Project, Volume 1: Capture and Separation of Carbon Dioxide form Combustion Sources, CO2 Capture Project, Elsevier Ltd UK 2005, ISBN 0080445748

[2] M. van den Broek, R. Hoefnagels, E. Rubin, W. Turkenburg, A. Faaij, Effects of technological learning on future cost and performance of power plants with CO2 capture, Progress in Energy and Combustion Science, Paper in Progress (2009) 1-24

[3] GeoForschungszentrum Potsdam, CO2-SINK - drilling project, description of the project PART 1 (2006) 1-39

[4] S. Nešic, "Key issues related to modelling of internal corrosion of oil and gas pipelines - A review", Corrosion Science 49 (2007) 4308-4338

[5] S. Hurter, Impact of Mutual Solubility of H2O and CO2 on Injection Operations for Geological Storage of CO2, International Conference of the Properties of Water and Steem ICPWS, Berlin, September 8-11

[6] L. Zhang, J. Yang, J.S. Sun, M. Lu, Effect of pressure on wet H2S/CO2 corrosion of pipeline steel, No. 09565, NACE Corrosion 2008 Conference and Expo, New Orleans, Louisiana, USA, March 16th - 20th, 2008

[7] L.J. Mu, W.Z. Zhao, Investigation on Carbon Dioxide Corrosion Behaviors of 13Cr Stainless Steel in Simulated Strum Water, Corrosion Science, Manuscript No. CORSCI-D-09-00353 (2009) 1-24

[8] M. Bonis, Weight loss corrosion with H2S: From facts to leading parameters and mechanisms, Paper No. 09564, NACE Corrosion 2008 Conference and Expo, New Orleans, Louisiana, USA, March 16th - 20th, 2008

[9] [IPCC] IPCC Carbon Dioxide Capture and Storage: Technical Summary (2005) 5. Geological Storage, p. 29, http://www.greenfacts.org/en/co2-capture-storage/figtableboxes/figure-7.htm http://arch.rivm.nl/env/int/ipcc/pages_media/SRCCSfinal/SRCCS_TechnicalSummary.pdf

[10] M. Seiersten, Material selection for separation, transportation and disposal of CO2, NACE Corrosion, Corrosion paper no. 01042 (2001)

[11] Z.D. Cui, S.L. Wu, S.L. Zhu, X.J. Yang, Study on corrosion properties of pipelines in simulated produced water saturated with supercritical CO2, Applied Surface Science 252 (2006) 2368-2374

[12] Pfennig, A., Kranzmann, A., Reliability of pipe steels with different amounts of C and Cr during onshore carbon dioxide injection, International Journal of Greenhouse Gas Control 5 (2011) 757-769

[13] J. Enerhaug, A study of localized corrosion in super martensitic stainless steel weldments, a thesis submitted to the Norwegen University of Science and Technology (NTNU), Trondheim 2002

[14] V. Neubert, Beanspruchung der Förderrohrtour durch korrosive Gase, VDI-Berichte Nr. 2026, 2008

[15] R. Kirchheiner, P. Wölpert, Qualifizierung metallischer Hochleistungs-werkstoffe für die Energieumwandlung in geothermischen Prozessen, VDI-Berichte Nr. 2026, 2008

[16] H. Zhang, Y.L Zhao, Z.D. Jiang, Effects of temperature on the corrosion behaviour of 13Cr martensitic stainless steel during exposure to CO2 and Cl- environment, Material Letters 59 (2005) 3370-3374

[17] J.N. Alhajji and M.R. Reda, The effect of alloying elements on the electrochemical corrosion of low residual carbon steels instagnant CO2-saturated brine, Corrosion Science, Vol. 34, No. 11 (1993) 1899-1911

[18] Y.-S. Choi and S. Neši?, Corrosion behaviour of carbon steel in supercritical CO2-water environments, No. 09256, NACE Corrosion 2008 Conference and Expo, New Orleans, Louisiana, USA, March 16th - 20th, 2008

[19] X. Jiang, S. Nešic, F. Huet, The Effect of Electrode Size on Electrochemical Noise Measurements and the Role of Chloride on Localized CO2 Corrosion of Mild Steel, Paper No. 09575, NACE Corrosion 2008 Conference and Expo, New Orleans, Louisiana, USA, March 16th - 20th, 2008

[20] Z. Ahmad, I.M. Allam, B.J. Abdul Aleem, Effect of environmental factors on the atmospheric corrosion of mild steel in aggressive sea coastal environment, Anti Corrosion Methods and Materials, 47 (2000) 215-225

[21] A. Pfennig, R. Bäßler, Effect of CO2 on the stability of steels with 1% and 13% Cr in saline water, Corrosion Science, Vol. 51, Issue 4 (2009) 931-940,

[22] A. Pfennig, A. Kranzmann, Influence of CO2 on the corrosion behaviour of 13Cr martensitic stainless steel AISI 420 and low-alloyed steel AISI 4140 exposed to saline aquifer water environment, Air Pollution XVII , Volume 123 (2009) 409-418, ISBN: 978-1-84564-195-5, ISSN: 1746-448X

[23] A. Pfennig, A. Kranzmann, The role of pit corrosion in engineering the carbon storage site Ketzin, Germany, WIT Transactions on Ecology and the Environment, Volume 126, (2010) 109-118, ISBN: 978-1-84564-450-5

[24] R. Nyborg, Controlling Internal Corrosion in Oil and Gas Pipelines, Business Briefing: Exploration & Production: The Oil & Gas Review, issue 2 (2005) 70-74

[25] D.S. Carvalho, C.J.B. Joia, O.R. Mattos, Corrosion rate of iron and iron-chromium alloys in CO2-medium, Corrosion Science 47 (2005) 2974-2986

[26] B.R. Linter, G.T Burstein, Reactions of pipeline steels in carbon dioxide solutions, Corrosion Science 41 (1999) 117-139

[27] S.L Wu, Z.D. Cui, G.X. Zhao, M.L. Yan, S.L. Zhu, X.J. Yang, EIS study of the surface film on the surface of carbon steel form supercritical carbon dioxide corrosion", Applied Surface Science 228 (2004) 17-25

[28] B. Brown, S. R. Parakala, S. Neši?, CO2 corrosion in the presence of trace amounts of H2S, Corrosion, paper no. 04736 (2004) 1-28

[29] http://www.standard.no/pronorm-3/data/f/0/01/36/9_10704_0/M-506d1r2.pdf, "CO2 corrosion rate calculation model"

[30] J. Han, Y. Yang, S. Nešic, B. N. Brown, Roles of passivation and galvanic effects in localized CO2 corrosion of mild steel, Paper No. 08332, NACE Corrosion 2008, New Orleans, Louisiana, USA, March 16th - 20th, 2008

[31] J. Banas, U. Lelek-Borkowska, B. Mazurkiewicz, W. Solarski, Effect of CO2 and H2S on the composition and stability of passive film on iron alloy in geothermal water, Electrochimica Acta 52 (2007) 5704-5714.

[32] P.D. Bilmes, C.L. Llorente, C.M. Méndez, C.A. Gervasi, Microstructure, heat treatment and pitting corrosion of 13CrNiMo plate and weld metals, Corrosion Science, ISSN: 0010-938X, Vol: 51, Issue: 4, (2009) 876-882

[33] S. Bülbül, Y Sun, Corrosion behaviours of high Cr-Ni cast steels in the HCl solution, Journal of Alloys and Compounds 598 [2010] 143-147

[34] B. Hou, Y. Li, Y. Li, J. Zhang, Effect of alloy elements on the anti-corrosion properties of low alloy steel, Bull. Mater. Sci, Vol. 23, No. 3 (2000) 189-192

[35] Z. Cvijovic and G. Radenkovic, Microstructure and pitting corrosion resistance of annealed duplex stainless stell, Corrosion Science 48 (2006) 3887-3906

[36] L. Zhang, W. Zhang, Y. Jiang, B. Deng, D. Sun, J. Li, Influence of annealing treatment on the corrosion resistance of lean duplex stainless steel 2101

[37] Y.-S. Choi, J.-G. Kim, Y.-S. Park, J.-Y. Park, Austenitzing treatment influence on the electrochemical corrosion behaviour of 0.3C-14Cr-3Mo martnesitic stainless steel, Materials Letters 61 (2007) 244-247

[38] A. N. Isfahany, H. Saghafian, G. Borhani, The effect of heat treatment on mechanical properties and corrosion behaviour of AISI420 martensitic stainless steel, Journal of Alloys and Compounds 509 (2011) 3931-3936

[39] J.-Y. Park, Y.-S. Park, The effects of heat-treatment parameters on corrosion resistance and phase transformation of 14Cr-3Mo martensitic stainless steel, Materials Science and Engineering A 449-451 [2007] 1131-1134

[40] D. Dyja, Z. Stradomski, A. Pirek, Microstructural and fracture analysis of aged cast duplex steel, Strength of Materials, Vol. 40, No. 1 [2008] 122-125

[41] M.A. Lucio-Garciaa, J.G. Gonzalez-Rodrigueza, , M. Casalesc, L. Martinezc, J.G. Chacon-Navaa, M.A. Neri-Floresa and A. Martinez-Villafañea, Effect of heat treatment on H2S corrosion of a micro-alloyed C-Mn steel, Corrosion Science 51 (2009) 2380-2386

[42] A. Förster, B. Norden, K. Zinck-Jørgensen, P. Frykman, J. Kulenkampff, E. Spangenberg, J. Erzinger, M. Zimmer, J. Kopp, G. Borm, C. Juhlin, C. Cosma, S. Hurter, 2006, Baseline characterization of the CO2SINK geological storage site at Ketzin, Germany: Environmental Geosciences, V. 13, No. 3 (September 2006), pp. 145-161.

[43] J. Schiz, Einfluss der Austenitisierungsparameter auf die Korrosionsbeständigkeit von Stählen in CCS-Umgebung, Bachelor-Thesis zur Erlangung des Grades B.Sc. im Fachbereich Maschinenbau an der HTW Berlin, 2011

[44] SW. Kraus and G. Nolze, POWDER CELL - a program for the representation and manipulation of crystal structures and calculation of the resulting X-ray powder patterns, J. Appl. Cryst. (1996), 29, 301-303

[45] A Pfennig, P. Zastrow, A. Kranzmann, Influence of heat treatment on the corrosion behaviour of stainless steels during CO_2-sequestration into saline aquifer, International Journal of Green House Gas Control 15 (2013) 213–224

Energy Technology 2014: Carbon Dioxide Management and Other Technologies
Edited by: Cong Wang, Jan de Bakker, Cynthia K. Belt, Animesh Jha, Neale R. Neelameggham,
Soobhankar Pati, Leon H. Prentice, Gabriella Tranell, and Kyle S. Brinkman
TMS (The Minerals, Metals & Materials Society), 2014

INFLUENCE OF VOLATILE FUNCTIONALITY ON PULVERIZED COAL EXPLOSIVITY

Qinghai Pang[1],

[1]School of Metallurgical and Ecological Engineering, University of Science and Technology
Beijing, Beijing 100083, PR China

Keywords: bituminous coal; explosivity; FT-IR; radical

Abstract

With the increasing proportion of bituminous coal in blend coal, the safety of blend coal during pulverization and transportation is gradually being paid more and more attention. Therefore, the explosivity of bituminous pulverized coal (PC) with different volatile content from 25 to 38% was studied by measuring the combustion flame length after its injection into a glass pipe with the heat source installed in the center. The results indicated that high volatile content bituminous PC is apt to have higher explosivity. By blending different bituminous PC with 10% anthracite, the explosivity of bituminous PC showed a positive correlation with the proportion –OH radical. However, different bituminous PC seemed to have the same explosivity when the addition of anthracite increased to 20%. With the further increasing proportion of anthracite to 30%, the explosivity of bituminous PC with high –OH proportion became negligible while a weak explosivity of high –CH3 proportion bituminous PC still remained.

Introduction

With the increasing iron ore prices and decreasing steel prices, it has become critical for iron and steel enterprises to control the cost in ironmaking and steelmaking process. Blast furnaces (BF) are the main consumer in ironmaking process, which occupies 70% fuel consumption of total steelmaking process [1]. In this circumstance, any means necessary is taken by BF operators to lower the cost of iron in production. Coke and iron ore are the main expense in this process, who takes up a very high proportion in total cost. Pulverized coal injection (PCI) is the most efficient way to reduce coke ratio, which replace coke with relatively cheaper PC.

At present, most of BF is injected with blend coal in the consideration of decent combustion property and quantity of heat. Simultaneously, bituminite and anthracite blending is an effective method to obtain blend coals with required composition and performance, such as volatile and ash content, etc. Nevertheless, a very huge gap was found between the price of bituminous coal and anthracite in some regions of China, in which the price of bituminite can be just 70% of the price of anthracite. Thus, iron enterprises in these regions attempt to possibly amplify the proportion of bituminite in blend coal. Increasing proportion of bituminite results in higher volatile content in blend coal, by which the safety of blend coal during pulverization or transportation was deteriorated.

Relatively, coals with higher volatile content usually possess stronger explosivity, which means that volatile and explosivity may related to some extent. However, the relationship between volatile and explosivity is not of certainty for some PC with high volatile content around 40% was found to be inexplosive in comparison to the coals with lower volatile content.

With the increasing proportion of bituminous coal in blend coal, investigations on the factors of PC explosivity have become more and more important. Therefore, bituminous coals with different volatile content were selected. The explosivity of blend coal and its relationship between volatile content and functionality was investigated in this research.

Experimental

2.1 Coal samples

Table 1. Proximate analysis and ultimate analysis of bituminite and anthracite samples.

Coal	Proximate analysis/%			Ultimate analysis/%				
	FCd	Vd	Ad	Cd	Hd	Nd	Od	Sd
A	55.42	38.30	6.28	61.21	2.91	0.89	15.68	0.21
B	56.97	35.66	7.37	67.92	4.11	0.78	17.44	0.36
C	58.32	34.67	7.01	68.35	4.28	0.80	17.11	0.32
D	56.55	33.50	9.95	68.52	4.10	0.98	14.38	0.26
E	59.83	33.44	6.73	74.30	4.22	1.07	12.36	0.22
F	62.97	31.73	5.30	75.10	4.34	0.89	12.50	0.44
G	63.54	28.77	7.69	73.38	3.98	0.87	12.23	0.22
H	60.46	28.45	11.09	69.02	4.07	0.74	11.67	0.56
I	66.17	25.64	8.19	69.45	3.46	0.89	14.13	0.36
J	89.03	7.40	3.57	89.81	3.53	0.79	2.05	0.12

FC, fixed carbon; V, volatile; A, ash; M, moisture; d, dried; Q, quantity of heat; gr, gross calorific value;

In order to study the influence of volatile content and functionality on the explosivity of PC, 9 bituminite and one anthracite samples with different volatile content from about 25% to 38% which are all applied for BF injection were selected as the specimens of this investigation. Coal samples were dried under 105°C for over 4h and grinded to -200 mesh for following experiments. The proximate analysis and ultimate analysis of coal samples were shown in Table 1.

2.2 FT-IR analysis

It can be observed from the proximate analysis of coal specimens that hydrogen content of coals selected range from about 3.0% to 4.5%. Hydrogen in coal combines with carbon and oxygen in forms of different radicals like –OH, CH3, aromatic hydrocarbon and other formations, which makes the volatile in coal of various properties[2]. Formations of radicals in coal are dominant in not only the decomposition process but also the products of decomposition[3]. As a result, the combustion and explosivity of coal under certain condition are of great variety. Consequently, radicals in coal are of great importance to the properties of coal. Therefore, the functional groups in coal specimens selected were analyzed with Nicolet Nexus 470 FT-IR type Fourier-transform infrared (FT-IR) system with a spectral resolution of 4 cm in the range 400 – 4000cm-1.

2.3 Thermogravimetric analysis

Thermogravimetric studies were carried out on WCT−2C type thermogravity analysis (TGA) device. In order to avoid the influence of heat and mass transfer limitations, a small amount of PC of 17.5mg was loaded into a corundum crucible and placed into the chamber of thermogravity analysis device for heating. The coal specimens were heated from ambient temperature to final temperature of 900°C at the constant heating rate of 20°C/min. Weight loss during combustion was recorded.

2.4 Explosivity measurement

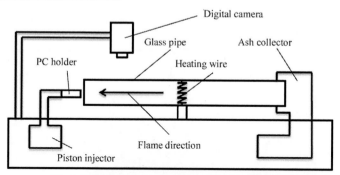

Fig. 1. Schematic diagram of PC combustion flame testing instrument

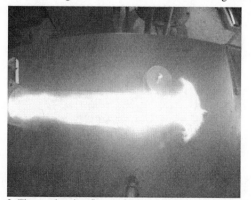

Fig .2. The combustion flame of PC observed by digital camera

1g pulverized coal was loaded in the PC holder connected to the piston injector and injected into a glass pipe at a very high speed and suspended in the chamber of the glass pipe. The dispersed PC particles spread into the center of the pipe and contact with a platinum heating wire in the center of the glass pipe which was preheated to 1500°C. Consequently, dispersed PC was ignited and combusted at a very high speed. The flame starts from the center of the pipe and moves towards the direction from which it was injected and finally vanished at a certain position along the pipe. The length of combustion flame of PC was visually recorded by a digital camera, the length of the flame was measured on the pictures recorded which was adopted to represent the explosivity of PC. The schematic diagram of the experimental apparatus was illustrated in Fig. 1 and the flame of coal combustion observed by the digital camera was shown in Fig. 2.

Discussion

3.1 FR-IR analysis

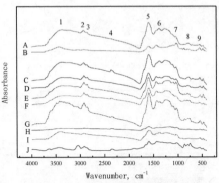

Fig. 3. FT-IR spectrums of coal specimens

Table 2 Classification of absorption peak in FT–IR spectra

Peak	Wavenumber / cm-1	Functional group
1	3300	-OH
	3030	-CH (aromatic ring)
2	2950 (shoulder)	-CH3
3	2920, 2860	-CH3 of cyclane or aliphatic hydrocarbon
4	2780 ~ 2350	-COOH
5	1610	Carbonyl substituted aromatic hydrocarbon
	1590 ~ 1470	Aromatic hydrocarbon
	1460	-CH2, -CH3, inorganic carbonate
6	1375	-CH3
7	1330 ~ 1110	C-O (phenol, alcohol, ether, ester)
	1040 ~ 900	Ash
8	860	CH (1,2,4-; 2,4,5; 1,2,3,4,5 substituted aromatic hydrocarbon)
	750	CH (1,2 substituted aromatic hydrocarbon)
	700	CH (Single substituted or 1,3 substituted aromatic hydrocarbon)
9	550	-SH

The decomposition of volatiles in coal takes place first when PC is heated, which follows the combustion of volatile and residual coke[4]. The explosion of PC occurs in a very short time after its contact with heat source, the intensity of which depends on the gas product produced by the decomposition of coal. Therefore, the decomposition of coal plays a very important role in this process, the initial temperature and product of which are decided by the formation and quantity of radicals. Different gas product will be produced by the oxidation and breach of different radicals, who has a great influence on the initial temperature and combustion process of coal. –COOH will be generated by the oxidation of –OH, with the breach of radicals like –CH3 produced methane. Radical in 10 coal specimens selected was analyzed by Fourier-transform infrared (FT-IR). The FT-IR spectrums of coal specimens are shown in Fig. 3 and the peaks of the spectrums were marked as 1 to 9 from the left to the right. The corresponding radicals of 7 characteristic peaks are illustrated in Tab. 2[5].

FT-IR spectrums of coal specimens in Fig. 3 indicated that volatiles in low rank coals tend to possess higher intensity in peak1, which reflects the higher content of –OH in PC. –OH is a very active radical. The oxidation of –OH is easily to take place at relatively lower temperature and its oxidation may lead to the breach of hydrocarbon structures, which means that coals with more – OH radicals are usually of higher activity[6]. Besides, it also can be observed that radicals of – OH were found to be decreased with the enhancing degree of coalification. Simultaneously, an obvious increase in the intensity of peak 3 and 6, which means the quantity of –CH3 was increased with the enhancing degree of coalification. –CH3 is a radicals of lower activity in comparison to –OH, so the oxidation –CH3 occurs under relatively higher temperatures in comparison to –OH.

3.2 Thermogravimetric analysis

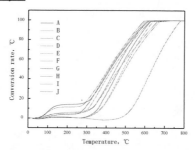

Fig. 4. Conversion rate curves of coal specimens

Conversion rate curves of different PC are shown in Fig. 4. It can be found that certain proportion of crystal water in A and D PC are 14% and 11% respectively, the contents of which are relatively higher than the other PC. Besides, the crystal water content in B, C, H and I is about 5%, while no crystal water is found in E, F and G PC. The curves in Tab.1 showed that a good relationship between conversion rate and the functionalities in PC. Weight loss rate of the PC with more –OH radical seems to be faster, while that of PC with high –CH3 is found to be relatively slower.

Table 3. Flame length of PC tested

Coal	Vhd/%	Hd/%	Flame length with anthracite, mm			
			0%	10%	20%	30%
A	38.30	2.91	785	575	98	23
B	35.66	4.11	608	343	125	93
C	34.67	3.79	800	608	113	37
D	33.50	4.10	773	719	86	33
E	33.44	4.22	800	586	102	65
F	31.73	4.34	800	462	136	103
G	28.77	3.98	790	575	117	16
H	28.45	4.07	711	298	103	76
I	25.64	3.46	580	167	115	89
J	7.40	3.53	–	–	–	–

Pure bituminous coal is usually found to be explosive and flame length of which often is over 800mm, which exceeds the measuring range of this device. Therefore, in order to enlarge the difference in explosivity of bituminous coals, the explosivity of mixture of bituminous coal and anthracite were investigated. One anthracite type with relatively lower volatile and ash content as well as negligible flame length was selected in this investigation. The explosivity of bituminous coal with anthracite blending proportion of 10%, 20% and 30% were observed.

It can be inferred from the data in Table 3 that coals with volatile content is apt to be explosive, while the combustion flame length of lower volatile content PC are found to be shorter. Meanwhile, crystal water may have a slight influence on the flame length of PC. It can be found that the flame length of A and D PC is a little shorter in comparison to PC with same or even lower volatile content, though these two coals have high –OH content and therefore active volatiles. The same phenomenon is also observed in B, C, H and I, which means that crystal water may possibly weaken the explosivity of PC. Many experiment results showed that no obvious combustion phenomenon of anthracites PC can be observed when PC is injected in to the glass pipe, which means that anthracite PC is relatively stable during the transportation in pipelines. Explosivity results showed that small proportion addition of anthracite PC in bituminous PC around 10% may significantly reduce the explosivity of PC with more –CH3 radicals. However, no obvious difference in explosivity of different bituminous PC can be noticed when the proportion of anthracite PC increases to 20%. The explosivity of bituminous coal with high –OH proportion became negligible with the further increase of anthracite proportion to 30%, while high –CH proportion bituminous coals still showed a weak explosivity.

Conclusion

A good relationship between weight loss rate and functionality was observed in the comparison of conversion rate curves. PC with high –OH content is found to have a faster conversion rate in comparison to PC with high –CH3. Therefore, FT-IR spectrum of PC is valid to predict its combustion property.

Bituminous PC with crystal water seemed to be less explosive, though the volatile contents of these PC were even higher. Besides, the higher the crystal water content in PC, the weaker the explosivity was observed. Therefore, the conclusion can be made that crystal water in PC may weaken the explosivity of PC.

Pure bituminous PC with higher –OH content is apt to has stronger explosivity in comparison to high –CH3 PC, while this difference was narrowed or even eliminated by anthracite addition to certain proportion. However, the high –CH3 PC was found to be more explosive with the further addition of anthracite.

References

[1] Y. S. Shen, B. Y Guo, A. B. Yu, P. Zulli, Model Study of the effects of coal properties and blast conditions on pulverized coal combustion, ISIJ. International 49 (2009) 819-826.

[2] M.G. Yu, Y.M. Zheng, C. Lu, H.L. Jia, Experimental research on coal spontaneous combustion characteristic by TG-FTIR, Journal of Henan Polytechnic University (natural science) (in Chinese) 28 (2009) 547-551.

[3] T.X. Chu, S.Q. Yang, Y. Sun, J.K. Sun, Z.P. Liu, Experimental study on low temperature oxidizaiton of coal and its infrared spectrum analysis, China Safety Science Journal (in Chinese) 18 (2008) 171 -176.

[4] J. Feng, W.Y. Li, K.C. Xie, Research on coal structure using FT-IR, Journal of China University of Mining & Technology (in Chinese) 31(2002) 362-366.

[5] J.S. Yu, Coal chemistry, Metallurgical Industry Press, 2006.

[6] I. K. Oikonomopoulos, M. Perraki, N. Tougiannidis, T. Perraki, M. J. Frey, P. Antoniadis, W. Ricken, A comparative study on structural differences of xylite and matrix lignite lithotypes by means of FT-IR, XRD, SEM and TGA analyses: An example from the Neogene Greek lignite deposits, International Journal of Coal Geology, 115 (2013) 1-12.

Energy Technology 2014: Carbon Dioxide Management and Other Technologies
Edited by: Cong Wang, Jan de Bakker, Cynthia K. Belt, Animesh Jha, Neale R. Neelameggham,
Soobhankar Pati, Leon H. Prentice, Gabriella Tranell, and Kyle S. Brinkman
TMS (The Minerals, Metals & Materials Society), 2014

Thermodynamic and Experimental Study on the Steam Reforming Processes of Bio-oil Compounds for Hydrogen Production

Huaqing Xie, QingboYu, Kun Wang, Xinhui Li, Qin Qin

School of Materials and Metallurgy, Northeastern University, No. 11, Lane 3, WenHua Road, HePing District, Shenyang, 110819, Liaoning, P.R. China.

Keywords: Bio-oil, Steam reforming, Hydrogen production, Thermodynamic analysis, Catalysts

Abstract

Three typical model compounds of bio-oil (ethanol, acetone and phenol) were studied in the steam reforming process for hydrogen production. In the thermodynamic analysis, the three componds showed the similar trends. The hydrogen yields increased with the increase of steam/carbon (S/C) ratios, and the temperatures of the maximum hydrogen yields moved afterward to low temperature zone. However, the hydrogen contents were just about 70%. For improving hydrogen yield, a new system for steam reforming of bio-oil with site CO_2 capture was put forward. This system required the catalst with the larger granularity having greater mechanical strength and higher abrasion resistance. So, six Ni-base catalysts loaded on Al_2O_3 particles prepared were used and among them, Mg-Ni/Co catalyst showed the best catalytic performance. For the simulated bio-oil (the mixture of the compounds), the hydrogen yield can attain 63% with Mg-Ni/Co catalyst with the good stability and regeneration ability.

Introduction

Currently the interest in the use of hydrogen as a clean energy carrier is increasing, and the main way for producing hydrogen is steam reforming of fossil fuels, containing steam reforming of nature gas, naphtha and coal. But this way caused substantial amounts of CO_2 emitted, and these sources of hydrogen which are non-renewable energy will eventually be used up [1, 2]. As a kind of renewable energy, biomass is recognized as an attractive alternative to fossil fuels, because of its neutral CO_2 emissions [3]. There are mainly two thermochemical processes using biomass to produce hydrogen, gasification and flash pyrolysis followed by steam reforming (SR) of the bio-oil produced in the flash pryolysis process [4]. The latter was known as one of the most promising and economically viable methods for hydrogen production [5]. In the last decades, the biomass flash pyrolysis technology was adequately developed and fairly mature [6]. But the steam reforming of bio-oil is still at the research stage and yet has been leading to more and more concerns by the researchers. In their research, the catalysts were widely studied and yet mainly were powdery[7-10], and the reactors manily were fixed bed reactors[11-13], readly causing the carbon deposition and the catalyst deactivation, against the continuous running. In this paper, a new systerm was put forward and could realize continuous production of hydrogen, yet needed catalysts with the larger granularity having greater mechanical strength and higher abrasion resistance, as well as powdery CO_2 sorbents which mainly calcium oxide based ones.

In this paper, the thermodynamic analysis of the steam reforming processes of three typical bio-oil model compounds (ethanol, acetone, phenol) was performed, and the effect of S/C ratio and temperature on the hydrogen yield and the compositions of gaseous products were investigated. Then, a new systerm for steam reforming of bio-oil with site CO_2 capture was put forward, and the catalysts with a certain granularity were prepared and studied.

Thermodynamic analysis of SR of bio-oil model compounds for hydrogen production

In this paper, ethanol, acetone and phenol, representative of the major classes with the highest concentrations of bio-oil were studied as typical model compounds. Their equilibrium compositions in SR processes were calculated with HSC Chemistry 5.0. For them (abbreviated to $C_nH_mO_k$), the steam reforming reaction is:

$$C_nH_mO_k + (n-k)H_2O ® nCO + (n+m/2-k)H_2 \tag{1}$$

followed by the water gas shift (WGS) reaction:

$$nCO + nH_2O ® nCO_2 + nH_2 \tag{2}$$

Given the Eq.(1) and Eq.(2) happening successively, the overall reaction can be represented:

$$C_nH_mO_k + (2n-k)H_2O ® nCO_2 + (2n+m/2-k)H_2 \tag{3}$$

In the SR process, the side reactions also took place, such as thermal decoposition reaction:

$$C_nH_mO_k ® C_xH_yO_z + gas(H_2, CO, CO_2, CH_4 \cdots) + coke \tag{4}$$

Figure 1 (a~c) showed the hydrogen yield as a function of temperature at different steam/carbon (S/C) ratios (Eq. (5)) for the three compounds. The hydrogen yield was defined as the ratio of the amount of hydrogen in the actual gaseous products to the theoretical amount of hydrogen that can be obtained when complete reforming to CO_2 and H_2 occurs through Eq. (3).

$$S/C = \text{Moles of water steam in the feed} / \text{Moles of carbon in the feed} \tag{5}$$

The trends of the hydrogen yields for the three compounds with the temperature under the different S/C ratios were similar: with the increase of the temperature, the hydrogen yield first increased to the maximum and then decreased slowly. It is mainly because the WGS reaction (Eq. (2)) was an exothermic one, and was inhtbited in high temperature. The hydrogen yields increased with the S/C ratio. The maximum hydrogen yields increased from just about 70% at the S/C ratios of the stoichiometric values to approximate 100% at the S/C ratios of 24, and the temperatures of the maximum hydrogen yields moved to low temperature zone, indicating adequate steam can observably accelerate the SR reaction progresses. However, the hydrogen yields at the S/C ratios over 6 were very close, especially under the temperature range corresponding to higher hydrogen yields, especially after the hydrogen yields reached the top. Figure 1 also showed the carbon yields as a function of temperature at differernt S/C ratios. The carbon yield to some extend can reflect the degree of the side reactions. With the increases of the temperature and the S/C ratio, the carbon yields of the three compounds decrease and finally vanish. When the S/C ratios passed a certain value (3 for ethanol, 3 for acetone and 6 for phenol), the carbon can completely vanish under the whole temperature range (400K~1300K).

As mentioned above, higher hydrogen yields of the three compounds can be obtained at the S/C ratio of over 6, but the hydrogen contents were lower, just about 70%, and the residual 30%

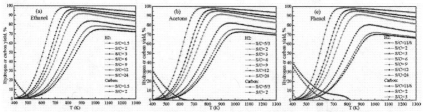

Figure 1. The hydrogen yields and carbon yields with the temperature at different S/C ratios.

242

almost consisted of carbon dioxide and carbon monoxide (shown in Table 1). If the kinds of gaseous products was applied to fuel cell, it will cause the poisoning of the fuel cell, which generally must be feed by a stream with the content of carbon monoxide lower than 20 ppm. So, the studies of novel catalysts with high performance to promote hydrogen yield and to decrease carbon deposit, and novel processes to decrease the contents of carbon dioxide and carbon monoxide, such as SR with site CO_2 adsorption [13-15], attracted more researchers' attentions.

Table 1. The component contents of dry gaseous products from steam reforming at 900K at the S/C ratio of 6.

Sample	H_2 (%)	CO_2 (%)	CO (%)	CH_4 (%)
Ethanol	73.38	20.94	5.41	0.27
Acetone	70.97	23.26	5.55	0.22
Phenol	68.09	26.07	5.66	0.18

Experiment of SR of bio-oil model compounds for hydrogen production

At the moment, the study on the SR with site CO_2 adsorption were still under the experimental stage. The mainly used CO_2 sorbents were CaO based ones [13, 14, 16]. Absorbing CO_2 in the SR site can in theory not only decrease the CO_2 content in the gaseous products, but also promote the water gas shift reaction, resulting improving the hydrogen yield and hydrogen purity. The used reactors presently were mainly fixed bed reactors, and the experiments showed the addition of CO_2 sorbents could obviously improve the hydrogen yield and content. However, the fixed amount of CO_2 sorbents limited the the continuity of the SR process with high efficiency. So, given keeping hydrogen with high yield and content produced continuously, our team put forward the following process system (see Figure 2), which mainly consisted of a catalytic reformer and CO_2 sorbent regenerator. In the catalytic reformer where SR catalysts were padded, the SR reaction of bio-oil inserted continuously took place. Meanwhile, the sorbent also was inserted and absorbed CO_2 produced in the SR process, and then was transferred by air into the regenerator, where the sorbent desporption reaction took place, with the combustion of a small amount of coke. To keep this system running reliably, it required the sorbent with samll granularity and the catalyst with larger granularity as well as greater mechanical strength and higher abrasion resistance to reduce the flow resistance. But, the catalysts currently applied to the SR of bio-oil to product hydrogen mostly were powdered. So, the new catalysts with the larger granularity (3~5mm) were prepared and their performances were examined in this paper.

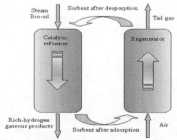

Figure 2. Schematic representation of the new system for reforming with site CO_2 capture.

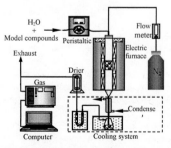

Figure 3. Schematic of the experimental setup used for reforming of model.

Preparation and selection of Ni-based catalysts with large granularity

Six nickel-based catalysts (Mg-Ni/Mg, Mg-Ni/Ce, Ce-Ni/Mg, Ce-Ni/Ce, Mg-Ni/Co and Ce-Ni/Co) loaded on the Al_2O_3 particles with 3~5 mm in diameter were prepared with the incipient wetness method. Mg, Ce and Co were as the auxiliary agent elements. The precursors used were $Ni(NO_3)_2$, $Mg(NO_3)_2$, $Ce(NO_3)_3$, and $Co(NO_3)_2$. Mg or Ce was firstly impregnated to improve thermostability of the carrier, and the mixed Ni and Mg or Ce or Co metals then impregnated. The nominal metal composition of the final catalysts was 5wt% for Mg and Ce in the first impreganting, and 10wt% for Ni, 5wt% for Mg, Ce and Co in the second impreganting. Ethanol, acetone and phenol were also as the feed stock to examine these catalysts. The schematic of the experimental setup was shown in Figure 3. The Al_2O_3 particles were used as blank samples. The experimental conditions were 1023K, the S/C ratio of 6, and the liquid hourly space velocity of $850m^3 \cdot h^{-1}$. Besides the actual hydrogen yield, the other two types hydrogen yields were studied. The second type of hydrogen yield (H_2(+CO))was defined as the sum of the actual hydrogen yield and the potential hydrogen yield from the complete reaction of carbon monoxide in the gaseous products through the WGS reaction; the thrid type of hydrogen yield (H_2(+CO+CH_4)) was defined as the sum of the second type hydrogen yield and the potential hydrogen yield from the complete reaction of methane in the gaseous products through the mathane steam reforming reaction (MSR) (Eq. (6)). Because the WGS reaction readily took place, the second hydrogen yield generally was studied. Although MSR is an endothermic reaction, the reaction will be promoted under the suitable conditions (such as efficiency catalyst, sufficent steam). What's more, the higher the third hydrogen yield is, the fewer carbon deposits is produced in SR process.

$$CH_4 + 2H_2O ® CO_2 + 4H_2 \tag{6}$$

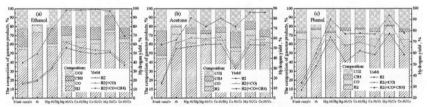

Figure 4. Compositions of gaseous products and hydrogen yields with the different catalysts.

Figure 4 (a~c) showed the compositions of gaseous products and the hydrogen yields with the different catalysts for three compounds. For the different catalysts, the concentrations of the same compositions were different. For ethanol and acetone, the hydrogen contents of the blank samples were obviously lower than those of the seven nickeliferous catalysts, with higher methane contents than those of the latters, to some extent indicating that the main reactions of ethanol and acetone in the absence of the nickel metal were thermodecomposition. However, for phenol, the products contentrations of the blank sample were broadly similar to those of the nickeliferous catalysts, probably because the gaseous products of phenol thermodecomposition were similar to those of phenol SR. For the hydrogen yields, the orders on the whole were consistent: nickeliferous catalysts with auxiliary agent elements > Ni catalyst > blank sample, suggesting Ni was obviously catalytic for SR of the compounds and the auxiliary agent elements were efficient to improve the catalytic effect of Ni. But for acetone, the actual and second hydrogen yields of Ce-Ni/Ce unexpectedly were lower than the those of Ni catalyst and even the blank sample, probably because the excessive Ce elements to some extent derease the dispersity of Ni on the carrier. Additionally, the third hydrogen yields of the nickeliferous catalysts with auxiliary agent elements were obviously higher than those of the other samples, verifying the auxiliary agent elements improved the anti-carbon deposition abilities of the catalysts. Notable, the third hydrogen yields of the compounds with Mg-Ni/Co catalyst were over 90%. For the three compounds, Mg-Ni/Co catalyst showed the best preformance among all the catalysts.

Catalystic performance of Mg-Ni/Co catalyst on simulated bio-oil

In this part, the three compounds with the equal masses were mixed as the simulated bio-oil and its steam reforming with Mg-Ni/Co catalyst was studied under the same conditions above.

Figure 5 (a) showed the hydrogen yields and the compositions of gaseous products with Mg-Ni/Co catalysts for the simulated bio-oil. The three hydrogen yields of the simulated bio-oil obtained through experiment were higher than the values obtained by the cumulation of the three compounds, indicating there was the synergistic effect between the three compounds in the SR processes of simulated bio-oil. The actual hydrogen yield of the simulated bio-oil could attain 63%. Although the H_2 concentration was lower than those of the three compounds (see Figure 4), the CO_2 concentration was obviously higher, to some extent indicating the SR reactions of the three compounds in simulated bio-oil were improved.

Long term experiments were carried out to study the stability of Mg-Ni/Co catalysts fresh and regenerative (the results shown in Figure 5 (b&c)). The regenerative catalysts derived from the ones after the first long term experiments. The regeneration process was: after the end of the first long term experiment, air was pumped into the reactor to make carbon deposits on the surface of the catalysts combust until the compositions of the tail gas were similar to those of air detected by the gas analyzer. For the fresh and regenerative catalysts, the compositions showed good stability and high coincidence in the whole term. Since with the peristaltic pump the delivery of the bio-oil and water was not strictly continuous (i.e. the drpolets were fed into the reactor every 3 seconds on average), the yields of the gaseous products showed some fluctuations with time. But the overall trends of the three types of hydrogen yields for the two kinds of catalysts were stable and the values of the same types of hydrogen yields were also similar, showing Mg-Ni/Co catalysts had the better stability and regeneration ability.

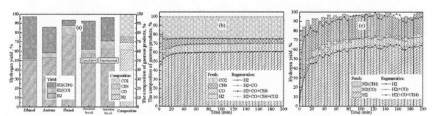

Figure 5. Gaseous products and hydrogen yields for the simulated bio-oil with Mg-Ni/Co catalyst.

Conclusions

The steam reforming process for hydrogen production of ethanol, acetone and phenol as three typical model compounds of bio-oil were studied through thermodynamic analysis and experiment. In the thermodynamic analysis, the three componds showed the similar trends. The hydrogen yields increased with the increase of S/C ratios, but when S/C ratios passes over 6, the hydrogen yields were very close. And with the increase of S/C ratios, the temperatures of the maximum hydrogen yields moved afterward to low temperature zone. However, the hydrogen contents were just about 70% with the the remaining gases mainly composed of CO_2. For improving hydrogen yield, a new system for SR of bio-oil with site CO_2 capture which can realize continuous running was put forward, requiring the catalyst with the larger granularity having greater mechanical strength and higher abrasion resistance, besides powdery CO_2 sorbents which mainly CaO based ones. So, six Ni-base catalysts loaded on Al_2O_3 particles prepared were studied and among them Mg-Ni/Co catalyst showed the best catalytic performance. For the simulated bio-oil (the mixture of the compounds), the hydrogen yield can attain 63% with Mg-Ni/Co catalyst, showing the good stability and regeneration ability.

References

1. M.A. Pena, J.P. Gomez and J.L.G. Fierro, "New catalytic routes for syngas and hydrogen production", *Applied Catalysis A: General*, 144 (1-2)(1996), 7-57.

2. E.C.Vagia and A.A. Lemonidou, "Thermodynamic analysis of hydrogen production via steam reforming of selected components of aqueous bio-oil fraction", *International Journal of Hydrogen Energy*, 32(2)(2007), 212-223.

3. X.D. Zhang et al., "Study on biomass pyrolysis kinetics", *Journal of Engineering for Gas Turbines and Power.* 128 (3)(2006), 493-496.

4. E.C. Vagia, and A.A. Lemonidou, "Investigations on the properties of ceria-zirconia-supported Ni and Rh catalysts and their performance in acetic acid steam reforming", *Journal of Catalysis*, 269(2) (2010), 388-396.

5. E.C. Vagia, and A.A. Lemonidou, "Thermodynamic analysis of hydrogen production via autothermal steam reforming of selected components of aqueous bio-oil fraction", *International Journal of Hydrogen Energy*, 33(10)(2008), 2489-2500.

6. D. Wang, S. Czernik and E. Chornet, "Production of hydrogen from biomass by catalytic steam reforming of fast pyrolysis oils", *Energy & Fuel*, 12(1998), 19-24.

7. W.J. Wang and Y.Q. Wang. "Steam reforming of ethanol to hydrogen over nickel metal catalysts", *International Journal of Energy Research*, 34(14) (2010), 1285-90.

8. X. Hu, and G. Lu, "Investigation of steam reforming of acetic acid to hydrogen over Ni-Co metal catalyst", *Journal of Molecular Catalysis a-Chemical*, 261(2007), 43-48.

9. Y. Cui et al., "Steam reforming of glycerol: The experimental activity of La1-xCexNiO3 catalyst in comparison to the thermodynamic reaction equilibrium", *Applied Catalysis B: Environmental*, 90(2009), 29-37.

10. E.C. Vagia, and A.A. Lemonidou, "Hydrogen production via steam reforming of bio-oil components over calcium aluminate supported nickel and noble metal catalysts", *Applied Catalysis A: General*, 351(2008), 111-121.

11. F. Seyedeyn-Azad et al., "Biomass to hydrogen via catalytic steam reforming of bio-oil over Ni-supported alumina catalysts", *Fuel Processing Technology*, 92(2011), 563-569.

12. C. Rioche et al., "Steam reforming of model compounds and fast pyrolysis bio-oil on supported noble metal catalysts", *Applied Catalysis B-Environmental*, 61 (2005), 130-139.

13. C.F. Yan, E.Y. Hu and C.L. Cai, "Hydrogen production from bio-oil aqueous fraction with in situ carbon dioxide capture", *International Journal of Hydrogen Energy*, 35 (2010), 2612-2616.

14. C. S. Martavaltzi and A. A. Lemonidou, "Development of new CaO based sorbent materials for CO2 removal at high temperature", *Microporous and Mesoporous Materials*, 110(2008), 119-127.

15. C. M. Kinoshita and S. Q. Turn, "Production of hydrogen from bio-oil using CaO as a CO2 sorbent", *International Journal of Hydrogen Energy*, 28 (2003), 1065-1071.

16. I. Zamboni, C. Courson and A. Kiennemann, "Synthesis of Fe/CaO active sorbent for CO2 absorption and tars removal in biomass gasification", *Catalysis Today*, 176(2011), 197-201.

Energy Technology 2014
Carbon Dioxide Management and Other Technologies

SYMPOSIUM:
High-Temperature Material Systems for Energy Conversion and Storage

Energy Technology 2014
Carbon Dioxide Management and Other Technologies

SYMPOSIUM: HIGH-TEMPERATURE
MATERIAL SYSTEMS FOR ENERGY
CONVERSION AND STORAGE

Solid Oxide Fuel Cells II

Session Chairs:

Xingbo Liu

JinhuaTong

Energy Technology 2014: Carbon Dioxide Management and Other Technologies
Edited by: Cong Wang, Jan de Bakker, Cynthia K. Belt, Animesh Jha, Neale R. Neelameggham,
Soobhankar Pati, Leon H. Prentice, Gabriella Tranell, and Kyle S. Brinkman
TMS (The Minerals, Metals & Materials Society), 2014

SURFACE SEGREGATION AND PHASE FORMATION IN THIN FILMS OF SOFC CATHODE MATERIALS

Jacob Davis[1], Yang Yu[1], Deniz Cetin[1], Karl Ludwig[1,3], Uday Pal[1,2], Srikanth Gopalan[1,2], and Soumendra Basu[1,2]

[1]Division of Materials Science and Engineering, Boston University,
15 St. Mary's Street, Brookline, MA 02446, USA
[2]Department of Mechanical Engineering, Boston University,
110 Cummington Street, Boston, MA 02215, USA
[3]Department of Physics, Boston University,
590 Commonwealth Avenue, Boston, MA 02215, USA

Keywords: LSM, LSCF, Surface segregation, Surface phase formation, TXRF, HAXPES

Abstract

A limiting factor for Solid Oxide Fuel Cell (SOFC) performance is the oxygen reduction reaction (ORR) at the cathode surface. Changes in the surface composition and structure on annealing at 800°C has been studied, since these can influence the ORR. Idealized single crystals of cathode materials $La_{0.8}Sr_{0.2}MnO_3$ (LSM) and $La_{1-x}Sr_xCo_yFe_{1-y}O_3$ (LSCF) were grown as heteroepitaxial thin films on lattice matched single crystal substrates. Changes upon heating the films to operating temperature and pressures were characterized using various synchrotron x-ray techniques. Total Reflection X-ray Fluorescence (TXRF) measurements, which probe compositional changes, were made at high temperature in real time. The LSM surfaces were found to develop manganese enrichment when heated. Highly strontium doped LSCF were found to develop strontium-rich surfaces. HArd X-ray PhotoElectron Spectroscopy (HAXPES) was used to investigate the electronic structure of the materials. Highly strontium doped LSCF precipitates a surface strontium phase that contains both oxide and carbonate contributions.

Introduction

Solid oxide fuel cells (SOFCs) are chemical to electrical energy conversion devices that could potentially reshape how energy demands are fulfilled. SOFCs have a potential efficiency over 70%, fuel flexibility, low emissions, and low degradation over long time scales. Commonly, SOFCs are operated at high temperatures (above 800°C). At these temperatures expensive housing is needed to contain an operating stack as well as coatings to contain the oxidation of the metallic interconnects. Lowering the temperature of an operating device would allow for more conventional materials to be used, thus lowering overall cost. Understanding the reaction kinetics of the oxygen reduction reaction at the surface of the cathode is vital to designing a system that will perform well at lower temperatures. A commercialization goal is to reduce operating temperatures without sacrificing performance.

In order to exploit the advantages of SOFCs, an understanding is needed of the surface chemical states of cations on the surface of the cathode and how this relates to the oxygen reduction reaction (ORR):

$$\frac{1}{2}O_{2(g)} + 2e' + V_{\ddot{o}} \rightarrow O_O^x \qquad (1)$$

where e' is an electron in the electrode, $V_{\ddot{o}}$ is an oxygen vacancy in the electrode and O_O^x is an oxygen on an oxygen site. Two different categories of cathodes are studied; an electronic conductor, strontium doped lanthanum manganite (LSM), and a mixed ionic electronic conductor, lanthanum strontium cobalt ferrite (LSCF). Both of these are well known SOFC cathode materials and have shown promising properties in working cells.

In LSM, surface instabilities have been reported in the form of strontium segregation to the surface leading to the formation of strontium enriched nanoparticles [1]. Yildiz et al. also reported an increase in the number of electronic La 5d-band vacancies in the near surface region of LSM [2]. In LSCF, the widely used composition $La_{0.6}Sr_{0.4}Co_{0.2}Fe_{0.8}O_3$ (LSCF-6428) exhibits surface instability resulting in surface strontium segregation. For both idealized thin films and grain surfaces of sintered cathodes LSCF-6428, surface strontium enrichment has been reported previously [3-5]. Oh et al. showed that precipitates form on grain surfaces upon annealing after 50 hours [6]. Previous studies attribute the surface strontium to oxide and hydroxide phases [4,7-9]. Segregation has been found for other ABO_3 perovskites including $La_{0.8}Sr_{0.2}CoO_{3\pm\delta}$. (LSC) [8,9]. Strontium oxide is an insulator [10]. High surface coverage of strontium oxide will reduce activity by blocking incorporation sites [6,11].

In this study, idealized thin films of LSM and LSCF were grown on single crystal lattice matched substrates. Thin films allow for a well-defined surface/gas interface with surfaces amiable to surface x-ray studies. A variety of x-ray fluorescence experiments were carried with the aim of measuring bulk and surface structure, local atomic environments after high temperature exposure.

Experimental

Epitaxial thin films of $(La_{0.8}Sr_{0.2})_{0.95}MnO_{3\pm\delta}$ (LSM-20) and $La_{0.6}Sr_{0.4}Co_{0.2}Fe_{0.8}O_{3\pm\delta}$ (LSCF-6428) deposited on lattice matched single crystal (100) $LaAlO_3$ (LAO) and (110) $NdGaO_3$ (NGO) substrates by pulsed laser deposition (PLD) at the Environmental Molecular Sciences Laboratory (EMSL) at Pacific Northwest National Laboratory (PNNL). In all cases, the films were nominally 250 nm in thickness, and were atomically smooth with surface roughness of the order of 0.5 nm. X-ray diffraction analysis showed excellent heteroepitaxy in all films, with LSM (100), LSCF (100) ∥ LAO (100) and LSM (100), LSCF (100) ∥ NGO (110).

The surface of the samples were examined by a variety of synchrotron x-ray, in the as deposited state and after annealing in air at 800°C for various times. Total reflection x-ray fluorescence (TXRF) experiments were carried out at beamline X23A2 at the National Synchrotron Light Source (NSLS) at Brookhaven National Laboratory (BNL). During a measurement, the incoming photon energy is set to be above the absorption edge energy of any lines that are to be measured, and the incident angle of the beam is scanned from a grazing angle to increasing values. In general, data taken at angles smaller than the critical angle, reflects the surface while at increasing incident angles above the critical angle interrogates the sample at increasing depths.

HArd X-ray PhotoElectron Spectroscopy (HAXPES) measurements were carried out at beamline X24A at the NSLS in BNL. Photoelectron spectroscopy makes use of the photoelectric effect. Incident x-rays of $h\nu$ collide with and eject electrons of a material. The emitted photoelectrons have a kinetic energy, E_K of [12]:

$$E_K = h\nu - \phi - E_B \tag{2}$$

where ϕ is the work function, and E_B is the binding energy of the electron in the material (before it was emitted). The binding energy is specific to the core-binding environment of each atomic element and oxidation state [12]. HAXPES offers some advantages over lower energy XPS techniques. Higher incident photon energies overcome surface contamination and result in improved signals without the need to heat in vacuum. Incident photon energy of 2140eV was used, and the energy was calibrated using the 3d binding energy measured from a silver foil.

Results and Discussion

LSM-20

Figure 1 shows TXRF data, the composition is plotted as (Sr+La)/Mn, which is the A/B ratio in LSM-20 thin films on LAO as a function of incident angle for samples in the as-deposited state, annealed in-situ for various times ranging for 1-8 hours in air at 800°C, and a sample annealed for 5 hours in air, quenched, and the data measured ex-situ.

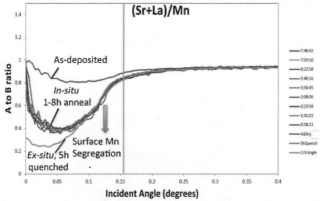

Figure 1. TXRF data showing A/B ratio as a function of incident angle for as deposited, in-situ annealed samples for 1-8 hours at 800°C in air, and a 5h annealed sample that was quenched, and measured ex-situ at room temperature.

The critical angle is marked at an incident angle of around 0.16 degrees. It is important to note that as the incident angle approaches zero, the beam hits the side of the sample, and the La signal from the substrate increases. Thus the best way to read this figure is to start at the critical angle and look at the signal from right to left since the increase in the plots close to 0 degrees is an artifact from substrate interference. The figure shows that there is a clear decrease in the A to B ratio as a result of annealing, and that decrease occurs within the first hour of annealing. Also, in the region of interest (just to the left of the critical angle), the ex-situ sample overlaps with the in-situ samples, indicating that ex-situ studies are valid in these materials. It should be noted that the ex-situ data has much smaller noise, since the data collection can occur for long times, without changing the effective annealing time of the samples. To remove the substrate interference, LSM-20 films on NGO were examined in the as-deposited, and after a 5 hour anneal and cool down, by ex-situ measurements. Figure 2 shows the data, plotted both as Sr/(Sr+La) and (Sr+La)/Mn. Also shown in the figures is a plot of the attenuation length, the depth from which information is being collected from the sample.

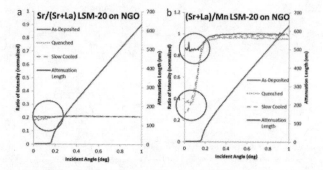

Figure 2. TXRF data of a) Sr/(Sr+La) and b) (Sr+La)/Mn for as-deposited and room temperature ex-situ measurements on samples cooled slowly and quenched after a 5h anneal in air at 800°C.

Figure 2 shows that the cooling rate does not have a major effect on the room temperature state of the samples after annealing. Figure 2a shows that no changes in the A-cation sublattice takes place on annealing, indicating that the reduction of (Sr+La)/Mn signal on annealing in Figure 2b is due to Mn enhancement at the surface. Figure 3 shows an AFM scan of a sample annealed for 12h at 800°C in air. The figure shows virtually no change in surface roughness, and more importantly, no second phase formation at the surface.

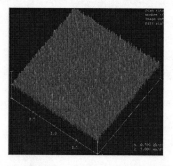

Figure 3. AFM scan of the surface of LSM-20 after a 12h anneal at 800°C in air, showing that there is no second phase formation at the surface.

Thus, it is evident that Mn surface segregation occurs with no second phase formation in LSM-20 with rapid kinetics at 800°C in air, where the surface composition reaches an equilibrium value within 1 hour. Figure 4 shows the effect of the Mn segregation on the point defects in LSM, based on a defect model, developed by Poulsen [13]. The figure shows that Mn surface segregation enhances the concentration of Mn^{+4} sites and oxygen vacancies, both of which should enhance the ORR reaction by increasing the sites of oxygen adsorption and oxygen incorporation at the surface [14]. Thus, to the extent, application of bulk thermodynamics is valid in the near surface region, the Mn segregation appears to be beneficial.

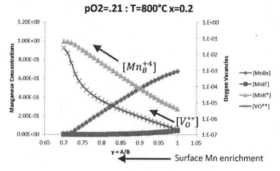

Figure 4. Effect of Mn surface segregation on defect concentrations based on defect model developed by Poulsen [13].

LSCF-6428

Figure 5 shows in-situ TXRF data, showing the evolution of the Sr/(Sr+La) signal ratio of LSCF-6428 measured as a function of time at 800°C in air. At 800°C the surface is still evolving, with strontium content increasing over eight hours.

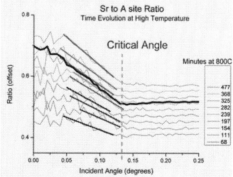

Figure 5. TXRF data of in-situ Sr/(Sr+La) ratio, as a function of incident angle. The plots are intentionally offset to highlight the change in the slope of the plots to the left of the critical angle, implying continued Sr surface enhancement. The black line is a 5h ex-situ data, showing good match to the in-situ data for a similar time.

A problem with *in-situ* high temperature measurements is limited ability to time-average scans as the surface is evolving. To get better statistics, an attempt was made to quench from high temperature and see if the high temperature state is preserved. Also plotted in Figure 5 is a thick black line is data taken for a sample quickly quenched to room temperature after 5 hours at 800°C. The *ex-situ* sample is consistent with high temperature measurements and allows for longer count times and better counting statistics. The match with the in-situ and ex-situ data is reasonable enough that ex-situ samples were used in more detailed studies of the surface segregation and changes to the surface bonding states.

The slope of the Sr/(Sr+La) plot just below the critical angle was taken to be an indicator of surface segregation. Figure 6 shows a plot of how the slope evolves for the in-situ and ex-situ measurements. Although the two plots do not completely overlap, they show similar trends. Also, the error bars of the ex-situ measurements are significantly smaller, a reflection of longer count times.

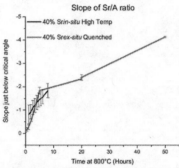

Figure 6. Evolution of the slope of the Sr/(Sr+La) plot just below the critical angle for an LSCF 6428 sample for both in-situ and ex-situ measurements.

To examine if this Sr segregation is accompanied by surface phase formation, the surface of the ex-situ samples were examined by HAXPES. Figure 7 shows a typical Sr 3d signal from a sample. Strontium 3d is a doublet from spin orbit splitting into $3d_{3/2}$ and $3d_{5/2}$ [4].

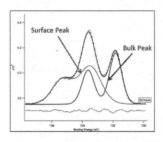

Figure 7. HAXPES Sr 3d signal from LSCF-6438, showing spin orbit splitting into $3d_{3/2}$ and $3d_{5/2}$ [4]. The signal can be fitted to two separate doublets, one identified with a surface phase, and another with the bulk perovskite phase.

Strontium 3d is a doublet from spin orbit splitting into $3d_{3/2}$ and $3d_{5/2}$. The data has been fit to two doublets. The peak at 131.8eV can be identified as the bulk perovskite phase. The higher energy peak is from the surface and evolves in both area and position with annealing time.

To compare the amount of surface strontium to bulk perovskite strontium the ratio of the integrated areas of the two peaks is plotted in Figure 8a. The contribution of surface strontium is increasing with anneal time and seems to reach equilibrium after 5 hours. In Figure 8b, the position of the surface and bulk peaks are plotted for the various annealing times. The bulk perovskite peak at lower energy is essentially unchanging. The surface peak falls between 133.0 and 134.0eV. These energies correspond to strontium oxide (SrO) and strontium carbonate (SrCO$_3$) respectively [4]. The increase in binding energy with time indicates an increase in strontium carbonate, which is thermodynamically expected to form due to the reaction between SrO and atmospheric CO$_2$ for the ex-situ samples. Due to the formation of the second phase, it is important to separate the composition of the surface perovskite phase, before point defect models can be applied. This is the focus of the ongoing research.

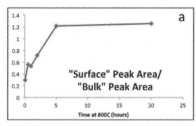 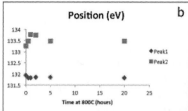

Figure 8. a) Plot of the area under the Sr 3d surface peak/area under the Sr 3d bulk peak as a function of annealing time at 800°C in air. b) Change in position of the Sr 3d surface and bulk peaks with annealing time at 800°C in air.

Conclusions

The surface of LSM and LSCF thin films undergo complex changes when they are exposed to conditions relevant to SOFC operation. Since it is at these cathode surfaces where oxygen is incorporated, such surface instability should effect oxygen incorporation. LSM-20 exhibits surface manganese enrichment at 800°C in air. This enrichment saturates within 1 hour and is irreversible upon cooling. Defect modeling suggests the surface manganese enrichment may enhance the ORR reaction by increasing the ratio of Mn^{4+} to Mn^{3+}, and the oxygen vacancy concentration at the cathode surface. LSCF-6428 develops surfaces phases as strontium segregates out of the perovskite phase at 800°C in air. Comparative TXRF measurements in real time at high temperature and of quenched samples show that quenching preserves the high temperature state of the surface. This segregation begins in the first hour at high temperature and continues to increase over at least 50 hours. The strontium rich surface phases are consistent with SrO and SrCO$_3$.

Acknowledgements

This work is supported through the DOE SECA program under Grant DEFC2612FE0009656. A portion of the research was performed at EMSL, a national scientific user facility sponsored by the Department of Energy's Office of Biological and Environmental Research and located at Pacific Northwest National Laboratory, and at the National Synchrotron Light Source, Brookhaven National Laboratory. The authors would like to acknowledge the contributions of K.E. Smith at Boston University, T.C. Kaspar at PNNL, and J.C. Woicik at NIST.

References

1. K.C. Chang, B. Ingram, B. Kavaipatti, B. Yildiz, D. Hennessy, P. Salvador, N. Leyarovska, and H. You, "In situ Synchrotron X-ray Studies of Dense Thin-Film Strontium-Doped Lanthanum Manganite Solid Oxide Fuel Cell Cathodes", *Solid State Ionics*, 1126 (2009), 27-32.

2. B. Yildiz, D.J. Myers, J.D. Carter, K.C. Chang, and H. You, "In Situ X-ray and Electrochemical Studies of Solid Oxide Fuel Cell/Electrolyzer Oxygen Electrodes", *Advances in Solid Oxide Fuel Cells*, 28 (2008), 153-164.

3. S.P. Simner, M.D. Anderson, M.H. Engelhard, and J.W. Stevenson. "Degradation Mechanisms of La–Sr–Co–Fe–O_3", *Electrochemical and Solid-State Letters*, 9 (2006), A478-A481.

4. P.A.W. van der Heide, "Systematic x-ray photoelectron spectroscopic study of $La_{1-x}Sr_x$-based perovskite-type oxides", *Surface and Interface Analysis*, 33 (2002), 414-425.

5. E. Bucher, and W. Sitte, "Long-term stability of the oxygen exchange properties of $(La,Sr)_{(1-z)}(Co,Fe)O_{3-\delta}$ in dry and wet atmospheres", *Solid State Ionics*, 192 (2011), 480-482.

6. D. Oh, D. Gostovic, and E.D. Wachsman, "Mechanism of $La_{0.6}Sr_{0.4}Co_{0.2}Fe_{0.8}O_3$ cathode degradation", *Journal of Materials Research*, 27 (2012), 1992-1999.

7. E.J. Crumlin, E. Mutoro, Z. Liu, M.E. Grass, M.D. Biegalski, Y.L. Lee, D. Morgan, H.M. Christen, H. Bluhm, and Y. Shao-Horn, "Surface strontium enrichment on highly active perovskites for oxygen", *Energy & Environmental Science*, 5 (2012), 6081-6088.

8. E. Mutoro, E.J. Crumlin, H. Popke, B. Luerssen, M. Amati, M.K. Abyaneh, M.D. Biegalski, H.M. Christen, L. Gregoratti, J. Janek, and Y. Shao-Horn, "Reversible Compositional Control of Oxide Surfaces by Electrochemical Potentials", *Journal of Physical Chemistry Letters*, 3 (2012), 40-44.

9. Z.H. Cai, M. Kubicek, J. fleig, and B. Yildiz, "Chemical Heterogeneities on $La_{0.6}Sr_{0.4}CoO_{3-\delta}$ Thin Films-Correlations to Cathode Surface Activity and Stability", *Chemistry of Materials*, 24, (2012), 1116-1127.

10. W.D. Copeland, and R.A. Swalin. "Sudies on Defect structure of Strontium Oxide", *Journal of Physics and Chemistry of Solids*, 29, (1968), 313-325.

11. E. Mutoro, E.J. Crumlin, M.D. Biegalski, H.M. Christen, and Y. Shao-Horn, "Enhanced oxygen reduction activity on surface-decorated perovskite thin films for solid oxide fuel cells", *Energy & Environmental Science*, 4 (2011), 3689-3696.

12. K. Siegbahn, "Preface to hard X-ray photo electron Spectroscopy (HAXPES)", *Nuclear Instruments & Methods in Physics Research Section A-Accerlaerators Spectrometers Detectors and Associated Equipment*, 547, (2005) 1-7.

13. F.W. Poulsen, "Defect chemistry modelling of oxygen-stoichiometry, vacancy concentrations, and conductivity of $(La_{1-x}Sr_x)_{(y)}MnO_{3+/-\delta}$", *Solid State Ionics*, 129 (2000), 145-162.

14. J.N. Davis, "Surface Phase Emergence and Evolution of Solid Oxide Fuel Cell Cathode Materials", Ph.D. Dissertation, Boston University (2013).

Energy Technology 2014: Carbon Dioxide Management and Other Technologies
Edited by: Cong Wang, Jan de Bakker, Cynthia K. Belt, Animesh Jha, Neale R. Neelameggham,
Soobhankar Pati, Leon H. Prentice, Gabriella Tranell, and Kyle S. Brinkman
TMS (The Minerals, Metals & Materials Society), 2014

AN INTERRUPTED IN-SITU METHOD FOR ELECTROCHEMICAL FORMATION OF Mg-Ni INTERMETALLICS

Fuat Erden[1,2], İshak Karakaya[3] and Metehan Erdoğan[4]

[1]Institute of Materials Research and Engineering (IMRE), Agency for Science Technology and Research (A*STAR); 3 Research Link, Singapore, 117602

[2]National University of Singapore; 21 Lower Kent Ridge Road, Singapore, 119077

[3]Middle East Technical University; Üniversiteler Mah. Dumlupınar Blv. No:1, 06800, Çankaya, Ankara/TÜRKİYE

[4]Yıldırım Beyazıt Üniversitesi; Çankırı Caddesi Çiçek Sokak No:3 Altındağ, Ulus, Ankara/TÜRKİYE

Keywords: Mg_2Ni, Electrochemical Reduction, Molten Salt Electrolysis, Electrodeoxidation

ABSTRACT
Mg-Ni system draws attention due to the promising hydrogen storage properties of Mg_2Ni. The proposed method involves the use of NiO and $MgCl_2$ in an electrochemical cell to from Mg-Ni intermetallics directly. The electrodeoxidation of NiO was followed by combination of Mg and Ni in molten salt electrolytes. The products were characterized by XRD analysis and SEM examinations. The quantitative results from Rigaku supported by Rietveld Refinement method conducted by Maud showed that more than 50 percent of the input Ni was converted to Mg-Ni intermetallics.

INTRODUCTION
It is well known that fossil fuels are the most common energy producing materials throughout the last century although, they are not renewable and their usage cause critical environmental problems [1, 2]. Hydrogen, on the other hand, is known as a clean, renewable energy carrier and instead of fossil fuels; usage of hydrogen for energy production will be the benefit of mankind. However, there are problems that hinder more common usage of hydrogen such as hydrogen storage [3]. Mg_2Ni is known as Mg based hydrogen storage material with high hydrogen storage capacity [3-6]. Common Mg_2Ni production methods are; mechanical alloying [8-11], vacuum induction melting [12, 13], vacuum arc melting [14], combustion synthesis [15], melt spinning [16], repetitive rolling [17] and isothermal evaporation casting [18] were also used. However, in all of these, pure Mg and Ni are required to be produced previously. Therefore, difficulties arise during handling, preparation and combination of Mg and Ni due to highly reactive nature of especially magnesium (i.e. oxidation of Mg) [11, 17, 19, 20]. Although hydrogen desorption dynamics can be significantly improved with usage of nanostructured Mg_2Ni [7], it is difficult and it requires very long processing durations to produce nanostructured Mg_2Ni by above processes.

Electrodeoxidation of MgO-NiO mixtures were attempted to form nanostructured Mg-Ni intermetallics [21]. The complete reduction of NiO and partial $MgNi_2$ formation were achieved but Mg_2Ni formation was not successful [21]. In another study, partial Mg_2Ni formation was obtained [22] when electrodeoxidation of NiO was followed by molten salt electrolysis of $MgCl_2$ in the same cell. The aim of this study is to enhance Mg_2Ni formation by interrupted combination of electrolysis and electrodeoxidation steps.

EXPERIMENTAL PROCEDURE

Figure 1 shows the schematic experimental setup used in electrochemistry experiments. 0.8 grams of NiO pellet was placed on a stainless steel spoon-like electrode, which was welded to a Kanthal current collector. The wire was covered by an insulating alumina tube near the bottom of the cell. A quartz insulator covered graphite rod (Alfa Aesar A10134) was used as anode. The electrolyte was initially CaCl2 but $MgCl_2$ and NaCl were added after the initial electrodeoxidation stage. Hygroscopic salts, CaCl2 and MgCl2 were subjected to a drying procedure [23], before they were used in the experiments. The crucible, containing the electrolyte, was positioned in a 5cm outside diameter, 60cm long quartz vessel. Anode lead and cathode lead were extended out of the cell vessel from the top and connected to positive and negative poles of Agilent N6700B Power Supply respectively.

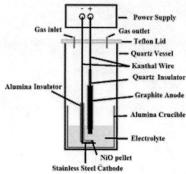

Figure 1. Schematic Representation of the Experimental Setup

Table 1 and Figure 2, which were used to determine the experiment conditions for the second part, show the potential requirement of possible reactions.

Table 1. Possible Cell Reactions, their standard Gibbs energy changes and standard potentials

No	Reaction	ΔG^o (J)	E^o (V)
(1)	NiO (s)+ C (s) = Ni + CO (g)	-37100.2 [24]	0.19
(2)	$MgCl_2$ = Mg + Cl_2 (g)	491922.4 [24]	-2.55
(3)	$MgCl_2$ + 2Ni = $MgNi_2$ + Cl_2 (g)	441108 [24]	-2.29
(4)	$MgCl_2$ +(1/2) Ni = (1/2)Mg_2Ni + Cl_2 (g)	469039 [24, 25]	-2.43
(5)	$3MgCl_2$ + $MgNi_2$ = $2Mg_2Ni$ + $3Cl_2$	1435048 [24, 25]	-2.48

260

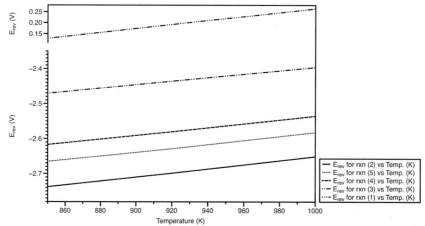

Figure 2. The reversible cell voltage vs temperature graphs of reactions (1), (2), (3), (4) and (5)

The electrochemical test procedure was divided to two parts. Initially, 60 grams of $CaCl_2$ was dried, and used as the electrolyte during the electrodeoxidation of 0.8g NiO at 3V applied potential at 825°C for 5 hours. In some of the experiments, NiO pellet was used directly in the experiments without sintering while it was sintered at 1200°C for 6 hours in some others. Ammonium bicarbonate was used to control the porosity when the pellet is sintered. After electrodeoxidation of NiO, the cell was cooled to room temperature and the electrolyte compositions were adjusted to form Mg-Ni intermetallics. Three different electrolyte compositions were prepared to examine the effect of $MgCl_2$ concentration on Mg-Ni intermetallic formation. 4.08, 8.16 or 16.32 grams of $MgCl_2$ together with 31.6 grams of NaCl and 3 grams of NH_4Cl were added to the $CaCl_2$ electrolyte. In the second part of experiments, voltages of 2.7 V or 2.9 V and temperatures of 645°C or 700°C were used. The duration for the second part of the experiments was 48 hours.

After the experiments, reaction products were washed in ultrasonic mixer by water, ethanol and methanol mixture for 2 hours. Reaction products were identified by x-ray diffraction analysis. Rigaku SA-HF-3 X-ray Generator and Diffractometer, Rigaku Geigerflex X-Ray Powder Diffractometer and Rigaku Ultima IV X-ray Diffractometer were used for this purpose. Quantitative analyses were conducted by "Rigaku" software to compare the test results. Byproducts were removed in quantitative analysis by normalization to show the Mg-Ni intermetallic formation results clearly. Results of "Rigaku" were checked by mass balance [26] and Rietveld Refinement Method which was conducted by "Maud".

RESULTS and DISCUSSION

At first, NiO electrodeoxidation was completed by applying 3.0 V at 825°C for 5 hours in molten $CaCl_2$ salt. Figure 3 shows the calculated accumulated charge vs. time graph obtained during one of the experiments for electrodeoxidation of a 0.8 gram unsintered NiO pellet.

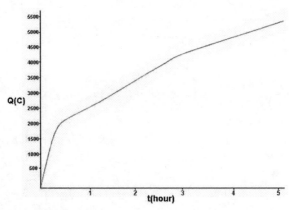

Figure 3. Calculated Total Charge passed versus time graph obtained for electrodeoxidation of unsintered NiO pellet

From Figure 3, it can be seen that accumulated charge passed was above 5000 Coulombs (C) after 5 hours. The calculations according to Faraday's Law show that around 2067 C was required for complete reduction of 0.8 g NiO when 100% current efficiency is assumed. It can be seen in Figure 3 that passage of 2067 C was achieved within about 45 minutes, which shows that most of the NiO powder was reduced in the first 45 minutes. Therefore, 5 hours of electrodeoxidation, which resulted in passage of about 5000 C of total charge, suggests that all of the NiO pellet was reduced to Ni, even if a current efficiency lower than 50% is assumed. As a result, the second part of the experiments were safely started after 5 hours of electrodeoxidation in the first step for both unsintered and sintered NiO pellets. It's known that the electrodeoxidation of NiO is expected to be faster in $CaCl_2$ when the pellet was sintered [27, 28].

The results of the second part of the experiments indicated partial Mg-Ni intermetallic formation. The results of some of the experiments are shown below to discuss the effects of applied potential, porosity, sintering, temperature and $MgCl_2$ amount on Mg-Ni intermetallics formation.

Effect of Applied Potential:
Table 2 summarizes the experimental parameters and sequences used for tests to determine the effect of applied potential in the second part on Mg-Ni intermetallic formation.

Table 2. Experimental Parameters used to study the effect of applied potential

Experiment 1	Experiment 2
Unsintered 0.8 gram NiO pellet	Unsintered 0.8 gram NiO pellet
Porosity: 43 %	Porosity: 43 %
First Step: NiO electrodeoxidation at 825°C for 5 hours by applying 3V	First Step: NiO electrodeoxidation at 825°C for 5 hours by applying 3V
Second Step: Electrolysis at **2.7** V for 48 hours at 645°C	Second Step: Electrolysis at **2.9** V for 48 hours at 645°C

In both experiments, X-ray diffraction results indicated formation of similar compounds. Table 3, which was obtained by the results of the quantitative analysis conducted by "Rigaku" software, shows the compositions of the cathode samples. Figure 4 shows the x-ray diffraction result of the sample obtained after experiment at 2.9 V applied potential.

Table 3. Compositions of the cathode samples obtained from tests used to study the effect of applied potential

Experiment 1 (wt. %)	Experiment 2 (wt. %)
Ni: 84	Ni: 31
MgO: 8	MgO: 23
MgNi$_2$: 5	Mg$_2$Ni: 41
Mg(OH)$_2$: 3	Mg(OH)$_2$: 5

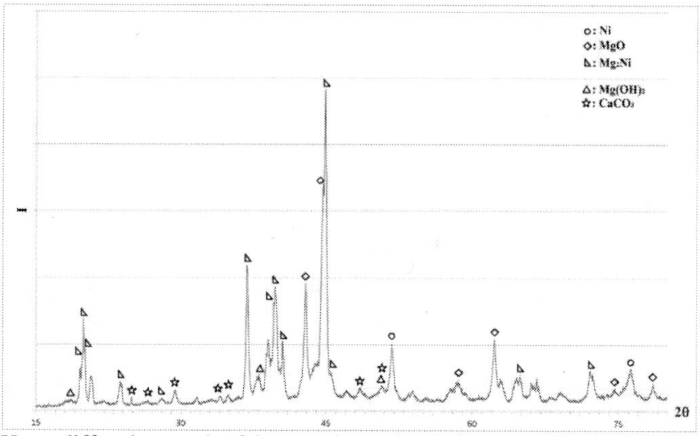

Figure 4. X-ray diffraction result of the sample obtained after the experiment at 2.9 V applied potential

Partial formation of MgNi$_2$ was observed in experiment at 2.7V applied potential without (or with very little) Mg$_2$Ni. On the other hand, about 40 % Mg$_2$Ni formation was observed when the applied potential was 2.9 V. It can be said that the applied potential of 2.7 V did not allow molten salt electrolysis of MgCl$_2$ when I-R drop and over-potentials are considered (see Table 1). However, the applied potential of 2.9 V was probably sufficient for all possible reactions in the cell to proceed. Moreover, better results in terms of higher Mg-Ni intermetallics amount at higher applied potential was probably the result of faster reactions at higher applied potentials.

It is important to note that Mg$_2$Ni formation was observed without or with very little MgNi$_2$ at 2.9 V applied potential. This can be explained by reaction (5) in Table 1, which shows that MgNi$_2$ produced could be converted to Mg$_2$Ni in the presence of excess MgCl$_2$.

In the following experiments, 2.9 V was applied as the cell potential because it yielded higher Mg-Ni intermetallics formation than 2.7 V. However, potentials higher than 2.9V were not studied in this study to avoid molten salt electrolysis of other components in the electrolyte.

Effect of Porosity & Sintering:
Table 4 summarizes the experimental parameters used for tests to determine the effect of sintering and porosity on Mg-Ni intermetallic formation.

Table 4. Experimental Parameters used to study the effect of sintering and porosity

Experiment 1	Experiment 2	Experiment 3
Unsintered 0.8 g NiO pellet	Sintered 0.8 g NiO pellet	Sintered 0.8 g NiO pellet
-	Porosity: 40 %	Porosity: 65 %
First Step: NiO electrodeoxidation at 825°C for 5 hours by applying 3V	First Step: NiO electrodeoxidation at 825°C for 5 hours by applying 3V	First Step: NiO electrodeoxidation at 825°C for 5 hours by applying 3V
Second Step: Electrolysis at 2.9 V for 48 hours at 645°C	Second Step: Electrolysis at 2.9V for 48 hours at 645°C	Second Step: Electrolysis at 2.9V for 48 hours at 645°C

Table 5 was prepared by the results of the quantitative analysis conducted by "Rigaku" software. Figure 5 shows the x-ray diffraction results of sample obtained after Experiment 3.

Table 5. Compositions of the cathode samples obtained from tests used to study the effect of sintering and porosity

Experiment 1 (wt. %)	Experiment 2 (wt. %)	Experiment 3 (wt. %)
Ni: 31 MgO: 23 Mg_2Ni: 41 $Mg(OH)_2$: 5	Ni: 68 MgO: 19 Mg_2Ni: 7 $MgNi_2$: 4 $Mg(OH)_2$: 2	Ni: 50 MgO: 19 Mg_2Ni: 23 $MgNi_2$: 7 $Mg(OH)_2$: 2

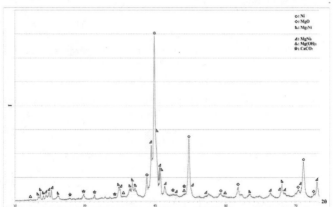

Figure 5. X-ray diffraction result of the sample obtained after Experiment 3

It is observed that the amounts of Mg_2Ni and $MgNi_2$ were less in experiments with sintered samples when compared to experiments with unsintered NiO. Sintered pellets were observed to remain in compact form after the experiments; whereas, unsintered ones were found to disperse on the cathode. The reason for higher Mg_2Ni content in experiments with unsintered starting materials can be explained by the creation of space due to dispersed pellet.

When experiments with sintered starting materials are compared, it can be seen that higher porosity yielded higher Mg-Ni intermetallic amount. This finding contributed to the high porosity requirement for Mg-Ni intermetallics formation. Calculations show that at least 64.2

% porosity was required to fully convert NiO to Mg_2Ni because the molar volume of Mg_2Ni is higher when compared to NiO. Although 65% porosity in Experiment 3 seemed to be sufficient to convert all NiO to Mg_2Ni, shrinkage is expected during electrodeoxidation of NiO to Ni. Moreover, as the atomic radius of Mg is larger than that of Ni, even with higher porosities, it is difficult to produce Mg_2Ni, when Mg starts diffusing to the inner lattice of the Ni particles. Besides, crystal structures of Mg and Mg_2Ni are hexagonal close packed (HCP) and that of Ni is face centered cubic (FCC). Both diffusion, being a kinetically slower process, and the requirement for the formation of a HCP crystal structure may impose limitations on the extent of formation of Mg_2Ni phase.

Effect of Temperature:

Table 6 summarizes the experimental parameters used for tests to determine the effect of temperature on Mg-Ni intermetallic formation.

Table 6. Experimental Parameters used to study the effect of temperature

Experiment 1	Experiment 2
Unsintered 0.8 g NiO pellet	Unsintered 0.8 g NiO pellet
Porosity: 43%	Porosity: 43%
First Step: NiO electrodeoxidation at 825°C for 5 hours by applying 3 V	First Step: NiO electrodeoxidation at 825°C for 5 hours by applying 3V
Second Step: Electrolysis at 2.9 V for 48 hours at **645°C**	Second Step: Electrolysis at 2.9 V for 48 hours at **700°C**

Table 7 was prepared by the results of the quantitative analysis conducted by "Rigaku" software:

Table 7. Compositions of cathode samples obtained from tests used to study the effect of temperature

Experiment 1 (wt. %)	Experiment 2 (wt. %)
Ni: 31 MgO: 23 Mg_2Ni: 41 $Mg(OH)_2$: 5	Ni: 44 MgO: 21 Mg_2Ni: 15 $Mg(OH)_2$: 8 $MgNi_2$: 12

It can be seen that 645°C yielded higher Mg-Ni intermetallic formation. It should be noted that Mg is solid at 645°C and liquid at 700°C. Therefore, one of the possible reasons for better result at 645°C could be the difficulty of remaining on the cathode for liquid magnesium. However, it is important to note that more experiments should be conducted to clearly observe the effect of temperature on Mg-Ni intermetallic formation.

Rietveld refinement was carried out by the computer program Maud [29] on the x-ray results obtained after experiment at 700°C to check the reliability of quantitative analysis results of Rigaku. Figure 6 shows the difference in the observed x-ray diffraction pattern and x-ray diffraction fit calculated by Maud. Table 8 was prepared to compare the quantitative analysis results of Rigaku and Maud.

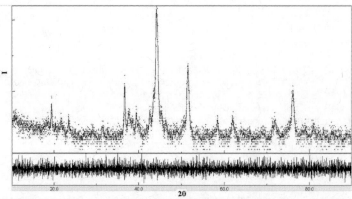

Figure 6. Rietveld Refinement conducted by Maud on the x-ray diffraction results of reduced sample obtained after experiment at 700°C

Table 8. Comparison of Quantitative Analysis Results of Rigaku and Maud

Results calculated by Rigaku (wt. %)	Results calculated by Maud (wt. %)
Ni: 44	Ni: 42.6
MgO: 21	MgO: 23.8
Mg_2Ni: 15	Mg_2Ni: 13.9
$Mg(OH)_2$: 8	$Mg(OH)_2$: 6.9
$MgNi_2$: 12	$MgNi_2$: 12.8

Blue circles in Figure 6 show the observed diffraction intensities and the black line is the best fit for the Rietveld Refinement. Rw value was calculated as 29.125 for the x-ray fit drawn by Maud. This value can be considered as acceptable when the instrumental errors, high noise on x-ray diffraction pattern and some unindexed phases are taken into account. In addition, the differences between quantitative analyses results performed by Maud and Rigaku are within acceptable limits as can be seen from Table 8. Therefore, it can be said that the Rietveld Refinement of Maud supports quantitative analysis results of Rigaku used in this study.

Effect of $MgCl_2$ amount in electrolyte:
Table 9 and 10 summarize the experimental parameters used for tests to determine the effect of $MgCl_2$ amount on Mg-Ni intermetallic formation and compositions of cathode samples obtained from these tests by "Rigaku" software, respectively.

Table 9. Experimental Parameters used to study the effect of $MgCl_2$ amount

Experiment 1	Experiment 2	Experiment 3
2 times excess $MgCl_2$ $MgCl_2$ amount: 4.08 g	4 times excess $MgCl_2$ $MgCl_2$ amount: 8.16 g	8 times excess $MgCl_2$ $MgCl_2$ amount: 16.32 g
Unsintered 0.8 g NiO pellet	Unsintered 0.8 g NiO pellet	Unsintered 0.8 g NiO pellet
Porosity: 43 %	Porosity: 43 %	Porosity: 43 %
First Step: NiO electrodeoxidation at 825°C for 5 hours by applying 3V	First Step: NiO electrodeoxidation at 825°C for 5 hours by applying 3V	First Step: NiO electrodeoxidation at 825°C for 5 hours by applying 3V
Second Step: Electrolysis at 2.9 V for 48 hours at 700°C	Second Step: Electrolysis at 2.9 V for 48 hours at 700°C	Second Step: Electrolysis at 2.9 V for 48 hours at 700°C

Table 10. Compositions of cathode samples obtained from of tests used to study the effect of MgCl$_2$ amount

Experiment 1	Experiment 2	Experiment 3
Ni: 78	Ni: 42	Ni: 24
MgO: 11	MgO: 19	MgO: 20
MgNi$_2$: 8	Mg$_2$Ni: 13	Mg$_2$Ni: 45
Mg(OH)$_2$: 3	Mg(OH)$_2$: 14	Mg(OH)$_2$: 5
	MgNi$_2$: 12	Mg$_6$Ni: 6

Mg$_2$Ni was not observed in reduction products when two times excess MgCl$_2$ was used as can be seen in Table 10. On the other hand; MgNi$_2$, which requires one mole of Mg per mole of compound formed, was observed to some extent. Moreover, MgO amount was lower when compared to higher MgCl$_2$ containing electrolytes. Since MgCl$_2$ amount was lower in Experiment 1, MgCl$_2$ activity in the electrolyte was smaller which results in higher potential requirements for reactions (2), (3), (4), and (5) given in Table 1. Another possible reason for less Mg containing compound formation in Experiment was the slow reaction kinetics because Mg amount in the electrolyte was less when compared to the other experiments.

Highest Mg-Ni intermetallic formation was observed after experiment with eight times excess MgCl$_2$. Stoichiometric calculations showed that about 52 % of the initial Ni was transformed to Mg-Ni intermetallics after Experiments 3.

From the results presented in Table 10, it can be proposed that the first Mg-Ni intermetallic to form during the experiments was MgNi$_2$. Mg$_2$Ni was probably formed from MgNi$_2$ and MgCl$_2$ according to reaction (5) as explained previously. The absences of Mg$_2$Ni in Experiment 1 and MgNi$_2$ in Experiment 3 add credibility to this proposal.

CONCLUSIONS

Electrochemical reduction of NiO particles to nickel powder was followed by in-situ formation of Mg-Ni intermetallics in molten salt electrolytes where MgCl$_2$ was introduced to the electrolyte after the reduction of NiO. Two step experiments revealed that higher potentials resulted in more Mg-Ni intermetallic formation. Although sintering and decrease in NiO porosity expected to affect NiO reduction positively, Mg-Ni intermetallic formation was affected negatively. It was also found that experiments conducted at 645°C yielded better results than 700°C. This result could be explained with reference to the fact that Mg is solid at 645°C and liquid at 700°C. Finally, increase in excess MgCl$_2$ addition to the electrolyte was found to yield better results for Mg$_2$Ni formation. However it is important to set a limit to MgCl$_2$ amount, because above a certain value, Mg$_6$Ni formation may be seen at the expense of produced Mg$_2$Ni.

References

[1]Momirlan, M., and Veziroglu, T.N. *Renewable and Sustainable Energy Reviews.* Current Status of Hydrogen Energy, 2002. 6: pp. 141-179.
[2]Veziroglu, T.N. *Hydrogen Technology for Energy Needs of Human Settlements.* International Journal of Hydrogen Energy, 1987. 12(2): pp. 99-129.

[3]Sakintuna, B., Lamari-Darkrim, F., and Hirscher, M. *Metal Hydride Materials For Solid Hydrogen Storage: A Review.* International Journal of Hydrogen Energy, 2007. 32: pp. 1121-1140.

[4]Vijay, R., Sundaresan, R., Maiya, M.P., and Srinivasa Murthy, S. *Comparative evaluation of Mg–Ni hydrogen absorbing materials prepared by mechanical alloying.* International Journal of Hydrogen Energy, 2005. 30: pp. 501-508

[5]Abdellaoui, M., Cracco, D., and Percheron-Guegan, A. *Structural characterization and reversible hydrogen absorption properties of Mg_2Ni rich nanocomposite materials synthesized by mechanical alloying.* Journal of Alloys and Compounds, 1998. 268: pp. 233–240.

[6]Iturbe Garcia, J. L., Lopez-Munoz, M. N., Basurto, R., and Millan, S. *Hydrogen Desorption Process in Mg_2Ni hdyrides.* Revista Mexicana De Fisica, 2006. 52(4): pp. 365-367.

[7]Jin, Y., Yang, X., Li, Y. and Zhang, W. *Ynthesis and Characterization of Nonstructural MG2NI with Replacement Diffusion Method.* Modern Applied Science, 2010, 4(8): pp. 114-118.

[8]Liang, G., Boily, S., Hout, J., Van Neste, A., and Schulz, R. *Mechanical alloying and hydrogen absorption properties of the Mg–Ni system.* Journal of Alloys and Compounds, 1998, 267: pp. 302–306.

[9]Song, M.Y. *Effects of mechanical alloying on the hydrogen storage characteristics of Mg– x wt% Ni (x = 0, 5, 10, 25 and 55) mixtures.* International Journal of Hydrogen Energy, 1995, 20: pp. 221–227.

[10]Aymard, L., Ichitsubo, M., Uchida, K., Sekreta, E., and Ikazaki, F. *Preparation of Mg_2Ni base alloy by the combination of mechanical alloying and heat treatment at low temperature.* Journal of Alloys and Compounds, 1997, 259: pp. L5–L8.

[11]Zaluska, A., Zaluski, L., and Ström-Olsen, J. O. *Synergy of hydrogen sorption in ball-milled hydrides of Mg and Mg_2Ni.* Journal of Alloys and Compounds, 1999, 289: pp. 197–206.

[12]Terashita, N., Takahashi, M., Kobayashi, K., Sasai, T., and Akiba, E. *Synthesis and hydriding dehydriding properties of amorphous $Mg_2Ni_{1.9}M_{0.1}$ alloys mechanically alloyed from $Mg_2Ni_{0.9}M_{0.1}$ (M=none Ni, Ca, La, Y, Al, Si, Cu and Mn) and Ni powder.* Journal of Alloys and Compounds, 1999, 293–295: pp. 541–545.

[13]Spassov, T., and Köster, U. *Hydrogenation of amorphous and nanocrystalline Mg-based alloys.* Journal of Alloys and Compounds, 1999, 287: pp. 243–250.

[14]Shao, H., Liu, T., Li, X., and Zhang, L. *Preparation of Mg_2Ni intermetallic compound from nanoparticles.* Scripta Materialia, 2003, 49: pp. 595–599.

[15]Li, L., Akiyama, T., and Yagi, J-i. *Hydrogen storage alloy of Mg_2NiH_4 hydride produced by hydriding combustion synthesis from powder of mixture metal.* Journal of Alloys and Compounds, 2000, 308: pp. 98–103.

[16]Friedlmeier, G., Arakawa, M., Hirai, T., and Akiba, E. *Preparation and structural, thermal and hydriding characteristics of melt-spun Mg–Ni alloys.* Journal of Alloys and Compounds, 1999, 292: pp. 107–117.

[17]Ueda, T. T., Tsukahara, M., Kamiya, Y. and Kikuchi, S. *Preparation and hydrogen storage properties of $Mg–Ni–Mg_2Ni$ laminate composites.* Journal of Alloys and Compounds, 2005, 386: pp. 253–257.

[18]Hsu, C., Lee, S., Jeng, R., and Lin, J. *Mass production of Mg_2Ni alloy bulk by isothermal evaporation casting process.* International Journal of Hydrogen Energy, 2007, 32: pp. 4907-4911.

[19]Aizawa, T., Kuji, T., and Nakano, H. *Synthesis of Mg_2Ni alloy by bulk mechanical alloying.* Journal of Alloys and Compounds, 1999, 291: pp. 248-253.

[20]Kodera, Y., Yamasaki, N., Yamamoto, T., Kawasaki, T., Ohyanagi, M., and Munir, Z. A. *Hydrogen storage Mg₂Ni alloy produced by introduction field activated combustion synthesis.* Journal of Alloys and Compounds, 2007, 446-447: pp. 138-141.

[21]Tan, S., Aydınol, K., Öztürk, T., and Karakaya, İ. *Direct synthesis of Mg-Ni compounds from their oxides.* Journal of Alloys and Compounds, 2010, 504: pp. 134-140.

[22]Erden, F., Karakaya, İ., and Erdoğan, M. *On the Preparation of Mg2Ni Alloy by a New Electrochemical Method,* High Temperature Electrochemistry: Electrochemistry and Materials Properties, "TMS2013",San Antonio 2013.

[23]Karakaya, İ., Thompson, W.T. W.T. *A Thermodynamic study of the system MgCl₂-NaCl-CaCl₂, "Can. Metall. Quart,* 1986, 25 (4): pp. 307-317.

[24]*FACT – Facility for the Analysis Chemical Thermodynamics.* [cited 2010 November]; Available from: http://www.crct.polymtl.ca/FACT/

[25]Smith, J. F., and Christian, J. L. Acta Metallurgica, 1960, 8: pp. 249–255.

[26]Erden, F. In-situ Formation of Mg-Ni Intermetallics. M. S. Thesis submitted to Middle East Technical University, 2013.

[27]Centeno-Sanchez, R. L., Fray, D. J., and Chen, G. Z. Study on the reduction of highly porous TiO₂ precursors and thin TiO₂ layers by the FFC Cambridge Process. Journal of Materials Science, 2007, 42: pp. 7494-7501.

[28]Ergül, E. Reduction of Silicon Dioxide by Electrochemical Deoxidation. M. S. Thesis submitted to Middle East Technical University, 2010.

[29]Lutterotti, L., Matthies, S., and Wenk, H. R. *MAUD: A friendly JAVA program for material analysis using diffraction.* IUCr: Newsletter of the CPD, 1999, 21: pp. 14-15.

Energy Technology 2014
Carbon Dioxide Management and Other Technologies

SYMPOSIUM:
Solar Cell Silicon

Energy Technology 2014
Carbon Dioxide Management and Other Technologies

SYMPOSIUM: SOLAR CELL SILICON

Silicon Production
and Solidification

Session Chairs:

Gabriella Tranell
Arjan Ciftja

Energy Technology 2014: Carbon Dioxide Management and Other Technologies
*Edited by: Cong Wang, Jan de Bakker, Cynthia K. Belt, Animesh Jha, Neale R. Neelameggham,
Soobhankar Pati, Leon H. Prentice, Gabriella Tranell, and Kyle S. Brinkman*
TMS (The Minerals, Metals & Materials Society), 2014

ELECTROCHEMICAL DEPOSITION OF HIGH PURITY SILICON FROM MOLTEN FLUORIDE ELECTROLYTES

Geir Martin Haarberg[1], Henrik Gudbrandsen[2], Karen S. Osen[2], Sverre Rolseth[2], and
Ana Maria Martinez[2]
[1]Department of Materials Technology, Norwegian University of Science and Technology, NO-7491 Trondheim, Norway
[2]SINTEF Materials and Chemistry, NO-7465 Trondheim, Norway

Keywords: Silicon, electrolysis, electrorefining, molten fluorides
geir.m.haarberg@material.ntnu.no

Abstract

The Several approaches were tried for developing an electrochemical route for producing high purity silicon from molten salts. K_2SiF_6 and metallurgical grade silicon were used as the source of silicon. Molten LiF-KF and pure KF with additions of K_2SiF_6 were used as electrolytes at temperatures from 550 - 900 °C. Silicon was cathodically deposited and anodically dissolved in the fluoride electrolytes. High electrorefining efficiency was obtained for many elements after electrolysis experiments at constant potential or constant current. The main challenge was to reduce the contents of boron and phosphorus. Another difficulty was to remove inclusions of solidified electrolyte. The lowest energy consumption was estimated to be ~ 3 kWh/kg Si, while the best current efficiency for Si deposition was found to be ~ 95 %.

Introduction

Silicon is the most important material for photovoltaic power generation devices because of high feasible efficiency, low ecological impact and slow degradation. Development of a new and less expensive route for producing silicon of the desired quality is necessary in order to make solar cells competitive. Electrochemical routes may be an alternative possibility, and many laboratory studies have been reported over the past few decades. Most of the studies have been performed in molten salts, chloride or fluoride, at moderate temperatures so that solid silicon was deposited at the cathode. The production of high purity silicon has shown an annual increase of about 25 %, due to the increasing demand of solar grade silicon (SoG-Si). For future increased use of solar energy it is important to develop alternative, more energy efficient and cheaper processes for SoG-Si production compared to the Siemens process.

Common silicon sources have been SiO_2 and K_2SiF_6 for electrowinning and MG-Si in the case of electrorefining. A serious challenge is to find an oxygen evolving anode for electrolytes containing dissolved oxides. Especially at moderate to high temperatures and in molten fluoride electrolytes this is a difficult task.

Many experimental results have been reported in molten chloride electrolytes. Nohira et al. [1-3] and Chen et al. [4] cathodically deposited silicon by electrodeoxidation of solid SiO_2 in molten $CaCl_2$ by potentiostatic electrolysis. An alternate approach is to use electrorefining of low-cost metallurgical grade silicon (MG-Si) to produce SOG-Si. In previous studies [5], an MG-Si anode was used, and silicon was successfully electrodeposited on a molybdenum wire in a $CaCl_2$ based melt. It was found that the anode was strongly passivated due to the precipitation of silicon.

Also, some deposits detached from the molybdenum wire because of the poor adherence to the cathode.
Considering the Cu-Si alloy which has a good segregation effect for some impurities, and should not passivate, could be a good candidate anode for electrorefining. [6].

The use of molten fluoride electrolytes may offer some advantages in the case of electrodeposition of high purity silicon by traditional electrowinning. Cohen and Huggins [7] reported successful electrodeposition of silicon from fluoride melts; molten KF-LiF (1:1) with 5 mol% K_2SiF_6. Later Elwell and coworkers [8-12] continued the studies of electrochemical deposition of Si from fluoride melts. They concluded that only molten KF-LiF and LiF-NaF-KF could be used to obtain good quality deposits. Boen and Bouteillon [13] and de Lepinay et al. [14] also published valuable results on Si deposition from fluoride electrolytes. Cathode substrates of Ag, W, and graphite were used. Pulsed current conditions gave good quality and adherent Si deposits on silver and graphite. A summary of electrochemical approaches to deposit high purity silicon from molten salts was recently published [15].

Experimental

Preliminary experiments were performed in molten LiF-KF-K_2SiF_6 (molar ratio 46:49:5) at 750 °C and in molten KF-K_2SiF_6 (molar ratio 97:3) at 850 °C. Metallurgical grade silicon was used as the anode material during electrolysis. The cathode was of high purity solar grade silicon, and it was rotating during electrolysis in order to enhance the mass transfer of dissolved Si containing species.

More recent experiments in fluoride electrolytes were carried out at 550 and 800 °C in a closed laboratory furnace with argon atmosphere. A glassy carbon crucible was used to contain the molten fluoride electrolyte consisting of eutectic KF-LiF with K_2SiF_6 (5 - 20 mol%). KF and LiF of p.a. grade and K_2SiF_6 were dried and stored at 200°C. Electrochemical measurements were performed by using wires of Ag and W and rods of glassy carbon as working electrodes. The counter electrode was a 3 mm (diameter) glassy carbon rod. A Pt wire placed in the electrolyte was used as a quasi reference electrode to avoid introducing impurities. Silicon was deposited on plates of Ag and high purity Si, mainly by galvanostatic electrolysis using a silicon dissolving anode. Deposited Si was examined by SEM/EDS. The current efficiency with respect to Si was determined by weighing the amount of deposited Si.

Results and discussion

The first electrorefining experiments in molten LiF-KF-K_2SiF_6 (molar ratio 45:50:5) were carried out in a closed furnace under argon atmosphere. The anodes were evenly consumed, and a brown powdery, partly hard adhering silicon containing deposit was obtained at the cathode. Cathode deposits from several experiments were collected and melted and subsequently washed in distilled water and in HF-HCl mixtures. Results from analyses by GDMS showed good electrorefining effects for boron, titanium, and aluminium. The electrorefining of phosphorus, boron, copper and iron was not satisfactory, which was mainly due to poor quality of chemicals and too high anodic current density. The current efficiency for silicon was determined to be about 50 %. At a constant current density of 0.38 A/cm^2 the cell voltage was about 2.5 V. The corresponding energy consumption can be estimated to be about 19 kWh/kg Si. Experiments were also carried out in molten KF-K$_2$SiF$_6$ (molar ratio 97:3). The washing of the deposits was more difficult in this electrolyte.

Recent studies in molten eutectic KF-LiF with K$_2$SiF$_6$ (5 - 20 mol%) gave more promising results. In these experiments more attention was given to using high purity materials and components such as electrolyte constituents, electrodes and container materials. Also the handling of the cell took place under argon atmosphere.

Electrochemical techniques were used to study the behaviour of dissolved silicon species. Cyclic voltammograms obtained on a silver electrode are shown in Fig. 1. In the pure molten KF-LiF electrolyte the cathodic potential window is limited by potassium deposition occurring at ~ -1.8 V versus the Pt reference electrode. A significant cathodic current increase at potentials less cathodic is due to the formation of dissolved potassium. After adding K$_2$SiF$_6$ a new red/ox couple appears at ~ -1.0 V vs Pt. The shape of the voltammograms indicates the cathodic formation of a metal and a subsequent anodic dissolution, which is likely to be the formation and dissolution of silicon. Similar results were reported by de Lepinay et al. [8]. Recording cyclic voltammograms at gradually increasing cathodic potential limit shows that Si deposition is the only cathodic reaction at potentials less cathodic than K deposition. Silver seems to be an inert cathode substrate for Si deposition. By varying the scan rate it was found that the cathodic peak current density was proportional to the square root of the scan rate, which indicates that the cathodic deposition of Si is controlled by diffusion of dissolved Si (IV) complexes in the electrolyte. Potentiostatic current transients confirmed the diffusion controlled nature of the silicon deposition process. On silver cathodes some effects of nucleation phenomena were observed.

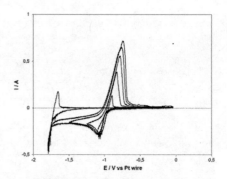

Figure 1. Cyclic voltammetry (200 mV/s) on a silver working electrode at 800 °C in pure molten KF-LiF and in molten KF-LiF with 5 mol% K$_2$SiF$_6$.

Fig. 2 shows typical microphotographs of deposited silicon on a silver cathode. The morphology is characterized by nodules similar to what has been reported by Boen and Bouteillon [7]. The structure seems to be uniform, and thus the nodular surface might be advantageous for use in a solar cell because of the increased area for sunlight absorption. The sample was washed in boiling water, but this was insufficient to remove all the salt remains, especially LiF which is the most difficult compound to remove because of the low solubility in aqueous solutions.

The apparent current efficiency with respect to silicon was determined by measuring the amount of silicon deposited during galvanostatic electrolysis. The current efficiency was consistently found to be in the range from 85 - 95 %. High temperature (800 °C), high current density (100 mA/cm^2) and the use of Ag cathode substrate were found to be beneficial for obtaining higher current efficiency for Si deposition. The obtained Si deposits were found to be of high purity as analysed by EDS.

Figure 2. SEM images of Si deposits obtained on Ag cathode substrate at 800 °C.

Conclusions

Silicon was found to be cathodically deposited and anodically dissolved in molten fluoride electrolytes. Pure silicon deposits with good adherence to silver cathode substrates were obtained by galvanostatic electrolysis. The current efficiency for Si deposition was generally higher than 90 %. The energy consumption was estimated to be ~ 3 kWh/kg Si. Such a low energy requirement suggests that electrorefining may be a promising way to produce silicon of solar grade quality.

References

1. T. Nohira, K. Yasuda, Y. Ito, Nat. Mater. 2 (2003) 397-401.

2. K. Yasuda, T. Nohira, K. Amezawa, Y.H. Ogata, Y. Ito, J. Electrochem. Soc. 152 (2005) 69-74.

3. K. Yasuda, T. Nohira, R. Hagiwara, Y.H. Ogata, Electrochim. Acta 53 (2007) 106-110.

4. X.B. Jin, P. Gao, D.H. Wang, X.H. Hu, G.Z. Chen. Angew. Chem. Int. Edit. 116 (2004) 751-754.

5. O.E. Kongstein, C. Wollan, S. Sultana, G.M. Haarberg, ECS Transactions 3 (2007) 357-361.

6. J.M. Olson, K.L. Carleton, J. Electrochem. Soc. 128 (1981) 2698-2699.

7. U. Cohen and R.A. Huggins, J. Electrochem. Soc. 123 (1976) 381-383.

8. D. Elwell and R.S. Feigelson, Solar Energy Mat 6 (1982) 123-145.

9. D. Elwell and G.M. Rao, J. Appl. Electrochemistry 18 (1988) 15-22.

10. G.M. Rao, D. Elwell, and R.S. Feigelson, J Electrochem. Soc, 127 (1980) 1940-1944.

11. D. Elwell and G.M. Rao, Electrochimica Acta 27 (1982) 673-676.

12. G.M. Rao, D. Elwell, and R.S. Feigelson, J Electrochem. Soc. 129 (1982) 2867-2868.

13. R. Boen and J. Bouteillon, J Applied electrochemistry 13 (1983) 277-288.

14. J. de Lepinay, J. Bouteillon, S. Traore, D. Renaud and M.J. Barbier, J. Appl. Electrochemistry, 17 (1987) 294-302

15. J. Xu and G.M. Haarberg, High Temp. Mater. Proc., 32 (2) (2013) 97-105.

16. P. Stemmermann, H.Pollmann, N.Jb.Miner.Mh., 7 (1992) 409.

Energy Technology 2014: Carbon Dioxide Management and Other Technologies
Edited by: Cong Wang, Jan de Bakker, Cynthia K. Belt, Animesh Jha, Neale R. Neelameggham,
Soobhankar Pati, Leon H. Prentice, Gabriella Tranell, and Kyle S. Brinkman
TMS (The Minerals, Metals & Materials Society), 2014

METALLURGICAL SILICON REFINING BY TRANSIENT DIRECTIONAL SOLIDIFICATION

Moyses L. LIMA[1,2], Marcelo A. MARTORANO[2], João B. F. NETO[1]

[1] Laboratory of Metallurgical Processes - Institute for Technological Research – IPT, Avenida Professor Almeida Prado 532, 05508-901, São Paulo/SP, Brazil
[2] Department of Metallurgical and Materials Engineering – University of São Paulo, Avenida Professor Mello Moraes 2463, São Paulo/SP, Brazil

Keywords: Metallurgical silicon, Silicon refining, Directional Solidification, Macrosegregation

Abstract

Directional solidification is an essential refining step to obtain solar grade silicon from metallurgical silicon. This step can be carried out in a Bridgman furnace, where nearly constant temperature gradients and solidification velocities are imposed on the solid-liquid interface. In the present work, this directional solidification was conducted in a static furnace, in which large temperature gradients and low solidification velocities were enforced to increase macrosegregation. The resulting ingots were analyzed regarding their macrostructures, microstructures and chemical composition. Using measured cooling curves in the ingot as boundary conditions, a mathematical model based on the concept of a stagnant liquid layer at the solid-liquid interface was implemented to predict the macrosegregation profiles. The chemical analyses of the ingots show macrosegregation of several impurities to the ingots top. The mathematical model indicates that liquid convection plays an important role in stabilizing the planar solid-liquid interface, increasing the macrosegregation of impurities.

Introduction

Most of silicon used for photovoltaic (PV) energy production is obtained by chemical processes, such as, Siemens and its derivations [1]. However, these chemical processes have drawbacks related to relatively high energy consumption and investment costs, and to safety and environmental problems. In order to overcome these difficulties and to fulfill the silicon demand for PV applications, several metallurgical routes are under development [2] to produce silicon of enough purity to PV applications (SoG-Si) from metallurgical grade silicon (MG-Si).

Metallurgical routes involve different steps among which the directional solidification is pointed out as a fundamental one [1]. Directional solidification can cause macrosegregation of impurities to the last part of the ingot to solidify. The impurity segregation during solidification is related to the low solute partition coefficient (k_0) between liquid and solid silicon. Most of the impurities in MG-Si have $k_0 \ll 1$, except for boron, phosphorus and oxygen, which require specific steps of the metallurgical routes to be eliminated. The equilibrium partition coefficient and the solubility limits for the most important impurities in MG-Si are shown in table 1.

Studies regarding the formation of macrosegregation during MG-Si directional solidification were done using methods like Czochralski [4], Bridgman [3], and electron beam furnaces [5]. In these studies, the influence of the solid-liquid interface velocity on the macrosegregation profile under constant thermal gradient was examined. The objective of the present work is to analyze the macrosegregation profiles of metallic impurities in silicon ingots obtained by transient directional solidification using a static solidification apparatus. Different experimental conditions were tested to study the effect of the solid-liquid interface velocity and of the axial temperature gradient on the macrosegregation profiles.

Table 1: Equilibrium partition coefficient (k_0), diffusion coefficient in the liquid (D_l), solute solubility in solid silicon (C_{sol}) and the concentration of the main metallic impurities in the MG-Si (C_0).

Element	k_0*	D_l (m²/s)*	C_{sol} (ppmw)*	C_0 (ppmw)**
Fe	8.0×10^{-6}	1.8×10^{-8}	0.3	1464
Ni	8.0×10^{-6}	1.8×10^{-8}	4.2	22
Ti	3.6×10^{-6}	1.5×10^{-8}	0.14	98
V	4.0×10^{-6}	1.6×10^{-8}	0.18	4
Mn	1.0×10^{-5}	1.6×10^{-8}	0.29	128
Cr	1.1×10^{-5}	1.6×10^{-8}	0.37	4
Cu	4.0×10^{-4}	1.7×10^{-8}	1.8	9
Al	2.0×10^{-3}	6.8×10^{-8}	20	534

*From reference [3];** Obtained by ICP-OES analyses

Experimental Methodology

A static electric resistance furnace was used to obtain silicon ingots by upward directional solidification. In this furnace there are two sets of resistive heaters ($MoSi_2$) with individual controls: one at the top of the internal chamber and another at the lateral wall. During melting, solidification, and cooling, an argon atmosphere was maintained over the melt surface to prevent excessive oxidation.

Cylindrical ingots were obtained using 8kg of MG-Si (composition in table 1) melted in crucibles coated with silicon nitride. After melting, the furnace was held for 40min. at 1600°C to homogenize the liquid silicon temperature at approximately 1550°C. The solidification step began by positioning a copper water-cooled block in contact with the crucible base and cooling down the furnace at a controlled rate by adjusting the temperature of the heaters. The different conditions for each experiment are given in table 2, considering GC-100 as a reference case.

Table 2: Experimental conditions: crucible base material, ingot length(mm) and the use of controlled cooling during the solidification step.

Experiment	Crucible base material	Ingot length (mm)	Controlled cooling
GC-100	Graphite-clay	100	Yes
GC-100_N	Graphite-clay	100	No
G-100	Graphite	100	Yes
GC-130	Graphite-clay	130	Yes

The silicon temperature was measured during melting, solidification, and cooling using two type-B thermocouples (Pt-6 pct Rh / Pt-30 pct Rh) located along the ingot axis. The furnace heaters were controlled to cool down the furnace in the same way in all experiments, except for the GC-100_N. In this experiment, the heaters were turned off at the beginning of the solidification step.

After cooling down the system, the ingots were cut in the longitudinal section. The central part was cut for macrostructure, microstructure, and chemical analysis by inductively coupled plasma mass spectrometry (ICP-MS).

Mathematical Model Description

A mathematical model was proposed and implemented to predict the formation of macrosegregation of the impurity elements during directional solidification of the ingots. To derive the model equations, the volume averaging method [6] and the stagnant liquid layer concept were used. The stagnant liquid layer concept was proposed by Burton et al. [7] to approximately consider the effect of liquid convection in the mass transport during solidification by a planar solid-liquid interface. In this concept, a liquid layer is assumed adjacent to the interface in which the solute transport takes place only by diffusion. The concentration of impurities in the remainder liquid is considered uniform. To correctly satisfy mass conservation at the interface between the stagnant liquid layer and homogeneous liquid, an equation proposed by Martorano et al. [3] was used. In the model, the following simplifying hypotheses were assumed: (1) unidirectional heat and mass transfer; (2) binary alloy (diluted solution); (3) constant thickness of stagnant liquid layer (δ_{BPS}); (4) the solute transport in the mushy zone and stagnant liquid layer takes place only by diffusion; (5) no solute diffusion in the solid; (6) the density, heat capacity and diffusion coefficient are constant; (6) local thermodynamic equilibrium at any solid-liquid interface; (7) the liquidus and solidus lines are straight. The equations, written for a rectangular coordinate system fixed at the ingot base, the initial and boundary conditions are summarized in table 3.

Table 3: Equations, initial and boundary conditions of the mathematical model.

Description	Equation	Initial condition	Boundary conditions
Constitutive relation	$\varepsilon_s + \varepsilon_l = 1$	-	-
Heat transport	$\rho C_P \dfrac{\partial T}{\partial t} = \dfrac{\partial}{\partial x}\left(K \dfrac{\partial T}{\partial x}\right)$ $+ \rho \Delta H_f \dfrac{\partial \varepsilon_s}{\partial t}$	$T = T_i$	$T = T_B(t)$ for $x = 0$ $T = T_T(t)$ for $x = L$
Solute conservation in the liquid	$\dfrac{\partial}{\partial t}(\rho\, \varepsilon_l C_l) = \dfrac{\partial}{\partial x}\left(D_l \rho \varepsilon_l \dfrac{\partial C_l}{\partial x}\right)$ $- kC_l \dfrac{\partial \varepsilon_s}{\partial t}$	$C_l = C_0$	$\dfrac{\partial C_l}{\partial x} = 0$ for $x = 0$ $C_l = C_l^\infty$ for $x = x_{BPS}(t)$
Solute conservation in the solid	$\dfrac{\partial}{\partial t}(\rho\, \varepsilon_s C_s) = \rho C_{sl} \dfrac{\partial \varepsilon_s}{\partial t}$	$C_s = kC_0$	-
Solute conservation in the stagnant liquid layer	$(L - x_t - \delta_{BPS}) \dfrac{dC_l^\infty}{dt}$ $= -D_l \dfrac{\partial C_l}{\partial x}\bigg|_{x=x_{BPS}}$	$C_l^\infty = C_0$	-

281

The symbols in the equations of table 3 are: ε_s and ε_l are the volume fractions of solid and liquid, respectively; ρ is the silicon density; C_0 is the initial solute concentration; C_l is the solute concentration in the liquid; C_l^{∞} is the solute concentration in the homogeneous liquid; C_S is the solute concentration in the solid; D_L is the solute diffusivity in the liquid; x_t is the solid-liquid interface position; ΔH_f is the silicon latent heat; T_B and T_T were the cooling curves measured at the ingot base and top, respectively, which were adopted as boundary conditions of the first kind; k is the solute partition coefficient; K is the silicon thermal conductivity; C_p is the silicon heat capacity; L is in the ingot length; x_{BPS} is the position of the end of the stagnant liquid layer; δ_{BPS} is the stagnant liquid layer thickness. The equations in table 3 were coupled by the equation of liquidus line $T_l = T_0 + m_l C_l$ where T_l is the liquidus temperature, m_l is the liquidus line slope, and T_0 is the MG-Si liquidus temperature.

The equations in table 3 were solved numerically using the implicit finite volume method using a mesh of at least 1000 volumes and a time step of 0.01s. The model was tested using several analytical models [8, 9] for solute redistribution, showing excellent agreement.

Results and discussion

Cooling Curves

Figure 1 shows the cooling curves measured in the silicon during cooling and solidification for experiments G-100 and GC-130, with the curves for the experiment GC-100. The initials points of the curves correspond to the time when the copper block was positioned at the crucible base. It was not possible to measure the temperature in experiment GC-100_N, because the thermocouple protection was disrupted after turning off the furnace. Figure 1-(b) shows that the cooling curves for experiments GC-100 and GC-130 are similar. Since the distance between the measuring points in experiment GC-130 is 15mm greater than that in the GC-100, the lower temperature gradient (G) in experiment GC-130 is evident.

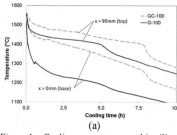
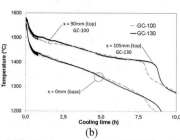

(a)   (b)

Figure 1 – Cooling curves measured in silicon during solidification in experiments: (a) GC-100 and G-100 and (b) GC-100 and GC-130.

Macro and Microstructures

The macrostructures of the longitudinal section of the ingots are shown in figure 2. It can be seen that the macrostructures are essentially formed by columnar grains parallel to the axial direction (solidification direction), as generally expected in directionally solidified ingots. In the GC-100_N experiment, columnar grains show side branches, while sharp boundaries are observed in the grains from other experiments. A progressive increase of grain size exists in the macrostructures of the ingots GC-100, GC-130, and G-100, possibly due to a competitive growth mechanism. A type of grain structure transition occurs in the GC-100 macrostructure at about 70mm from the ingot base, above which the directional growth is not evident. Analogous transitions also exist in the GC-130 and G-100 ingot macrostructures.

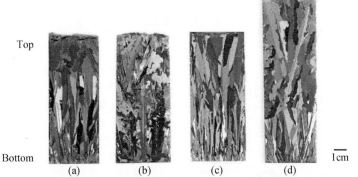

Figure 2 – Macrographs of the longitudinal section of the ingots (central portion) obtained in experiments: (a) GC-100, (b) GC-100_N, (c) G-100 and (d) GC-130.

Representative micrographs are shown in figure 3 for two different positions along the axial direction. In all ingots except GA-100_N, it is possible to observe the formation of a region free of intermetallic compounds. This type of region is an evidence of the reduction in the local impurity concentration [3]. These refined regions extend from the ingot base up to 70mm for GC-100, 40mm for GC-130, and throughout the ingot for G-100.

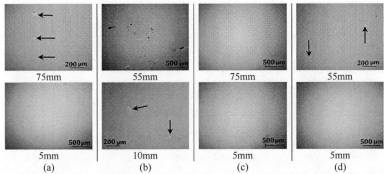

Figure 3 – Micrographs at two different positions relative to the ingot base along the axial direction: (a) GC-100, (b) GC-100_N, (c) G-100 and (d) GC-130. Below each micrograph is indicated the distance from the ingot base. The arrows point to small intermetallic compounds in the microstructure.

Macrosegregation Profiles

Profiles of Fe, Al, Mn, and Ti concentration relative to the average concentration (C_{imp}/C_0) along the ingot axis as a function of the relative position in the axial direction (x/L) are given in figure 4. The quantitative limits (QL) of the chemical analysis technique are indicated in the profiles. No evidence of important macrosegregation was seen in the micrographs of ingot GC-100_N, consequently, macrosegregation profiles were not measured. The concentration profiles show macrosegregation towards the ingot top ($x/L = 1$) for experiments GC-100 and GC-130. The position in which the first intermetallic compounds were found in the microstructure of experiments GC-100 and GC-130 coincides with the points in which the concentration of impurities increases abruptly in the profile (Figure 4) and also coincides with the transition in grain structure observed in the macrographs (Figure 2(a) and (d)), indicating that the increase in impurity concentration, the formation of intermetallic compounds, and the change in grain morphology are all related. These observations are in agreement with the results of Yuge et al. [5], who related the macrostructure transition and intermetallic compounds to the interface morphology transition from planar to cellular/dendritic. For experiment G-100, all concentrations in the profile are below the initial average concentration (C_0), which is consistent with the absence of intermetallic compounds in all micrographs. The impurities were segregated to the regions near the lateral surface, which was not analyzed.

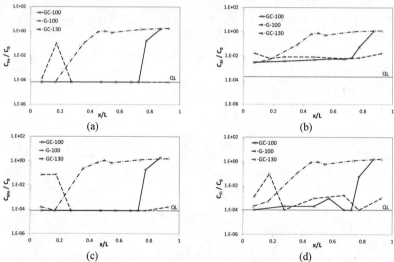

Figure 4 – Profiles of relative concentration (C_{imp}/C_0) measured by ICP-MS along the axial direction at the ingot center for (a) Fe, (b) Al, (c) Mn, and (d) Ti as a function of the relative position, x/L.

Mathematical model

In Figure 5(a) and (b), calculations with the present mathematical model show that the velocity of the solid-liquid interface (V) and the temperature gradient (G) at this interface were essentially equal ($V < 5.0 \times 10^{-6}$ m s^{-1} and G ~800 Km^{-1}) in experiments GC-100 and GC-130 up to ~0.03m from the ingot base. However, both the velocity and temperature gradient in experiment G-100 were larger (V~ 1.5 x 10^{-5} m s^{-1}). These results are consistent with the cooling curves presented in figure 1 and are probably related to the larger thermal diffusivity of graphite (7.3 x 10^{-5} ms^{-2}), which is the material of the crucible base in experiment G-100, in comparison with the conductivity of graphite-clay (1.8 x 10^{-6} ms^{-2}), the material of the crucible base for the GC-100 and GC-130 experiments. After solidification of this part of the ingots (~0,03m), the interface velocity for experiment GC-130 becomes larger and the temperature gradient lower than those for GC-100. These results are probably related to the longer ingot length in experiment GC-130 (Table 2). The decrease in temperature gradient and the increase in interface velocity for experiment GC-130 might have destabilized the planar solid-liquid interface, which might have become cellular/dendritic at about 0.04m from the ingot base, causing the precipitation of intermetallic compounds and the increase in impurity concentrations observed at this position.

The combination of larger solid-liquid interface velocities and temperature gradients in the ingots of experiment G-100 might be responsible for the severe macrosegregation and larger purification effect in comparison with the other experiments.

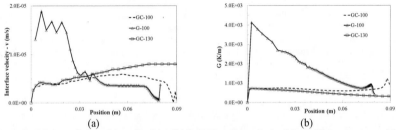

Figure 5 – Mathematical model results: (a) solid-liquid interface velocity (v) and (b) temperature gradient (G) as a function of the distance relative to the ingot base along the axial direction.

Figure 6 shows the concentration profiles of Fe calculated with the mathematical model (with and without convection) and that measured in experiments GC-100 and GC-130. The calculated profiles show an abrupt increase in concentration similar to that seen in the measured profiles. Nevertheless, this abrupt increase occurs earlier (nearer the ingot base) for the calculations without convection, i.e., only diffusive solute transport in the liquid. When convection is considered by adopting a stagnant liquid layer of thickness 45 x 10^{-3} m for GC-100 and 3.7 x 10^{-3} m for GC-130, the calculated and measured abrupt concentration increase occurs at approximately the same ingot position. For GC-100, the stagnant layer was an order of magnitude thinner than that in GC-130, possibly as a result of more vigorous convection in GC-100. Note that it is not possible to analyze the quality of the model results when the measured concentrations are below the quantification limit (QL), representing most of the measured profile in experiment GC-100. The calculated profile with convection in experiment GC-100 coincides with that given by the Scheil model up to ~ 0.06m, indicating that convection was very efficient in homogenizing the liquid composition. In all, the results in figure 6 indicate the fundamental role played by convection in the formation of macrosegregation during the directional solidification experiments in the present work. During the upwards directional solidification, convective currents might have originated from radial temperature gradients [10].

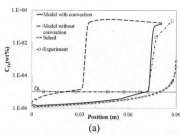
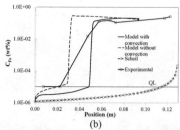

| (a) | (b) |

Figure 6 – Concentration profiles of Fe calculated using the mathematical model (with and without convection), measured in the experiments, and calculated with Scheil model: (a) GC-100; (b) GC-130.

Conclusions

Directionally solidified ingots were obtained from metallurgical grade silicon under transient conditions using a static electric resistance furnace with special control of heaters. The furnace heaters imposed relatively low solid-liquid interface velocities and relatively high temperature gradients. The grain macrostructure in most of the ingots consisted of columnar grains and the microstructure showed intermetallic compounds at ingot positions with larger impurity concentrations. An abrupt increase in the profile of impurity concentration coincided with the appearance of intermetallic compounds and a change in the columnar grain morphology. This transition seems to be related to a probable change in the solid-liquid interface morphology from planar to cellular / dendritic. The profiles of impurity concentrations showed that macrosegregation was more intense for the experiment with the lowest solid-liquid interface velocity and the largest temperature gradient. A mathematical model indicates that convective transport of solute played a fundamental role in increasing macrosegregation and in extending the region of a stable planar solid-liquid interface in the present work experiments.

Acknowledgements

Authors thank the support from Companhia Ferro-Ligas Minas Gerais (Minas Ligas) and from the National Bank for Social and Economic Development – BNDES.

References

1. J. Safarian, G. Tranell and M. Tangstad, Energy Procedia, vol. 20, pp. 88-97, 2012.
2. S. Pizzini, Solar Energy Materials & Solar Cells, vol. 94, pp. 1528-1533, 2010.
3. M. A. Martorano et al., Metallurgical and Materials Transactions A, vol. 42A, 2011.
4. E. Kuroda and T. Saitoh, Journal of Crystal Growth, vol. 47, pp. 251-260, 1979.
5. N. Yuge, K. Hanazawa and Y. Kato, Materials Transactions, vol. 45, pp. 850-857, 2004.
6. J. Ni and C. Beckermann, Metallurgical Transactions B, vol. 22B, pp. 349-361, June 1991.
7. J. A. Burton, R. C. Prim and W. P. Slitcher, J. of Chem. Phys., vol. 21, pp. 1987 - 1991, 1953.
8. J. J. Favier, Acta Metall., vol. 29, pp. 205-214, 1981.
9. M. C. Flemings, Solidification Processing, McGraw-Hill, 1974.
10. C. Beckermann, International Materials Reviews, vol. 47, pp. 243-261, 2002.

Energy Technology 2014: Carbon Dioxide Management and Other Technologies
Edited by: Cong Wang, Jan de Bakker, Cynthia K. Belt, Animesh Jha, Neale R. Neelameggham,
Soobhankar Pati, Leon H. Prentice, Gabriella Tranell, and Kyle S. Brinkman
TMS (The Minerals, Metals & Materials Society), 2014

NEW APPLICATIONS OF SHEET CASTING OF SILICON AND SILICON COMPOSITES

B. Kraaijveld[1], P.Y. Pichon[1], A. Schönecker[1], Y. Meteleva-Fischer[2]

[1]RGS Development B.V., Bijlestaal 54a, 1721 PW Broek op Langedijk, Netherlands
[2]Materials innovation institute (M2i), Mekelweg 2, 2628 CD Delft, Netherlands

Keywords: silicon, metal silicide, rapid crystallization, nanocomposites

Abstract
Ribbon-growth-on-substrate is a casting method, which offers a possibility to control heat flow, nucleation and subsequent crystal growth at the liquid-substrate interface. This technology allows for low-cost and high throughput shaping of silicon sheets with controlled crystal grain size and orientation for use in solar cells. In this work, sheets of silicon- metal silicide composite materials were produced and their microstructure was studied. It was shown that depending on the relevant phase diagram and casting conditions, a micro/nano-sized eutectic structure of the casted sheet can be formed. Such composite materials possess properties promising for various applications in energy generation, storage and conservation. The advantages, potential and challenges of the ribbon-growth-on-substrate technique for production of composite sheet materials as well as novel areas of application for such materials are discussed.

Introduction
The Ribbon Growth on Substrates (RGS) process is well known in the solar industry as a process to produce multicrystalline silicon wafers. An industrial size pilot production line for continuous casting of silicon wafer was developed by RGS Development B.V. in the Netherlands (Figure 1). A comprehensive review of this technology and properties of the produced silicon wafers were reported by Schönecker et al [1].

Figure 1. Industrial size pilot production line of the Ribbon Growth on Substrates process at RGS Development B.V. in the Netherlands

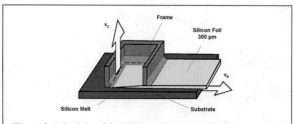

Figure 2. Principle of the Ribbon Growth on Substrate process
for silicon-based wafer

The principle of the RGS process is explained in Figure 2. The relatively cold substrate extracts the crystallization heat from the liquid silicon in the casting frame. This results in the growth of a silicon ribbon (Vc) in contact with the substrate. When the substrate (Vp) is transported from the casting zone into the cooling section, the crystal growth stops, the wafer can be removed and the substrate can be re-used. Last few years large work was done on the industrialization and market development of RGS wafers for the solar industry. Part of this work was discussed by Schönecker et al [2]. Although there is still a good perspective on the middle and long term solar market for the RGS wafers, the current market situation was found to be unfavorable for new entrants.

In addition to the mid-term development of competitive solar wafers produced by the RGS technology, alternative applications of the RGS process were investigated. One potential possibility is the rapid solidification of metal silicide and metal silicide-silicon composites thin sheets. The solidification velocity in conventional batch casting processes (Bridgman, CZ) is limited by thermal conduction of the solidification heat through the already solidified material. In such processes solidification is relatively close to thermodynamic equilibrium. In the RGS process larger deviations from thermodynamical equilibrium can be achieved by rapid solidification (in the order of 1mm/sec) and nucleation effects, where heat transfer and nucleation are controlled by the substrates material properties. The thermal material properties of the substrate and the quality of the thermal contact, together with the melt composition define the growth speed of the solid sheet and its microstructure. The effect of this rapid solidification process on casting eutectic and non-eutectic can result in a well-defined homogeneous microstructure and, with the proper process conditions, a nano-sized structure can be made. Such a nano structure can be used in various applications, like anodes for Li-ion batteries and thermo electric devices. The experimental results of two binary systems: Cr – Si and Mn – Si will now be reported.

Experimental

Using RGS process a large number of experiments was performed to cast metal silicide composites with composition listed in Table 1. Solar grade silicon (>7N, Wacker), chrome and manganese (>4N, Sigma Aldrich) were used. Melting of these materials resulted in a homogeneous melt. In spite of our best efforts, some oxidation of the manganese during handling and filling the melting crucible could not be prevented. During the melting of manganese and silicon some slag was formed. The amount of slag was clearly related with manganese oxidation. A considerable part of the slag consisted of SiC. Gas phase reactions taking place during the melting of silicon were studied by Meteleva-Fischer et al [3]. The process settings were adapted for the different compositions and their respective melting points. The casting speed was kept constant for all experiments at 6.5m/min. Various sheet thicknesses were casted ranging from 0.1mm till 1.0mm.

Obtained samples were analyzed using optical microscopy and SEM EDX (Energy-Dispersive X-ray spectroscopy). In order to see the microstructure of the composites, the chrome and manganese were etched from the composite with HF acid solution; silicon is resistant to HF and reveal the sample microstructure. Samples were studied from the top/bottom side and their cross section were prepared and analyzed as well.

Table 1: Material descriptions and their different compositions

Code	Description	Overall composition		Composition of phases	
C1	Eutectic Si CrSi$_2$	Si = 85mol%	Cr = 15mol%	Si = 52wt%	CrSi$_2$ = 48wt%
C2	"pure" CrSi$_2$	Si = 67mol%	Cr = 33mol%	Si =0wt %	CrSi$_2$ = 100wt%
M1	Eutectic Si MnSi$_{1.74}$	Si = 67mol%	Mn = 33mol%	Si = 7wt%	MnSi$_{1.74}$ = 93wt%
M2	"pure" MnSi$_{1.74}$	Si = 63mol%	MN = 37mol%	Si = 0wt%	MnSi$_{1.74}$ = 100wt%

A numerical model of the rapid solidification in RGS process was developped by RGS Development B.V together with TNO. The model was used to explain the microstructure development.

Discussion of results

The analysis of the optical images of the etched samples showed structure of material of different composition. Figure 3 presents optical images for substrate side of casted sheets. Visible are the nucleation areas, grains and grain boundaries. C1 and M1 show a fine sized eutectic structure and C2 and M2 show silicide crystals.

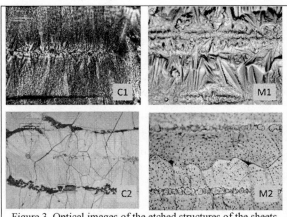

Figure 3. Optical images of the etched structures of the sheets from the substrate side.

Looking at the crystallographic structure of the produced sheet material (Figure 3) the following can be noticed: each sheet has nucleation areas more or less located on straight lines. This line pattern is induced by the design of the RGS substrates used for these experiments. Several solidification fronts start growing from the nucleation areas and distinct grain boundaries can be seen were the solidification fronts meet. There are also clear differences in the grain size depending on the initial melt composition: C2 and M2 show large grains of silicide material. In comparison, C1 shows a much finer microstructure, consisting of silicon and remained CrSi2 phases. M1 showed a very fine structure around the nucleation areas but further away clearly a larger structure.

On the Figure 4 the optical images of the cross section are presented. The silicide component was etched. The white part in the picture is silicon, the dark part are voids. Close to the nucleation point where the solidification speed is high, the typical size of the eutectic structure is 500nm. At the top of the sheet where the solidification speed is lower the structure size is typical 1500nm

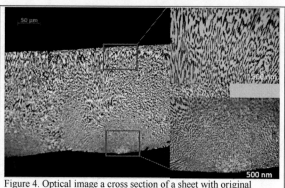

Figure 4. Optical image a cross section of a sheet with original composition C1

Looking at the cross section of C1 the following can be noted: at the bottom of the sheet and at the nucleation point the grain size of the eutectic structure is very small (in the order of 500nm). At the top of the sheet the structure is somewhat coarser (in the order of 1 to 2 μm). When comparing the images of the Figure 4 and the results of the crystal growth simulation model (developed by RGS Development and TNO) in Figure5, we conclude that the microstructure is oriented perpendicular to the solidification direction. These observations are in line with the lamellar eutectic growth model described by Jackson and Hunt [4]. This model predicts a dependence of the microstructure scale (lamellar spacing L) with the solidification velocity V according to:

$L = \frac{K}{V^2}$ with K depending in the material properties and the phase diagram.

In particular the smaller microstructure scale around the nucleation sites can be explained by a higher solidification velocity there. Moreover, compared to conventional casting processes, the larger average solidification velocity of the RGS process allows a finer microstructure to be formed.

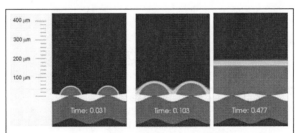

Figure 5. Snapshots of the numerical modelling of the RGS process: The red color indicates the solidified material, the blue is the liquid meltThe grey area is the substrate material

Based on these results, further work will be done on exploring the potential and limitations of the RGS process in producing smaller nano structures in a controlled way. In addition other metal-silicon systems will be explored such as for example Mo-Si and Ti-Si at different compositions. Initial melt compositions with more than 2 elements are also considered.

Potential Applications

The properties of the material prepared in these experiments are well suited for a range of very interesting applications. Initial property measurements in relation with some of these applications will be discussed (it is not intended to discuss applications in detail, the purpose is to demonstrate the application potential of these materials):

Mechanical properties: First of all for most applications the mechanical properties of the sheets are important. Silicon crystalline material is quite brittle it breaks easily. First results from tensile tests show for sample C1 (Si CrSi$_2$ eutectic composition) a much lower Young's modulus (more than a factor 2 compared with Silicon) and approximately the same high tensile strength as Silicon. However, it should be noted that the casting process itself and the cooling down afterwards, cause high thermal stresses, deformation and even cracks in the material. This is most likely due to the differences of thermal expansion coefficients between silicon and metal silicides. The RGS process provides a well-defined cooling body through the substrates which reduces this effect considerably.

Nano structure: For most applications it is desirable to produce a nano-structured material with as little as possible variation of the phase geometry. RGS material has demonstrated its ability to produce a relatively homogeneous material with geometry varying between 500 nm and 1500 nm. Due to the well-defined crystallization in contact with the substrate, such a non-steady state process is controllable and shows high reproducibility. Application examples of such nano-structured material are for example Lee [5] describes the application of Ti-Si composites in anodes for Li-ion batteries and Sano [7] describes the application of silicides in thermal electrical devices.

Porous structure: Besides the nano sized structure, sheets casted with the RGS process have a lamellar structure such that one component of the eutectic structure can be fully etched away (depending on the etchant this can be the silicide or the silicon). This results in a nano sized porous structure. Material with such structure has a very high surface area which could be applied in chemical processes or membranes. Föll et.all. [6] describe the application of porous silicon in anodes for Li-ion batteries.

Thermal electrical: As described by Sano[7] silicides have interesting thermal electrical properties. These properties were measured of some of the sheet material produced with the RGS process. The Seebeck coefficient of C1 (Si CrSi$_2$ eutectic composition) was found to be 453μV/K and for M1 (Si MnSi$_{1.74}$ eutectic composition) was found to be 138μV/K. The application of these materials was described in more detail by Schönecker [9] Compared to other thermoelectric materials, silicides are relatively cheap non pollutant and abundantly available

Conclusions

It was demonstrated that the RGS production platform is well suited for producing metal silicide and silicon-metal silicide composite sheet materials. These materials have very interesting properties, well suited for applications like anodes for Li-ion batteries or thermal electrical devices. In particular the controlled and reproducible production of nano sized structures on an industrial scale will increase the time to market of these applications. (This work is currently going on at RGS Development B.V.).
The field of applications is quite extensive and certainly not fully explored. For example recent developments on self-healing thermal barrier coatings, as described by Kochubey [8] poses again an interesting application for silicides.
The relatively low cost RGS technology, capable of producing at high material throughput rates for industrial solar silicon wafer manufacturing, in combination with the ability to produce structured silicon- metal silicide materials enables the wide-scale use of new silicon based materials in many application fields.

References

[1] A. Schönecker, A. Gutjahr, A. Burgers, G. Hahn, "Compatibility of the RGS Silicon Wafer with Industrial Type Solar Cell Processing", Proc. COMMAD (Conference on Optoelectronics and Microelectronic Materials and Devices), Sydney 2008

[2] A. Schönecker, S. Seren, L. Laas, M den Heijer, B. Kraaijveld, "Silicon Ribbon Materials: Entering the Market or Eternal Promise?", Proceedings 26th EU PVSEC, September 5-9, 2010, Hamburg, Germany

[3] Y. Meteleva-Fischer, S. Böttger, W. Sloof, B. Kraaijveld, "Gas Phase Interactions as Sources of Contamination in Solar Silicon. TMS 2014 Solar Cell Silicon Symposium

[4] K. A. Jackson, J. D. Hunt, "Lamellar and Roth Eitectic growth", Transaction of the methalurgical society, vol. 236, August 1966

[5] Yoo-Sung Lee, Jong-Hyuk Lee, Yeon-Wook Kim, Yang-Kook Sun, Sung-Man Lee, "Rapidly solidified Ti–Si alloys/carbon composites as anode for Li-ion batteries" Electrochimica Acta 52 (2006), p. 1523–1526

[6] Föll, H. Hartz, E. Ossei-Wusu, J. Carstensen and O. Riemenschneider "Si nanowire arrays as anodes in Li ion batteries" PHYSICA STATUS SOLIDI (RRL) Volume 4, Issue 1-2, February 2010, Pages: 4–6,

[7] Seijiro Sano, Hiroyuki Mizukami, Hiromasa Kaibe, "Development of High-Efficiency Thermoelectric. Power Generation System". Komatsu Technical report 2003 VOL.49 NO.152

[8] V.Kochubey and W.G.Sloof, "Self Healing Mechanism in Thermal Barrier coatings," in ITSC proceedings.

[9] A. Schönecker, J. König, M. Jägle, B. Kraaijveld, "A Novel Process for the Cost Efficient Synthesis of Metal Silicide Nanostructures on Industrial Scale", to be presented 11th European Conference on Thermoelectrics ECT 2013

Energy Technology 2014
Carbon Dioxide Management and Other Technologies

SYMPOSIUM: SOLAR CELL SILICON

Silicon Refining I

Session Chairs:

Yulia Meteleva-Fischer
Gabriella Tranell

Energy Technology 2014: Carbon Dioxide Management and Other Technologies
Edited by: Cong Wang, Jan de Bakker, Cynthia K. Belt, Animesh Jha, Neale R. Neelameggham,
Soobhankar Pati, Leon H. Prentice, Gabriella Tranell, and Kyle S. Brinkman
TMS (The Minerals, Metals & Materials Society), 2014

Thermodynamic Behavior and Morphology of Impurities in Its Solidification from a Si-Al Melt during the Refining of Silicon

Panpan Wang[1], Huimin Lu[1*], Shilai Yuan[1], Zhijiang Gao[1]

[1]Beihang University, School of Materials Science and Engineering;
37 Xueyuan Road; Beijing 100191, China
*Corresponding Author: lhm0862002@aliyun.com

Keywords: Al-Si melts, Solar grade Si, Purification, Impurities, thermodynamics.

Abstract

Solvent refining of silicon through using Si-Al melts is a valid method to remove the impurities from metallurgical grade silicon (MG-Si) melt. The thermodynamic behavior of impurities Fe, Al, Ca, Ti, Cu, C, B and P in MG-Si during the solidification of MG-Si was studied based on the available thermodynamic parameters and experimental data. The morphology of inclusions was investigated by Scanning Electron Microscope (SEM). The chemical composition of inclusions in MG-Si and Al-Si melt was analyzed by EPMA. It is found that the amount of white inclusion reduces and most impurities are solved into the Al-Si melts. After solidification of Si–Al alloy using induction heating and acid leaching, impurities can be effectively removed. The solidification of silicon from the Si–Al melt is found to be more effective for purification than general silicon solidification.

Introduction

Photovoltaic energy is one of promising energy resources in a sustainable society because of its inexhaustibility and cleanliness. Although the production and installation of solar cells have increased significantly during the last decade, the cost of solar cell installation is still high. Multicrystalline silicon has now become the main material in the photovoltaic market because of its low production cost and because of the relative high conversion efficiency of solar cells made from this material, which has been studied for a long time [1-3].

Electronic grade (EG) silicon (99.9999999% Si) is commonly used as the raw material to produce SOG-Si (99.9999% Si). Silicon produced from the chemical routes restricts the development of solar cells due to environment threaten, high energy consumption and low productivity. Purification of metallurgical grade silicon (MG-Si) is a low cost, environment friendly method. Although several metallurgical processes such as directional solidification, vacuum refining and electron beam melting [4], have been developed, the cost reduction and refining efficiencies for manufacturing silicon is still have to be further optimized.

In the purification process, it is difficult to remove boron and phosphorus by the conventional metal refining method. The refining efficiency depends on the segregation behavior of different elements in solid silicon and the liquid phase. The segregation ratios of all investigated impurities were found to be much smaller than the segregation coefficients between solid and liquid Al-Si melts [5]. For instance, the segregation coefficient of B in Si is 0.8 at the silicon melting point (1687K), whereas it is around 0.2 when silicon is solidified from Si-Al melt at 1273K. For phosphorus the segregation coefficient is 0.06 when silicon is solidified from Si-Al melt at 973K°C, which is much less than 0.35 for recrystallization from liquid silicon. Since the segregation coefficient of boron in silicon is 0.8, it cannot be removed by directional solidification.

Intensive research into solvent refining has been conducted recently, mainly with the Si-Al solvent [5-7]. This paper reviews the solvent processes and emphasizes their advantages and the key points governing the growth of silicon. Compared with preceding method, this process allows us to collect primary silicon crystals more effectively and avoid the considerable loss of aluminum and reduce acid waste.

In this work, the conditions to agglomerate solidified silicon were discussed, aiming at optimizing the refining process of silicon. A relationship that reflects the processing of the primary silicon agglomeration with the temperature has been established. Crystal growth of silicon was also studied during solidification of an Al-Si melt using induction heating. Furthermore, the difference between in silicon and Al-Si melts was discussed experimentally and theoretically. Although it was difficult to measure metallic impurities in extremely small contents in solid silicon, the thermodynamic considerations and evaluation of the metallic impurities was thus useful to understand the impurities formation and distribution. Finally, the low-cost overall optimized process for producing SOG-Si was proposed.

Experimental

The experiment is described as follows. When the vacuum of the furnace chamber was pumped up to 5×10^{-2} Pa, the samples were solidified in a mid-frequency induction furnace. After heating to and holding at 1300K, Al-50%Si was melted and then cooled to 1000K under a constant temperature gradient, which was measured to about 200 K/cm. The cooling rate was about 274K/s. Needle-shaped silicon crystals were successfully agglomerated at the top part of the sample as shown in Figure 1. To reveal clearly the phase morphologies and some other growth features of the solidified Al-Si alloy, the ingots were sectioned, polished, and etched.

Acid leaching is useful to remove the most metallic impurities which are on the surface of the grains of Si. SiO_2 layer and the impurities in it were removed by Fluor hydric acid. The separated was preprocessing by HCl (5%, 2h). It is possible to remove 95% of the metallic impurities after further leaching with HCl and HF (8:1, 360K, 2h).

The morphologies of the solidified Al-Si alloy were observed using scanning electron microscope (SEM). The chemical composition of inclusions in MG-Si and Al-Si melt was analyzed by EPMA. The chemical composition of the solidified sample after acid leaching was analyzed by inductively coupled plasma atomic emission spectrometry (ICP-AES).

Results and discussion

Principle for the liquid phase migration and silicon solidification in Al-Si melts.

MG-Si is initially alloyed with aluminum which is used as solvent metal to trap impurities, and later the refined silicon is recovered at a constant temperature gradient. Alloying MG-Si with Al is a process of redistribution of impurities in silicon by taking advantage of the thermodynamic instability of impurity elements in solid silicon at lower temperature, and the segregation ratios of impurity elements between solid Si and Al-Si melt are less than those between solid Si and liquid Si [8].

Therefore, the mass fractions of impurities in primary Si are less than those in MG-Si. In order to obtain primary Si crystal, the proportion of Al-Si should be selected in the hypereutectic region. The primary Si crystal would initially precipitate from liquid alloy when the temperature of Al-Si melt decreases. Similar to the previous work, the Si crystals solidified from the Si-Al melt tended to be uniformly distributed in the alloy melt. Under appropriate growth conditions, however, it was found that that bulk Si crystals can be obtained from the Si-Al melt. To intensify the separation, temperature gradient is introduced into the Al-Si refining process. The refined silicon is obtained after acid leaching.

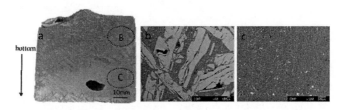

Fig. 1. Cross section of partial solidified Al-50%Si alloys; (b), (c) enlarged images of zones B and C.

Furthermore, the temperature gradient was of particular importance for considering the relationship between the convection and solidification of the melt. During the cooling process, the dendrite-like silicon was initially sedimented at the top along the direction of temperature increase. With the increasing temperature, the movement of primary silicon was restrained and changed into long needle-like primary silicon in the lower part of the alloy, as shown in Figure 1(a).

We have thus investigated the principle of the agglomeration of silicon crystals during solidification of the Al-Si alloy inserted in a graphite crucible. The heating theory of the intermediate frequency is the surface effect. Because of the graphite crucible, the magnetic field is shielded in medium frequency induction melting furnace. Therefore, compared with the research that the magnetic field has impact on the silicon solidification [8], we draw a different conclusion. It suggests that the solidified primary silicon distribution is not due to the magnetic field, but due to the temperature gradients. It has been found that the temperature distribution and particularly changes in temperature gradient are the main factors governing the growth of silicon.

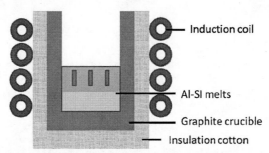

Fig. 2. Illustration of the sample during the solidification of silicon from the Al-Si melts.

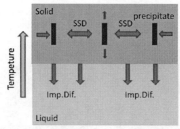

Fig. 3. Mechanisms involving solid state diffusion and solute
segregation during the silicon solidification of Al-Si melt.
SSD: Solid state diffusion (SSD). Imp. Dif.: Impurity diffusion.

During the cooling process, the solidified silicon is agglomerated at the top of the sample along
the direction of temperature increase. Many metallic elements such as Fe, Al, Ti, Ca, Mn, V and
Ni have high segregation coefficients in silicon. Thus, during solidification of silicon in Al-Si
melts, a major portion of these impurities should concentrate in the Ai-Si melts and a small
portion of these precipitated in primary silicon. Hence, acid leaching of separated silicon
dissolves the impurities located at the grain boundaries of silicon, while the silicon matrix is not
dissolved. Therefore, the silicon purity can be improved by leaching process. Additionally, the
contents of boron and phosphorus can be reduced less than 1 ppmw.

Table 1 The impurity contents of purified silicon, collected by acid leaching
of the solidified alloy [ppmw]; Si-1: metallurgical silicon. Si-2: metallurgical
silicon after acid leaching. Si-3: solidified silicon from Al-Si melt after acid leaching

Silicon source	Fe	Ca	Al	B	P
Si-1	4700	260	640	14	35
Si-2	15(99.7%)	3.1(98%)	450	6.8(51.4%)	4.5(87%)
Si-3	14(99.7%)	2.4(99%)	550	0.93(92.3%)	0.55(97%)

Thermodynamic behavior and morphology of impurities in its solidification from Al-Si melt

As mentioned previously, silicon is a very common element, although its production as a
metallic form is more difficult than extracting rare gold. This is because silicon can easily
combine with other elements chemically. Thermodynamic evaluation of the Gibbs energy change
of compounds during the solidification of silicon was considered in Al-Si melts, as shown in
Figure 4-7.
Diagram for the Gibbs energy change of boride formation with temperature is shown in Figure 4.
This diagram shows the temperature dependence of the Gibbs free energy change for 1 mole of
pure metallic substance reacting with boron to form borides. The Gibbs energy change of boride
is drawn in the lower part of Figure 4, which means that TiB is relatively stable. Diagram for the
Gibbs energy change of phosphide formation with temperature is shown in Figure 5, which
means that AlP is more stable than SiP.
Diagram for the Gibbs energy change of intermetallic compound formation with temperature is
shown in Figure 6. Diagram for the Gibbs energy change of silicide formation with temperature
is shown in Figure 7. It can be draw the conclusion that silicon tend to react with vanadium to
from Si_3V_5 is relatively stable.

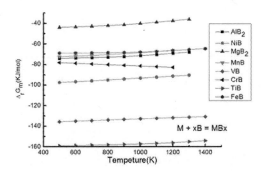

Fig. 4. The Gibbs energy change of boride formation with temperature.

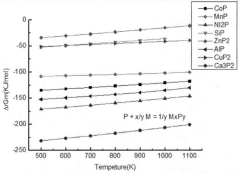

Fig. 5. The Gibbs energy change of phosphide formation with temperature.

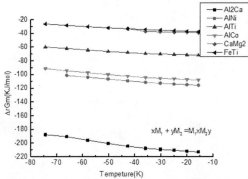

Fig. 6. The Gibbs energy change of intermetallic compound formation with temperature.

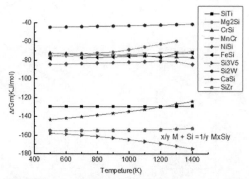

Fig. 7. The Gibbs energy change of silicide formation with temperature.

Optimized technological process of the refining of silicon

Figure 8 shows the scheme of optimized technical process of the refining of silicon. The process may be explained as follows: first the sample was kept in the coil and melted in 1300K under the constant temperature gradient; after cooling and holding at 1000K, the primary silicon was solidified in the Al-Si melts, needle-shaped silicon crystals were successfully agglomerated at the top part of the sample; Pouring the solidified silicon was to separate primary silicon from the Al-Si melts; then the separate solidified primary silicon were crashed into the particles, the refined silicon was collected after acid leaching and was subjected to the chemical analysis; additionally, through directional solidification most metallic impurities could be removed.

The segregation ratios of impurities between solid silicon and the Si-Al melt were much smaller, indicating the effective purification after the low-temperature solidification of silicon. This solidification allowed us to collect silicon crystals selectively from the alloy and to reduce the loss of aluminum and silicon considerably during the acid leaching. Pouring the solidified silicon was the effective method to separate primary silicon from Al-Si melts, it could be further to avoid the waste of acid and the loss of aluminum and most of metal concentrations were left in the Al-Si melts. The concentrations of Fe, Ca can be significantly reduced.

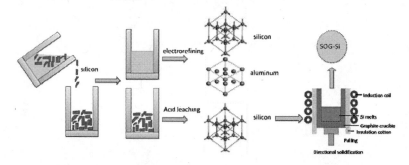

Fig. 8. The scheme of the optimized technical process of the refining of silicon.

The results show that the removal effect in the Al-Si melts is better than in other system. The boron content in MG-Si is successfully reduced from 34ppm to less than 1ppm during the process. Meanwhile, Al, Ca, and Fe elements in MG-Si are also well removed. Many metallic elements such as Fe, Al, Ti, Ca, Mn, V, Zn and Ni have high segregation coefficients in silicon. The unidirectional solidification method is effective to remove metallic impurities and a cost-effective technique for large-scale production of multi- crystalline silicon material. Additionally, thermodynamic evaluation for impurities was evaluated. Finally, the low-cost overall process for producing SOG-Si was proposed, as shown in Figure 8. These processes are used in combination to convert MG-Si into SOG-Si.

Conclusions

Characteristics of silicon morphology during solidification process and the thermodynamic properties of impurities in silicon were considered. From these considerations, we further understand the impurities distribution and the principle for refining of silicon in Al-Si melts.

The test refining was carried out by the solidification of silicon from a Si-Al solvent under the mid-frequency induction heating. A removal fraction over 99% for iron and 95% for phosphorus and boron were obtained. After the acid leaching, the best result of the concentration of B and P is respectively 1ppm and 0.55ppm. And it is possible to remove calcium to 2.4ppm.

The optimized technical process for SOG-Si production was evaluated. The most suitable process to fabricate SOG-Si from MG-Si and will be the combination of solvent refining, acid leaching, electrorefining and solidification refining in sequence.

References

[1] A.F.B. Braga et al., "New processes for the production of solar-grade polycrystalline silicon: A review," *Solar Energy Materials and Solar Cells*, 92 (2008), 418-424.
[2] Woosoon Lee, Wooyoung Yoona, and Choonghwan Park, "Purification of metallurgical-grade silicon in fractional melting process," *Journal of crystal growth*, 312 (2009), 146-148.
[3] Ekyo Kuroda, Hirotsugu Kozuka, and Yukio Takano, "The effect of temperature oscillations at the growth interface on crystal perfection," *Journal of Crystal Growth*, 47 (1979), 251-260.
[4] Tong Liu et al., "Large scale purification of metallurgical silicon for solar cell by using electron beam melting" *Journal of crystal growth*, 351 (2012), 19-22.
[5] Kazuki Morita, Takeshi YOSHIKAWA, "Thermodynamic evaluation of new metallurgical refining processes for SOG-silicon production," *Trans. Nonferrous Met. Soc. China*, 21 (2011), 685-690.
[6] R.K. Dawless et al, "Production of extreme-purity aluminum and silicon by fractional crystallization processing," *Journal of Crystal Growth*, 89 (1) (1988), 68-74.
[7] Yoshikawa T et al., "Thermodynamics of impurity elements in solid silicon," *Journal of Alloys and Compounds*, 490 (2010), 31-41.
[8] Yoshikawa T, Kazuki Morita, "Refining of silicon during its solidification from a Si–Al melt," *Journal of Crystal Growth*, 311 (3) (2009), 776-779.
[9] W.G. Pfann, "Zone melting," *International Materials Reviews*, 2 (1) (1957), 29-76.
[10] Yoshikawa T, Morita K, "Thermodynamics on the solidification refining of silicon with Si-Al melts" (Paper presented at the 134th Annual Meeting, San Francisco, EPD Congress 2005), 549.

Energy Technology 2014
Carbon Dioxide Management and Other Technologies

SYMPOSIUM: SOLAR CELL SILICON

Silicon Refining II

Session Chairs:

Yulia Meteleva-Fischer
Shadia Jamil Ikhmayies

Energy Technology 2014: Carbon Dioxide Management and Other Technologies
Edited by: Cong Wang, Jan de Bakker, Cynthia K. Belt, Animesh Jha, Neale R. Neelameggham,
Soobhankar Pati, Leon H. Prentice, Gabriella Tranell, and Kyle S. Brinkman
TMS (The Minerals, Metals & Materials Society), 2014

Research on the Forecast Model of the Boron Removal from Metallurgical Grade Silicon by Slag Refining Based on GA-BP Neural Network

Shilai Yuan[1], Huimin Lu[1,a], Zhijiang Gao[1], Liyuan Zhao[1]

[1]Beihang Univ., School of Materials Sci. & Eng.; 37 Xueyuan Road; Beijing 100191, China
[a]Corresponding Author: 13331151800@126.com

Keywords: Metallurgical grade silicon; Purification; Slag system; GA-BP Neural Network

Abstract

A purification process was developed to removal impurity element boron from metallurgical grade silicon by electromagnetic induction slag melting. Vacuum melting furnace was used to purify boron in both Al_2O_3-MgO-CaO-SiO_2 slag system and Al_2O_3-CaO-SiO_2 slag system. The relationship between different slag chemistry and the removal of boron in silicon were studied using Back Propagation (BP) Neural Network model. The best slag chemistry for the removal of boron was predicted by Genetic Algorithm (GA) contributed by the use of Matlab. The results show that the mass fraction of Boron in silicon is reduced from 11.7496×10^{-6} to 2.3259×10^{-6} after slag melting in $28.96\%Al_2O_3$-$3.43\%MgO$-$36.24\%CaO$-$31.37\%SiO_2$ slag system. The relative error obtained with GA-BP Neural Network model was below 0.35%.

Introduction

Solar energy, due to its low green house gas emissions, has been paid great attention. Accordingly, the demand for solar grade silicon (SoG-Si) has been growing rapidly. The quest for developing a low-cost method of producing SoG-Si has been moving forward in the last decade. Refining of metallurgical-grade silicon (MG-Si) attracts research interest due to its low material and energy costs, and more environmentally friendly technology in comparison to the traditional Simens process. As such, impurity elements in MG-Si must be removed to low levels for the photovoltaics (PV) cell to operate at optimum efficiency. Most impurities in metallurgical-grade silicon (MG-Si), especially metal elements, can be separated by directional solidification and the zone refining plasma-arc method [1, 2]. The main focus in refining MG-Si to SoG-Si is on the impurities which are the most difficult to remove. Boron has relatively high segregation coefficients (K_B=0.8, concentration of Boron in solid silicon divided by its concentration in liquid silicon) [3] and higher vapor tension (the saturated vapor pressure of boron and silicon are 6.78×10^{-7} Pa and 0.4 Pa, respectively). The vacuum melting and directional solidification methods hence have no effects on boron removal.

Slag refining involves melting MG-Si in the presence of a flux to produce a slag phase that can take up the impurity elements. The mechanism of slag refining is considered to include two main steps: the oxidation of boron and the absorption of boron oxide by slag materials. The oxidation of boron, described by the oxygen potential (P_{O2}), resulting from the equilibrium Si and SiO_2. The absorption of boron oxide, described as the slag basicity, is interpreted as expected to facilitate their extraction to alkali and alkali-earth oxides in the slag phase. Thus, effective removal may be obtained for a certain slag composition. Researchers [4-5] examined the effect of boron removal in $CaO-SiO_2$ based ternary systems and showed that the addition of excess CaO can decrease the activity of SiO_2. Researchers [6-8] studied the boron behaviors in $CaO-SiO_2-Al_2O_3$, $CaO-SiO_2-Na_2O$, $Al_2O_3-BaO-SiO_2$, $CaO-SiO_2-Li_2O$ and $Al_2O_3-CaO-MgO-SiO_2$ slag systems by adding boron in the silicon to a degree. Li_2O, Na_2O, MgO and BaO are thought to be associated with B_2O_3.

Artificial neural networks, due to their excellent ability of non-linear mapping, generalization, self-organization and self-learning, have been proved to be of widespread utility in engineering. The BP Neural Network (BPNN) is a kind of typical 'feed-forward, back propagation' neural network, trained by back-propagation of errors and a sample training model of the input layers, hidden layers and output layer structure is received [9]. The BP Neural Network can get proper weights and bias of a fixed network structure by fusing genetic algorithms (GA) avoiding being trapped in a local minimum when adjusted by BP algorithm [10].

Despite the expected complexity associated with optimizing slag chemistry, comprehensive data on the behavior of impurities in terms of fundamental variables such as temperature and slag composition are quite limited. In order to accurately predict the investment trends of the removal effect of boron in silicon ingot, this paper proposes one forecast model based on the GA-BP neural network. In this work, flux-refining of MG-Si is examined in detail using $SiO_2-CaO-MgO-Al_2O_3$ slags. The slag compositions were arranged using the uniform design method. Then, the relationship between slag compositions and boron removal were built using BP neural network model.

Experimental Procedure

SiO_2, CaO, Al_2O_3, and MgO powders were used in these experiments. SiO_2, CaO, Al_2O_3 and MgO are classified as fine powders. The MG-Si ingot with the boron concentration of 11.7496ppmw was prepared. The raw powders used in the preparation of the slag were fully blended and then dried at 100°C for one hour in order to completely removed humidity, then weighed and mixed thoroughly in the desired ratio for each particular experiment. The slag compositions were designed using the uniform design method, considering building a GA-BP neural network. The detail slag compositions were shown in Table I.

Table I Composition of slags before equilibrium experiments. All values are in wt%.

SiO₂	CaO	MgO	Al₂O₃
45	30	5	25
30	30	5	35
35	30	10	25
45	25	0	30
25	45	0	30
55	15	0	30

The slag powder mixtures were pressed into a pellet. Typically, 150g Si and 75g of fluxes were used in each experiment. The slag pellet and slag ingot were added to a graphite crucible inside an intermediate frequency furnace. The graphite crucible was used to melt slag mixture and silicon ingot in the induction furnace. The graphite crucible was conductive and during the melting process, the crucible itself was heated, which results in the heating and melting of the powder. The atmosphere in the furnace was 0.01Pa before the raw mixtures were heated up. The temperature was heated up to 1550°C and the temperature of the hot zone was measured with an infrared thermometer. A schematic diagram of the intermediate frequency furnace was shown in Fig. 1.

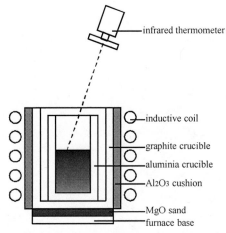

Fig. 1 The schematic diagram of the intermediate frequency furnace

The slag and the metal phases were physically separated from the crucible. The test samples were cut from the center of each ingot, then ground to a powder for digestion and analysis. The boron content in silicon was analyzed by inductively coupled plasma atomic emission spectroscopy (ICP-AES) and inductively coupled plasma mass spectrometry (ICP-MS), respectively.

Results and Discussion

<u>GA-BP Neural Network Training Process</u>

Considering the inputs number (SiO_2 proportion, CaO proportion, Al_2O_3 proportion, MgO proportion) and the output number (boron content), we choose the BP Neural Network with three layers to construct the non-linear reflection model, as shown in Fig. 2. Since the weights and the threshold values of BP Neural Network are both randomly generated, the predicted results may have large error, and the training process is of low efficiency, when the training epochs and training goal are 1000 and 1E-8, respectively. To increase the predicted accuracy and training efficiency, the GA was used to optimize BP neural network structure, transfer function, weights and the threshold values by choosing high fitness chromosome, crossing with their father generation and mutation. The experimental data was used to train the neural network. Then, the neural network was used to predict the boron content in specific slag composition. The predicted output is shown in Fig. 3 and the error percentage is shown in Fig. 4. The largest predicted error percentage is 0.35%, indicating that this model is feasible to predict. Based on this GA-BP Neural Network model, the corresponding process of GA optimization is programmed and run with MATLAB.

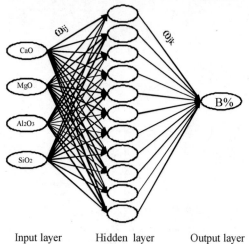

Input layer   Hidden layer   Output layer

Fig. 2 The common artificial three-layer neural network

The parameters are set as following: the initial population number N=100, the cross probability Pc=0.8, the mutation probability Pm=0.08, and the error e=1E-8. The GA-BP Neural Network prediction results show that the mass fraction of Boron in silicon is reduced from 11.7496×10^{-6} to 2.3259×10^{-6} after slag melting in 28.96%Al_2O_3-3.43%MgO-36.24%CaO-31.37%SiO_2 slag system slag system. The experimental data show that the boron content can be decreased below 2.35×10^{-6} using the prediction slag composition. This can also verify the valid of the GA-BP neural network.

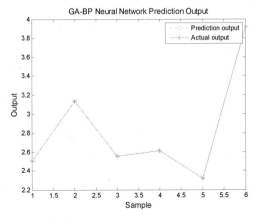

Fig. 3 The forecast output and actual output of GA-BP Neural Network mode

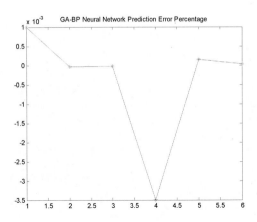

Fig. 4 The GA-BP Neural Network Prediction Error Percentage

<u>Basicity Analysis</u>

The effect of slag basicity on boron removal from silicon was calculated by varying the addition of CaO and SiO_2 with Al_2O_3 fixed at 30%. The boron removal effects are shown in Fig.5. It is found that the boron content in silicon ingot decreases with the rise of CaO proportion in slag and reaches a minimum value of 2.35 ppmw when the CaO proportion is between 0.20 and 0.25. However, the boron content in silicon ingot increases rapidly when the CaO proportion exceed 0.30. Although boron oxide is a strong acidic oxide, boron has weak affinity for a basic slag. When CaO was added to the slag, the $(BO_3)^{3-}$ combines with CaO, and then enters into the slag phase. Since CaO has a strong affinity for silica, an excess amout of CaO will give a high basic

311

slag at the expense of lowering the oxidization of SiO_2. Increasing the basicity too far may then have the effect of impeding the removal of boron as P_{O2} is in turn reduced.

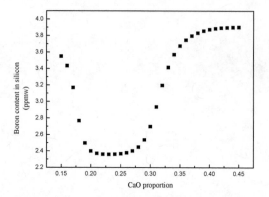

Fig. 5 Effects of CaO proportion on boron removal in $CaO-SiO_2-Al_2O_3$ salg system

The effects of MgO proportion on boron removal in $SiO_2-CaO-MgO-Al_2O_3$ slag system was calculated with CaO and Al_2O_3 fixed at 20% and 30%, respectively. The boron removal effects are shown as in Fig. 6. As the MgO proportion increasing , the boron content in silicon slightly decreases to a minimum value of 2.35 ppmw and then increased to a maximum value of 2.63ppmw. MgO can decrease the basicity of slags dramatically, which contribute to the boron removal from silicon ingot to slag phase. The minimum value when MgO proportion is around 3% can explained by the slag basicity. But the increasing part of the curve and the comparisons with larger proportion part need to be further studied, which may changed the slag phase structure and composition.

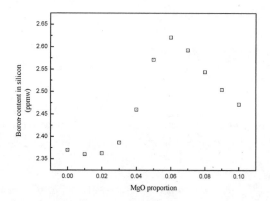

Fig. 6 Effects of MgO proportion on boron removal in $SiO_2-CaO-MgO-Al_2O_3$ slag system

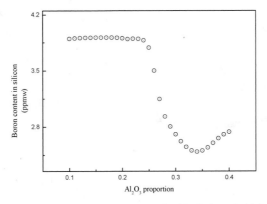

Fig. 7 Effects of Al_2O_3 proportion on boron removal in SiO_2-CaO-MgO-Al_2O_3 slag system CaO and MgO were fixed at 30% and 5%, respectively

Oxygen Potential Analysis

The effect of oxygen potential (P_{O2}) of the slag was calculated by varying the Al_2O_3 proportion with the proportion of CaO and MgO were fixed at 30% and 5%, respectively. The removal effects are shown in Fig. 7. The boron content in silicon ingot decreases with the rise of Al_2O_3 proportion in slag and reaches a minimum value of 2.49ppmw when the Al_2O_3 proportion is around 35%. However, the boron concentration in silicon ingot increases when the Al_2O_3 proportion exceed 35%. For the decreasing part in Fig. 7, the ability of the slag oxidizing boron into B_2O_3 increase with the increasing SiO_2 proportion, and there are enough O^{2-} ions to combine with B_2O_3 to form $(BO_3)^{3-}$ ions and enter into slag phase. For the uplifting part in Fig. 7, the decreasing CaO proportion results in the low contention of O^{2-} ions in slag phase, which hinder the B_2O_3 in silicon molten from entering into the slag phase and result in the increased boron concentration in silicon ingot. Alumina can also combine with O^{2-} ions to form AlO_4^{5-} tetrahedral in slag phase. It is expected that Ca^{2+} acts as the charge balancing cation. Then, the proportion of CaO present as part of the $Ca_{0.5}AlO_4^{4-}$ unit also increases. This CaO is then not free to donate oxygen to the slag, which decrease the boron removal effect.

Conclusion

The GA-BP Neural Network was chosen to construct model to study relationships between slag compositions and boron content in SiO_2-CaO-MgO-Al_2O_3 slag at 1550°C. The largest predicted error percentage in the prediction process was below 0.35%.

The boron concentration in silicon ingot decreases remarkably as the addition of CaO and MgO to slag, while the addition of SiO_2 has slight influence on boron removal.

The increase of basicity in slag could improve the boron removal effect. The addition of CaO has affinity to $(BO_3)^{3-}$. The increase of oxygen potential (P_{O2}) of the SiO_2-CaO-MgO-Al_2O_3 slag by varying the SiO_2 and Al_2O_3 proportion can also contribute to the boron removal in silicon ingot.

The GA-BP Neural Network prediction results show that the mass fraction of Boron in silicon is reduced from 11.7496×10^{-6} to 2.3259×10^{-6} after slag melting in $28.96\%Al_2O_3$-$3.43\%MgO$-$36.24\%CaO$-$31.37\%SiO_2$ slag system slag system. The experimental data shows that the boron content can be decreased below 0.5×10^{-6} using the prediction slag composition.

References

1. M.A. Martorano, J.B. Ferreira Neto, T.S. Oliveira and T.O. Tsubaki, "Refining of metallurgical silicon by directional solidification," *Mterials Science and Engineering B*, 176 (2011), 217-228.

2. Kazuhiro H, Noriyoshi Y, and Yoshiei K, "Evaporation of phosphorus in molten silicon by an electron beam irradiation method," *Mater Trans*, 45(2004), 844-849.

3. L Teixeira, and K Morita, "Removal of boron from molten silicon using CaO-SiO2 based slag," *ISIJ Internationa*, 49(2009), 783-787..

4. Leandro A, Viana T, Yomei T, Yoshinobu Y, and Kazuki M, "Behavior and state of boron in CaO-SiO_2 slag during refining of solar grade silicon," *ISIJ Internationa.* 49(1990), 777-782.

5. M.D. Johnston, and M Barati., "Distribution of impurity elements in slag-silicon equilibria for oxidative refining of metallurgical silicon for solar cell applications," *Sol Energ Mat Sol C*, 94(2010), 2085-2090.

6. J.J.Wu, W.H.Ma, B.J. Jia, Yang B, D.C.Liu, and Y.N. Dai, "Boron removal from metallurgical grade silicon using a CaO-Li_2O-SiO_2 molten slag refining technique," *J Non-Cryst Solid.* 358 (2012), 3079-3083.

7. J. Cai, J.T. Li, W.H. Chen, C Chen, and X.T. Luo, "Boron removal from metallurgical silicon using CaO-SiO_2-CaF_2 slags," *Trans Nonferrous Met Soc China*, 21(2011), 1402-1406.

8. D.W. Luo, N Liu, Y.P. Lu, G.L. Zhang, T.J. Li, "Removal of boron from metallurgical grade silicon by electromagnetic induction slag melting," *Trans Nonferrous Met Soc China*, 21(2011), 1178-1184.

9. K Liu, W.Y. Guo, X.L. Shen, and Z.F. Tan, "Research on the forecast model of electricity power industry loan based on GA-BP neural network," *Energy Procedia*, 14(2012), 1918-1924.

10. S.W. Yu, K.J. Zhu, F.Q. Diao, "A dynamic all parameters adaptive BP neural networks model and its application on oil reservoir prediction," *Appl Math Comput.*, 195(2008), 66-75.

Energy Technology 2014: Carbon Dioxide Management and Other Technologies
Edited by: Cong Wang, Jan de Bakker, Cynthia K. Belt, Animesh Jha, Neale R. Neelameggham,
Soobhankar Pati, Leon H. Prentice, Gabriella Tranell, and Kyle S. Brinkman
TMS (The Minerals, Metals & Materials Society), 2014

THEORETICAL ASPECTS ON PUSHING AND ENGULFMENT OF SiC PARTICLES DURING DIRECTIONAL SOLIDIFICATION EXPERIMENTS WITH MOLTEN SILICON

Arjan Ciftja[1], Orion Ciftja[2]

[1]SINTEF Materials and Chemistry, Trondheim, N-7465, Norway
[2]Department of Physics, Prairie View A&M University, Texas 77446, USA

Keywords: Silicon carbide, Solidification, Engulfment

Abstract

The directional solidification method is routinely used to produce polycrystalline silicon, an important material for the photovoltaic (PV) industry. The form of the solid-liquid interface during solidification has considerable influence on the material quality. The dissolved impurities and the solid inclusions above the solidification front affect the interface and may cause interface breakdown. We investigate the effects of pushing and engulfment of silicon carbide (SiC) particles by the solidification front during directional solidification of silicon. We employ a 12kg directional-crystallization furnace to cast the molten silicon. We use upgraded metallurgical grade silicon with low concentrations of metallic impurities (< 5ppm) and a high content of carbon (700 ppm) present as SiC particles. Samples from the cast polycrystalline silicon ingots are investigated by light microscope. The expected interaction of impurities with the solidification front is investigated theoretically using empirical models. The data obtained from the models are in good agreement with the experimental results.

Introduction

Directional solidification is a well-known method for producing multi-crystalline silicon for the PV industry. In this method a planar front, achieved during solidification, results in a clearly defined interface between the solid and the liquid phase. The solubility of major impurities is higher in the liquid than in the solid phase. Therefore, directional solidification will work as a purification process where the impurities are retained in the liquid phase [1]. Apart from the process of removing the dissolved elements in molten silicon, the directional solidification plays also another role. It will interact with the solid particles that may be present in the melt. When a liquid containing an insoluble particle is solidified, three distinct interaction phenomena are observed: engulfment of particles, continuous pushing, and entrapment [2]. The term "engulfment" is used to describe incorporation of a particle by a planar interface, and the term "entrapment" is used for particle incorporation by a cellular or dendritic interface (see Figure 1) [3].

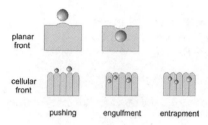

Figure 1: Schematic representation of particle-interface interaction as a function of interface morphology [3].

The form of the solid-liquid interface during solidification of silicon has considerable influence on the multi-crystalline silicon ingot quality. Impurities will build up in the boundary layer close to the solidification front and with increased concentration of impurities chances for interface breakdown will increase too. If the interface breaks up and columnar cells start to grow, both grain size and impurity distribution will be largely influenced [4]. Here we report the results from two directional solidification experiments on the behavior of SiC particles in molten silicon.

Materials and Methods

Materials

Two parallel tests were performed in a directional solidification furnace Crystalox DC250 at SINTEF Materials and Chemistry. The material used was UMG silicon with very little content of metallic impurities (all metals < 5 mass ppm) but with high content of carbon (~700 mass ppm) which is present as SiC particles. The experimental procedure is described elsewhere [5]. Ingots after solidification were named FS-4A and FS-4B. They were round with a height of about 100 mm and diameter of 250 mm.

Sampling and Analyzing Techniques

After solidification, the silicon cast ingots were cut in half. Samples were taken close to the center of the ingot from the top to the bottom. To analyze quantitatively the inclusion content from the bottom to the top, the ingot was divided into 7 regions from bottom to the top of the ingot. Analysis of the inclusion number distribution along the whole height of the ingot is carried out by means of light microscopy. Microscopic analysis consists of measuring the surface area of the inclusions and then calculating the equivalent particle diameter [6]. First, a surface area of 1 cm^2 is observed in each region. All the particles found inside this area were counted and their surface area was measured. Then, the samples were ground to a depth of 0.5 cm in the horizontal direction, and a surface area of 1 cm^2 was analyzed again in each region. This procedure was repeated twice thus, allowing us to obtain information from three different areas in each region.

Results and Discussion

Only particles within the size range of 10 - 30 μm from the middle (17 – 76 mm) of the ingots are considered for theoretical analysis. This range of particle size is selected since secondary inclusions smaller than 10 μm may be present after directional solidification. The bottom of the Si ingot is contaminated with Si_3N_4 particles from the coating layer which interfere with our results and therefore is neglected in our calculations. All the particles counted and measured in the middle of the ingot (17 – 76 mm from the bottom of the ingot) are considered to be engulfed particles. All the particles counted and measured in the top of the ingot (76 – 100 mm from the bottom of the ingot) are considered as pushed by the S/L interface.

Table I and II give the number size distribution of SiC particles in the input silicon and in the middle of the cast ingots.

Table I: Number size distribution of particles in 3 cm^2 in the input silicon and in the middle of ingot FS-4A.

d [μm]	Av. before	17–32 mm	32–46 mm	46–63 mm	63–76 mm	Av. after
12.5	128	18	34	15	16	21
17.5	101	13	7	3	13	9
22.5	51	7	9	4	9	7
27.5	32	8	3	2	3	4

Table II: Number size distribution of particles in 3 cm^2 in the input silicon and in the middle of ingot FS-4B.

d [μm]	Av. before	17–32 mm	32–46 mm	46–63 mm	63–76 mm	Av. after
12.5	128	7	4	1	7	5
17.5	101	4	3	2	3	2
22.5	51	3	1	0	1	1
27.5	32	1	0	0	0	0

The number size distribution of the particles at 76 – 100 mm from the bottom of the ingots in 2 cm^2 observed area are given in Table III and IV for the ingot FS-4A and FS-4B respectively. The summarized data from 3 cm^2 of the samples before and after solidifications are given in Table V and VI.

Table III: Number size distribution of pushed particles in 3 cm^2 of the upper part in the ingot FS-4A.

d [μm]	76 – 89 mm	89 – 100 mm	Average after
12.5	33	16	25
17.5	11	9	10
22.5	12	6	9
27.5	4	5	5

Table IV: Number size distribution of pushed particles in 3 cm^2 of the upper part in the ingot FS-4B.

d [μm]	76 − 89 mm	89 − 100 mm	Average after
12.5	13	13	13
17.5	5	5	5
22.5	8	4	6
27.5	2	1	2

Table V: Number size distribution of particles before and after refining in ingot FS-4A.

d (range) [μm]	Before Exp.	Engulfed	Pushed
12.5 (10 − 15)	128	21	25
17.5 (15 − 20)	101	9	10
22.5 (20 − 25)	51	7	9
27.5 (25 − 30)	32	4	5

Table VI: Number size distribution of particles before and after refining in ingot FS-4B.

d (range) [μm]	Before Exp.	Engulfed	Pushed
12.5 (10 − 15)	128	2	13
17.5 (15 − 20)	101	2	5
22.5 (20 − 25)	51	1	6
27.5 (25 − 30)	32	0	2

Tables V and VI indicate that pushing is more important than engulfment for the system studied. The silicon was kept in molten state for about two hours before solidification took place. Therefore, settling of SiC particles happened first. The difference of the particles before solidification and the particles pushed + engulfed by the S/L interface gives the settled particles. Taking into consideration settling, pushing and engulfment, a mass balance of the SiC particles could be formulated. Collision and growth of particles are neglected.

At this point, we note that: Mass of the particles before refining = Mass of settled particles + Mass of pushed particles (found in the top) + Mass of the engulfed particles (found in the middle of the ingot). In terms of particle size distribution we have:

$$f_N(d) = f_{N,\text{settling}}(d) + f_{N,\text{pushing}}(d) + f_{N,\text{ engulfment}}(d) \qquad (1)$$

Numerous models for particle behavior at the S/L interface have been proposed. The basic assumptions of most kinetic models assume spherical and inert particles and a macroscopically planar S/L interface [7]. All the models agree that there exists a critical velocity of the planar S/L interface, below which particles are pushed ahead of the advancing interface and above which particle engulfment occurs. This velocity is a function of the solidification velocity and some material parameters.

Søiland [8] calculated the critical velocity for pushing / engulfment, u_c, as a function of particle radius for the system Si-SiC based on the model of Stefanescu et al. [9]. According to this model the critical velocity for pushing / engulfment of particles is:

$$u_c = \frac{\Delta\sigma_0 d_0}{6(n-1)\mu R}\left(2 - \frac{\lambda_p}{\lambda_l}\right) \qquad (2)$$

where $\Delta\sigma_0$ is the surface tension difference, $\Delta\sigma_0 = \sigma_{ps} - \sigma_{pl} - \sigma_{ls}$, d_0 is the interatomic distance in the melt, μ is viscosity of the melt, R is the particle radius, λ_p is the thermal conductivity of the particle and λ_l is the thermal conductivity of the melt and, n is equal to 2. According to calculation by Søiland [8], at a growth velocity of 5.55 µm/s engulfment will occur for a particle with radius 1296 µm or above.

When a solidification front intercepts a solid particle, it can either push it or engulf it. Engulfment occurs through growth of the solid over and around the particle, followed by enclosure of the particle in the solid. As suggested in calculations from Søiland [8] (Section 2.3.3), SiC particles with size smaller than 1.2×10^{-3} m are pushed by the planar solidification moving with a velocity of 5.5×10^{-6} m s^{-1}. In the Equation (2) formulated by Stefanescu *et al.* [9] and used by Søiland [8] to calculate pushing of SiC particles by a planar S/L front, the critical velocity, u_c, is derived from the interaction of two forces, the repulsive force between a particle and a S/L interface, F_r, and the drag force of a fluid on the particle, F_d. Thus, the Stokes force exerted on the particles and the lift force produced by the liquid flow parallel to the interface are not considered. The model is valid for non-terrestrial experiments under zero gravity and for very pure metals. Terrestrial experiments give rise to three additional forces on the particle which often masks the influence of those two fundamental forces, F_r and F_d [10]. First, the gravitational (buoyancy) force on the particle is determined by the difference in density between the particle and the melt, $\Delta\rho$ and the g-level. Secondly, the particle experiences a lift known as the Saffman force, F_S, due to the difference in melt velocity in the gap between the particle and the interface and that on the opposite side of the particle. Finally, the particle can experience an additional lift force due to particle rotation. This force is defined as the Magnus force, F_M [10]. These forces are not taken into consideration in the model of Stefanescu *et al.* [9].

Shangguan *et al.* [11] argues that under the influence of gravity forces, the velocity of the particle itself does not significantly affect the critical velocity, because the particle velocity is very small compared to the critical interface velocity. The settling velocities for SiC particles within the size range of 1.0×10^{-6} - 4.0×10^{-5} m (1 – 40 µm) are from 5×10^{-7} - 9×10^{-4} m s^{-1}. Comparing these velocities with the respective critical interface velocities, u_c, as calculated by Søiland [8], we conclude that for particles larger than 15 µm the particle velocity affects the critical velocity of the pushing / engulfment process. Therefore, the critical velocity is lower than what the model predicts.

The difficulty in verifying and applying the models by experiments lies in the fact that these models are valid for pure metals. Predictions made by the models fail in the case of impure matrices as the critical velocity suffers an extreme reduction even for impurities in the ppm region [12].

When the concentration of impurities ahead of the S/L interface reaches a certain level, engulfment of particles will occur. The situation will resemble that of an alloy where the critical interface velocity is much lower than for the pure melts [13]. The difference in thermal conductivity, which is the driving force for pushing of particles in pure melts, will not have any effect in pushing of the particles in alloys because it is much higher than the diffusivity of the solute in the liquid. According to Schvezov and Weinberg [14] the thermal gradient at the S/L interface has no measurable effect on the critical velocity when the particle size is smaller than 5×10^{-4} m (500 µm).

Conclusions

Directional solidification of silicon pushes the particles to the top of the ingot. Even though the calculated critical velocity for engulfment of SiC particles in silicon is higher than the S/L interface velocity, some engulfment of the large particles takes place. This may explain the presence of SiC particles in the middle of the ingot.

Results from two ingots differ from each other by a factor 5 to 10, indicating the need for further experiments. Investigation of removal by settling should be studied separately from the pushing / engulfment process that takes place during casting.

Better theoretical models that can explain the pushing and engulfment of SiC particles by the solidification front during directional solidification of silicon are needed.

Acknowledgments

The research of one of the authors (O. C) was supported in part by ARO Grant No. W911NF-13-1-0139 and NSF Grant No. DMR-1104795.

References

1. T. A. Engh, Principle of Metal Refining, Oxford University Press, Oxford, New York, (1992) 375.
2. A. Catalina, S. Mukherjee, D. Stefanescu, Metallurgical and Materials Transactions A, 31(10) (2000) 2559.
3. F. Juretzko, D. Stefanescu, B. Dhindaw, S. Sen, and P. Curreri, Metallurgical and Materials Transactions A, 29(6) (1998) 1691.
4. R. Kvande, PhD thesis, NTNU, Trondheim, (2008).
5. A. Ciftja, M. Tangstad, T. A. Engh, JOM, Vol. 11 (2009).
6. A. Ciftja, Solar Silicon Refining: Inclusions, settling, filtration, wetting, LAP Lambert Academic Publishing, New York (2010).
7. B. Dhindaw, Current Science, 79(3) (2000) 341.
8. A. K. Søiland, Silicon for Solar Cells, NTNU, Trondheim, (2004).
9. D. Stefanescu, B. Dhindaw, S. Kacar, A. Moitra, Metallurgical and Materials Transactions A, 19(11) (1988) 2847.
10. A. V. Bune, S. Sen, S. Mukherjee, A. Catalina, D. M. Stefanescu, Journal of Crystal Growth, 211(1-4) (2000) 446.
11. D. Shangguan, S. Ahuja, D. Stefanescu, Metallurgical and Materials Transactions A, 23(2) (1992) 669.
12. U. Hecht, S. Rex, Metallurgical and Materials Transactions A, 28(3) (1997) 867.
13. J. K. Kim, P. K. Rohatgi, Materials Science and Engineering A, 244(2) (1998) 168.
14. C. E. Schvezov, F. Weinberg, Metallurgical Transactions B, 16(2) (1985) 367.

Energy Technology 2014: Carbon Dioxide Management and Other Technologies
*Edited by: Cong Wang, Jan de Bakker, Cynthia K. Belt, Animesh Jha, Neale R. Neelameggham,
Soobhankar Pati, Leon H. Prentice, Gabriella Tranell, and Kyle S. Brinkman*
TMS (The Minerals, Metals & Materials Society), 2014

Separation of Si and SiC Microparticles of Solar Grade Silicon Cutting Slurry by Micropore Membrane

Suning Liu, Kai Huang, Hongmin Zhu*

*State Key Laboratory of Advanced Metallurgy, University of Science and Technology
Beijing, Beijing 100083, PR China*
Corresponding author: E-mail address: hzhu@metall.ustb.edu.cn

Key words: silicon; micropore membrane; silicon cutting slurry

Abstract

Recovery of silicon particles from the cutting slurry for the manufacturing of solar grade silicon is very important to recycle this kind of special resource. The challenge is to separate the silicon particles from the mixed abrasive particles which contain SiC. Various methods such as sedimentation, centrifugation, and flotation have been conducted, while in present work, a simple and effective novel method was proposed to study the separation of Si and coexisting SiC particles in the cutting slurry. The new process is based on the size difference between the two kinds of microparticles. And the membranes with the different pore sizes of 1μm, 2μm, 3μm and 5μm were used to test the pass ratio and separation efficiency of the suspension. Results shows that the effective separation of Si and SiC microparticles was obtained under the optimal conditions, and indicating great potential of practical enrichment of Si from the coexisting abrasive particles.

Introduction

With the development of the world economy, energy shortage and environment problem is becoming more and more seriously. Solar energy, which is regarded as one of the most important new energy in the word, has received much attention and interest from the scientists [1, 2]. More than 80% solar cells are manufactured by the high purity polysilicon as the raw material. The main process of producing silicon wafer is by multi-wire slicing. In this slice processing, large amount of slurry is produced. The slurry contains more than 30% wt silicon powder, silicon carbide powder as the abrasive, scraps of metal chips, as well as cutting abrasive suspension polyethylene glycol (PEG). The high-purified silicon powders are worth recycling and reusing as the materials for preparing solar-grade silicon though it is a big challenge [3].

Most of the current studies were focusing on the recycling of the silicon carbide and PEG, while work on the recycling of the high-purified silicon is not so been focused since there is lacking of very effective separation and recovery technologies till up now. Nishijima utilized the method of superconducting magnetic to recycle the abrasive [4]. Wang applied a combinational process of chemical treatment, high-gravity centrifugation, high-temperature precipitation technique and directional solidification to enrich the silicon [5]. And other purification methods, such as centrifugation [6], froth flotation technologies [7], electrical filed separation [8] and phase-transfer process [9] have been employed. The disadvantages such as using poisonous media or complex equipments decrease the feasibility and efficiency of the above mentioned techniques.

The aim of this study is to develop a novel process to separate the silicon and silicon carbide particles in the cutting slurry. This process is based on the different particle sizes and the surface charges of the two kinds of particles, and utilized the millipore filtration as the basic separation method.

Experimental

Pre-dispersion treatment of the silicon cutting slurry

The silicon cutting slurry usually contains PEG, silicon and silicon carbide particles and scraps of metal. The PEG can be removed by the deionized water and metal purities dissolved by hydrochloride acid firstly. After filtrating and drying the mixture, the filtered cake was ball milled to disperse into small fragments or particles.

The surface charge of Si particles and SiC particles is quite different at pH range of 2-10, and the electrostatic repulsion force between the particles is quite strong between 8 and 9. So it is useful for the sufficient suspension and dispersion of the mixed microparticles of Si and SiC in the water at pH equal to 8.5. Then the mixture was treated by an ultrasonic wave to make the particles dispersed sufficiently.

Analysis methods

The particle size distribution was measured using LMS-30 Laser Particle Analyzer (JAPAN Seishin). X-ray diffraction measurements of the samples were measured by a Rigaku D/max-RB X-ray diffractometer. Scanning electron microscopy (SEM) instrument (Cambridge S-360) was used for observing the size and surface morphology of the particle samples.

Results and discussion

The SEM image of the mixed powder was shown in Fig. 1(a). Obviously, the mixing powder is made up of two sizes of particles. It can be proved by SEM that the particles with larger size are mainly composed of SiC and the smaller ones are mainly Si. It is clear that the average particle size of SiC is around 10 μm and Si is 1μm. The Fig. 1(b) also shows the two peaks in the particle size distribution curve. The XRD patterns of the mixture are presented in Fig. 1(c). Since the size of SiC particles and Si particles so drastically different, and the idea of separating these two particles by microfiltration with the membrane pore size between 1μm and 5μm was put forward based on our above findings.

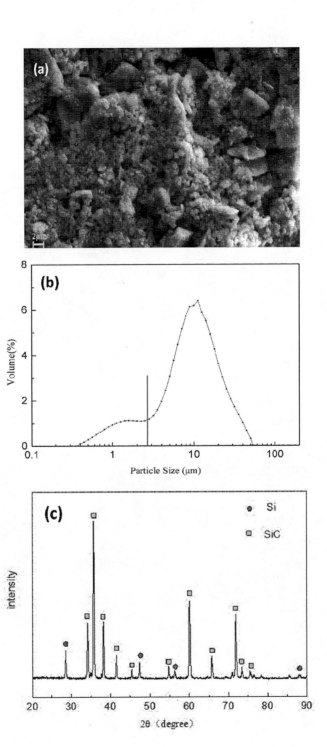

Figure 1. SEM photograph (a), size distribution curve (b)
and XRD pattern (c) of the crude slurry particles

To investigate the microfiltration separation effectiveness of the SiC and Si particles, four membranes with different pore sizes of 1 μm, 2 μm, 3 μm and 5 μm respectively were used in the experiments. The SEM images of the microfiltration membrane with the pore size of 2 μm and 5 μm are shown in the Fig. 2, which shows that the membranes used in present study have the uniform pore sizes.

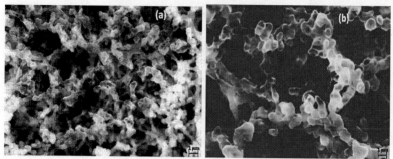

Figure 2. SEM photographs of the membranes with different pore size (a)2 μm; and (b) 5μm.

The powder obtained by the microfiltration mainly was made up of the particles with the size of 1 μm (Fig. 3). And these samples observed by scanning electron microscope are made of uniform and no extraordinarily large particles. From the obtained results, it appears that all the membranes with four different pore sizes can separate silicon and silicon carbide microparticles effectively.

324

Figure 3. SEM photographs of the Si-rich particles collected after filtering through the membranes with different pore size (a) 1μm; (b) 2μm; (c) 3μm; and (d) 5μm.

Fig. 4 shows the particle size distribution curve of the obtained Si-rich powder by different pore sizes microfiltration. The results indicated that the average particle size of the particles was smaller and the distribution range becomes narrow with membrane pore size decreasing. But there are still some large particles, especially for the powders obtained by 5 μm membrane microfiltration which is between 0.6-10 μm. There are two factors that might lead to these results, i.e., one hand is the pore sizes of micromembrane not absolutely uniform, and the other hand is error analysis of the laser particle size analyzer.

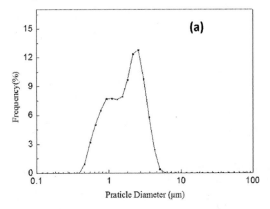

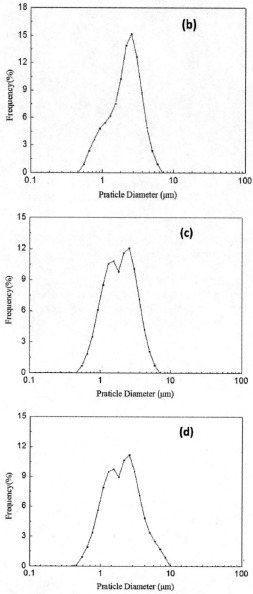

Figure 4. Size distribution curves of the Si-rich particles collected after filtering through the membranes with different pore size (a) 1μm; (b) 2μm; (c) 3μm; and (b) 5μm.

XRD patterns of the obtained powders by the microfiltration with different pore size membrane are presented in Fig. 5. The results showed that the silicon powder can be initially separated by this method. And the main constituent of the obtained powders is silicon while in the raw materials is silicon carbide. Above results demonstrate that the Si and SiC in the cutting slurry can be separate effectively.

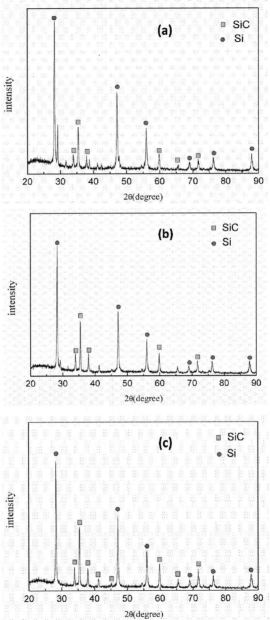

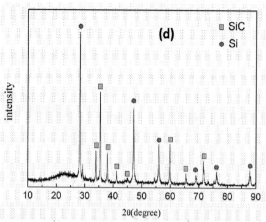

Figure 5. XRD patterns of the Si-rich particles collected after filtering through the membranes with different pore size (a) 1μm; (b) 2μm; (c) 3μm; and (d) 5μm.

As listed in Table. 1, it can be found that with the decrease of the membrane pore size, content of silicon in the Si-rich powders increased though the recovery ratio decreased. When the pore size of micropore membrane is 3 μm, the content of obtained Si and recovery ratio are both can be accepted.

Table. 1. The recovery ratio and content of Si in the Si-rich powder obtained by the microfiltration method.

Pore size	1 μm	2μm	3μm	5μm
Recovery ratio (%)	57.1	69.5	76.2	80.5
Content of obtained Si (%)	81.2	70.6	68.5	59.4

Conclusion

The silicon and silicon carbide from the cutting slurry can be separated effectively by a microfiltration process. This method is very simple and low-cost. With the pore size of membrane is 3 μm, the silicon content and the recovery ratio of the collected Si-rich powder can reach 68.5% and 76.2%, respectively.

Acknowledgement

This work was supported by National Science Foundation of China (No. 50934001, 21071014, 51102015 and 51004008), National High Technology Research and Development Program of China (863 Program, No. 2012AA062302).

References:

1. D. Sarti, R. Einhaus, "Silicon feedstock for the multi-crystalline photovoltaic industry", *Solar Energy Material & Solar Cells*, 72 (2002) 27–40.

2. A. Lee, H.H. Chen and H. Kang, "A model to analyze strategic products for photovoltaic silicon thin-film solar cell power industry", *Journal of Renewable and Sustainable Energy*, 15 (2011) 1271–1283.

3. A. Trabelsi, "Internal quantum efficiency improvement of polysilicon solar cells with porous silicon emitter", *Renewable Energy*, 50 (2013) 441-448.

4. S.Nishijima et al., "Recycling of abrasives from wasted slurry by superconducting magnetic separation", *IEEE Transactions on applied superconductivity*, 13(2003) 1596-1599.

5. H.Y. Wang et al., "Removal of silicon carbide from kerf loss slurry by Al-Si alloying process", *Separation and Purification Technology*, 89 (2012) 91-93.

6. S.N. Liu, K. Huang and H.M. Zhu, "Recovery of silicon powder from silicon wiresawing slurries by tuning the particle surface potential combined with centrifugation", *Separation and Purification Technology*, 118 (2013) 448-454.

7. M.L. Huang et al., "Froth flotation technology for separation of Si and SiC powders in wire saw slurry from silicon wafering process", *Electronic Components and Materials*, 29 (2010) 74-77.

8. Y.F. Wu, Y.M. Chen, "Separation of silicon and silicon carbide using an electrical field", *Separation and Purification Technology*, 68 (2009) 70-74.

9. Y.C. Lin, C.Y. Tai, "Recovery of silicon powder from kerfs loss slurry using phase-transfer separation", *Separation and Purification Technology*, 74(2010) 170-177.

AUTHOR INDEX
Energy Technology 2014

SUBJECT INDEX
Energy Technology 2014